PARADOXES AND INCONSISTENT MATHEMATICS

Logical paradoxes – like the Liar, Russell's, and the Sorites – are notorious. But in *Paradoxes and Inconsistent Mathematics*, it is argued that they are only the noisiest of many. Contradictions arise every day, from the smallest points to the widest boundaries. In this book, Zach Weber uses *dialetheic paraconsistency* – a formal framework where some contradictions can be true without absurdity – as the basis for developing this idea rigorously, from mathematical foundations up. In doing so, Weber directly addresses a longstanding open question: how much standard mathematics can paraconsistency capture? The guiding focus is on a more basic question, of why there are paradoxes. Details underscore a simple philosophical claim: that paradoxes are found in the ordinary, and that is what makes them so extraordinary.

ZACH WEBER is Associate Professor of Philosophy at the University of Otago, New Zealand.

PARADOXES AND INCONSISTENT MATHEMATICS

ZACH WEBER

University of Otago

CAMBRIDGE
UNIVERSITY PRESS

Shaftesbury Road, Cambridge CB2 8EA, United Kingdom

One Liberty Plaza, 20th Floor, New York, NY 10006, USA

477 Williamstown Road, Port Melbourne, VIC 3207, Australia

314–321, 3rd Floor, Plot 3, Splendor Forum, Jasola District Centre, New Delhi – 110025, India

103 Penang Road, #05–06/07, Visioncrest Commercial, Singapore 238467

Cambridge University Press is part of Cambridge University Press & Assessment, a department of the University of Cambridge.

We share the University's mission to contribute to society through the pursuit of education, learning and research at the highest international levels of excellence.

www.cambridge.org
Information on this title: www.cambridge.org/9781108995009

DOI: 10.1017/9781108993135

First published 2021
First paperback edition 2024

A catalogue record for this publication is available from the British Library

ISBN   978-1-108-83441-4   Hardback
ISBN   978-1-108-99500-9   Paperback

A paradox
    is only the truth
    standing
    on
    its
    head to get attention.

       — Oscar Wilde

Contents

Preface   *page* ix

Part I   What Are the Paradoxes?

Introduction to an Inconsistent World   3
   0.1   The Problem   3
   0.2   The Choices   9
   0.3   Prospectus: Fixed Points   22

1   Paradoxes; *or*, "Here in the Presence of an Absurdity"   28
   1.1   Sets   28
   1.2   Vagueness   40
   1.3   Boundaries   50
   1.4   Conclusion   59

Part II   How to Face the Paradoxes?

2   In Search of a Uniform Solution   65
   2.1   In Search of an Explanation   65
   2.2   Two Schemas   66
   2.3   Stepping Back from the Limits of Thought   79

3   Metatheory and Naive Theory   84
   3.1   The Myth of Metatheory   84
   3.2   Classical Recapture   96
   3.3   Naive Theory   102

4   Prolegomena to Any Future Inconsistent Mathematics   110
   4.1   Curry's Paradox   110
   4.2   Grišin's Paradox and Identity   120
   4.3   Logic   129
   Appendix: BCK and DKQ   145

Part III   Where Are the Paradoxes?

5   Set Theory   151
    5.1   Elements   151
    5.2   A Sketch of the Universe   164
    5.3   Order   180
    Excursus: Partitions, Equivalence Classes, and Cardinality   185

6   Arithmetic   189
    6.1   Thither Paraconsistent Arithmetic!   189
    6.2   Addition, Multiplication, and Order   193
    Excursus: Number Theory   201
    6.3   Descent: Inconsistency and Irrationality   207

7   Algebra   212
    7.1   Algebra for Inconsistent Mathematics: A Triviality Problem   212
    7.2   Vectors   216
    7.3   Groups, Rings, and Fields   221
    7.4   A Short Conclusion to a Short Chapter   229

8   Real Analysis   230
    8.1   Into the Labyrinth: Real Numbers   230
    8.2   Dedekind Cuts   238
    8.3   Continuity; *or*, "Amongst the Ghosts of Departed Quantities"   246
    8.4   Out of the Labyrinth: The Topology of a Point   255

9   Topology   256
    9.1   Closure Spaces   256
    Excursus: Consequence as Closure   264
    9.2   Boundaries and Connected Space   265
    9.3   Continuity   272

Part IV   Why Are There Paradoxes?

10   Ordinary Paradox   285
    10.1   Dividing the Universe   285
    10.2   The Last Horizon   297
    10.3   A Fixed Point Where None Can Be   300

Bibliography   303
Index   319

Preface

Why are there paradoxes? That there *are* logical paradoxes is not in question. A notorious bunch of them besets core parts of our understanding the world:

- The liar paradox in theories of truth
- Russell's paradox in theories of sets and properties
- The sorites paradox in any theory with vagueness

Many proposals have been offered that address aspects of these problems, from seminal work [Russell, 1905b; Tarski, 1944; Kripke, 1975] to the more recent [Priest, 2006b; Field, 2008; Beall, 2009]. Modern logic has given us unprecedented insight into the mechanics, so to speak, of paradoxes – we know *how* they arise, and how, at least temporarily, they can be evaded. The question investigated in this book is more basic: asking for an explanation of *why* there are paradoxes at all.

Paradoxes can look like they are unusual accidents, exceptional borderline cases, self-referential anomalies at the edge of the world. And paradoxes certainly are found at these dramatic and distant limits – but not only there. The famous paradoxes are only the noisiest of many. In the pages ahead, I rethink the paradoxes as much smaller in scale, appearing everywhere as innocuous parts of everyday objects. They are found at the edge of the universe but also at the edge of a coffee cup. Shifting our thinking to the local level demystifies the problem. Rather than fixating on bizarre things in bizarre places, we can begin to appreciate that paradoxes are found in the ordinary: *that* is what makes them so extraordinary.

The paradoxes are deductive arguments that end in contradictions; they appear to be *proofs* of contradictions. According to prevailing views, though, a contradiction can only be a mistake. So according to prevailing views, the paradoxes simply cannot be what they appear to be. But this classical approach rules out a priori the simplest explanation of the paradoxes: that *they are exactly what they appear to be*.

The framework in this book, then, uses *paraconsistent logic*, in the *dialetheic* tradition: the framework that allows for some contradictions to be true, without everything whatsoever being true.[1] New understanding of the paradoxes becomes possible – and precise – by situating them where they first arise, in foundational logico-mathematics, but now described with formalism specially designed to handle inconsistency. Central chapters develop elementary axiomatic theories in paraconsistent logic, beginning with Frege–Cantor naive set theory, and going on to describe rudiments of arithmetic, algebra, real analysis, and topology, with a focus throughout on paradoxes as they arise at *boundaries*. In doing so, this work thus addresses a longstanding open question, sometimes phrased as an objection: "Can paraconsistency capture any standard mathematics? If so, how much?" The (qualified) answer provided is by direct demonstration.

Crucially, I argue that the entire presentation ought to be purely paraconsistent, from the object level to the "meta"-level, without any recourse to classical resources (including right now). What is the picture of the world that emerges when we describe it in a fully nonclassical language? I provide a sketch, emphasizing qualitative aspects that localize the paradoxes at geometric *points*. With an appropriate paraconsistent framework, we can understand paradoxes as bona fide mathematical and metaphysical objects–explained, rather than explained away.

* * *

Here is the **argument in a nutshell**:

   I There are true contradictions, both in the foundations of logic and mathematics, and in the everyday world.[2]

  II If the world is inconsistent but not absurd, then the logic underlying our theory of the world, including all of logic and mathematics, ought to be paraconsistent.

 III Paraconsistent logic then must, and can, show that it supports some ordinary reasoning. A minimum requirement is that the logic be able to reestablish the motivating paradoxes – *proving* the contradictions, on their own terms, in elementary mathematics.

 IV In proving the paradoxes paraconsistently, the basic components of a nonclassical picture come into view (including nonstandard descriptions of identity, boundaries, and points). Then we are finally positioned to (re)address the question of why there are paradoxes.

For (I), I largely follow standard arguments for dialetheism, and in this obviously owe an enormous debt to Graham Priest, Richard Routley/Sylvan, and other pioneers[3] in

---

[1] The term "paraconsistent" was invented by Miró Quesada in the mid-1970s. The term "dialetheic" was invented in the late 1970s by Graham Priest and Richard Sylvan (Routley at the time) [Priest et al., 1989, p. xx], to replace the term "dialectic," which had been in use from Hegel. Dialetheism is simply the thesis that there are true contradictions; etymologically, the word is intended to evoke something like "two-way truth," and is sometimes spelled "dialethism." The (somewhat) less esoteric term "glut theory" means the same thing, and is used too, e.g., by Beall [Beall, 2009].

[2] *Being contradictory* is a property of sentences (or propositions), so literally speaking, a coffee cup can't be contradictory or "have a contradiction on it"; but if there is a *true description* of the coffee cup that is contradictory, then I will say that the coffee cup is too. Here and throughout, to say that *the world is inconsistent*, or that there are true contradictions in the world, is an evocative shorthand to mean that some glutty theory of the world is true (but without thereby committing some elementary category mistake). See Section 0.2.2 and [Priest, 2006b, pp. 299–302].

[3] Some key works by Priest and Roultey are [Routley, 1977; Priest and Routley, 1983; Priest, 2006b]. Other important contributors (in varying ways) include Arruda [Arruda and Batens, 1982], Asenjo [Asenjo, 1966, 1975], Beall [Beall, 2009], Brady [Brady, 2006], Colyvan [Colyvan, 2008a], da Costa [da Costa, 1974], Dunn [Dunn, 1980], Meyer [Meyer, 1976], Mortensen [Mortensen, 1995], Restall [Restall, 1992], Slaney [Slaney, 1982], and others.

paraconsistency and *inconsistent mathematics*: the application of paraconsistent logic to the study of contradictory abstract structures. In the Introduction and Chapter 1, I will offer various examples and considerations that I think motivate dialetheism; I don't simply assume that there are true contradictions, and I am almost certain that you don't. (Please write to me if you do.) Nevertheless, while the thesis that there are truth gluts remains contentious, I think to advance from here does not require further rehearsal of abstract polemics about the law of noncontradiction or the like. I don't try to defend my approach point-for-point against other options, because what is most important for the project now is to show what it can *do*. Dialetheic paraconsistency is a *motive* and a *method*; I argue that it both independently supports some genuine mathematical reasoning and, in doing so, offers new insights into the paradoxes. My plan is mainly just to show what a committed paraconsistent picture of the world drawn with precise tools might look like, taking a "deep dive" into the substructure of inconsistent space.

So in taking proposition (I) to be largely defended elsewhere, it is in the scope and depth of propositions (II) and (III) – the question of how mathematically revisionary dialetheic paraconsistency might be, and the "classical recapture" – that I am predominantly attempting new advances, with (IV) a matter of setting targets – a "paradox recapture" – and following out our commitments to their logico-mathematical end. I am sketching, to put it colorfully, a possible world in which the reaction to the paradoxes circa 1900 was quite different, and subsequently the mathematics that developed in the twentieth century had a different tenor, preserving obvious and intuitive principles (such as "collections are sets") by treading more carefully with our logic. This is not *Principia Mathematica Paraconsistenta* – the mathematical chapters are not strictly cumulative – but perhaps indicates how such a tome could eventually (or why it never will) be written. This is not a textbook. I hope it might invite others to use it as a stepping stone.

If some of the claims in this book seem extreme (the word "unhinged" may cross your mind – it has mine), they are nevertheless motivated by very traditional, almost Socratic sorts of commitments: the world can be made sense of, even (especially) when it seems senseless. Or at least, it is incumbent upon us to try. The "argument in a nutshell" has an unstated Premise 0: the world is ultimately and profoundly *intelligible*, in something like the mode of the Enlightenment project and the principle of sufficient reason, up to Hilbert's program.

Is this axiom of the solvability of every problem ... a general law inherent in the nature of mind, that all questions which it asks must be answerable? ... We hear within us the perpetual call: There is the problem. Seek the solution. You can find it by pure reason, for ... there is no *ignorabimus* ... *[Hilbert, 1902b]*

The spirit of this work is with rationalists and button-down logicists, give or take a disjunctive syllogism or two. I think that the challenge posed by the paradoxes is a challenge to reason itself, and that we are called to respond: that even (especially) when reason gives out, that is exactly when not to abandon reason. I ultimately believe that the paradoxes exist because *the world has no gaps*, that as Leibniz said, "*La nature ne fait jamais de sauts*."[4]

---

[4] E.g., in the Preface to the *New Essays*, [Leibniz, 1951, p. 378].

(If the world has gaps, those gaps are part of the all-inclusive world, too.) I am transfixed by Leibniz's beautiful statement from the *Monadology* [Leibniz, 1714, §47] that all objects are generated by the continual flashes of silent lightning [*fulgurations continuelles*] ...

By the end, I hope to have discovered a little of what this could mean.

In 1931, Gödel famously dashed Hilbert's hopes, if not the hopes of the entire Enlightenment project, by using a version of the liar paradox to prove that any complete, axiomatic theory of the world will be inconsistent. I read this result as an invitation. Welcome to our inconsistent world.

* * *

Acknowledgments

This work draws on papers published over my career to date. In putting this all together, everything is either extensively rewritten, remixed, or entirely new; but descendants of previously published work (or, published work that descended from this book) include the following. Chapter 1§2 draws from [Weber, 2010c], and 1§3 from [Weber and Cotnoir, 2015]. Chapter 2 and bits of Chapter 3 are modified from [Weber, 2019]. Some of Chapter 3§1 and 3§2 draws from from [Weber et al., 2016]. Some of Chapter 4 began in [Badia and Weber, 2019]. Thank you to my coauthors, and thank you to the original publishers for permission to draw on these articles. Additionally, sections throughout owe debts to (at least) the following: [Weber, 2009; 2010d; Weber and Colyvan, 2010; Weber, 2010d; McKubre-Jordens and Weber, 2012; Caret and Weber, 2015; Meadows and Weber, 2016; Girard and Weber, 2019; Omori and Weber, 2019].

This book has been a very long time coming. Many, many people contributed to it, directly or indirectly, in many countries over many years. I owe at least the following people, and surely others too, an enormous thanks. Thank you to Guillermo Badia, Jc Beall, Ross Brady, Colin Caret, Petr Cintula, Mark Colyvan, Aaron Cotnoir, Eric Dietrich, Patrick Girard, Lloyd Humberstone, Erik Istre, Franci Mangraviti, Maarten McKubre-Jordens, Toby Meadows, Hitoshi Omori, Graham Priest, Greg Restall, Dave Ripley and Shawn Standefer. For any errors, small and large, I am responsible.

Thank you to Hilary Gaskin at Cambridge University Press for supporting this project.

The writing of this book was supported by a grant from the Marsden Fund, Royal Society of New Zealand. Earlier than that, I had support from the Australian Research Council. Thank you to both agencies for supporting this sort of research.

Thank you to my parents. And thank you, finally, to my wife, and to my children.

Part I

What Are the Paradoxes?

There are whole mathematical cities that have been closed off and partially abandoned because of the outbreak of isolated contradictions . . .

They have become like modern restorations of ancient cities, mostly just patched up ruins visited by tourists [Routley, 1977, p. 927].

Introduction to an Inconsistent World

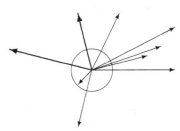

Look! The sun is rising on the horizon, where the earth meets the sky. The earth is not the sky and the sky is not the earth, but they touch, and these together are the world. The horizon: sky and not sky, earth and not earth, while the dawn chorus of birds sings. It is just an ordinary day, with paradoxes right in front of you.

0.1 The Problem

0.1.1 There Are Contradictions Inside Truisms

A *paradox* is a seemingly sound argument to a seemingly false conclusion.[1] A paradox is *genuine* when things are as they seem: a genuine paradox *is* a valid argument with true premises and a false conclusion. Since a valid argument preserves truth from premises to conclusion, the conclusion of a paradox is also true. The most striking paradoxes present themselves as logical-mathematical *proofs* of propositions that are both true and false. The conclusion of a genuine paradox, if there is any such thing, is a *dialetheia*, or *glut*, a true contradiction.

The paradoxes I am concerned with are very easy to state. Their contradictory conclusions are derived with a minimum of logical resources. They follow from principles that seem like they could not fail to be true, things we know to a certainty:

- A proposition p is true iff it is the case that p.
- An object a is in the set of φs iff it is the case that a is φ.
- Sometimes it is raining, and sometimes it is not.

---

[1] Following standard usage, following [Sainsbury, 1995], following Quine; see [Lycan, 2010].

And yet these ineluctable truths, these banalities, enclose inconsistencies – respectively, the liar paradox, the Russell paradox, and the sorites paradox.[2] Concentrated efforts to make it otherwise have foundered, despite their ingenuity and sophistication. This is fascinating: *vast* energies have been expended on showing that the paradoxes are not genuine, that things *could not be* as they seem. Through it the paradoxes have remained; they have *flourished*.

One good indicator of truth is that it persists, especially through attempts to repress or deny it. And so a good indicator of a genuine paradox is that it is somehow irrepressible – or as we have learned to say, it exhibits *revenge*.[3] Upon solution of a paradox, a new paradox arises that is essentially the same as the original paradox. To use a kind of geometric metaphor, the genuine paradoxes look to be *invariants* of certain spaces under solution-transformations. They do not go away.

This is not a pessimistic induction on the failure of logicians to solve some problem. It is a considered reevaluation of the paradoxes as having been unsolvable *for very good reason*: there is nothing to "solve." The idea is that logicians have been like apocryphal Pythagoreans attempting to "solve" the existence of irrational magnitudes such as $\sqrt{2}$ (cf. Chapter 6), whereas the better route forward is to see the rational numbers as only some among many.[4]

Paradoxes seem like a trick, except you can't figure out how the illusion works. And despite your persistence, you *never* figure it out. Every time you think you've got it, explained the deception, the magician repeats the trick, now with your attempted explanation in plain sight. "You see? No self-reference up my sleeve." After long enough, with successively more prolix attempts to expose the trick failing one by one, it becomes more reasonable to consider whether it is somehow not a trick at all.[5] The most amazement a magician can generate, after all, is for the viewer to realize that it isn't an illusion: the magician has done the impossible for real.[6]

Here's the trick. You were once a baby. Now you are not a baby. Because of the nature of time, there must have been a last moment you were a baby, or a first moment you were not. If a change occurred – and it did – then it must have occurred at some point, some *instant*. Even if the change was gradual, then the *beginning* of the gradual change itself still must have been at some precise moment. Or the beginning of the beginning ... It has to be! But as anyone can tell you, there is no one exact instant when a baby stops being a baby. So it looks like you changed in a profound way, without the change occurring anywhere. It's like escaping from a locked box without ever passing through its surfaces. Teleportation!

This problem – the *sorites paradox* – is so hard that it has led some philosophers to deny the existence of babies,[7] among other solutions. But perhaps the vagaries of terms such

---

[2] For *Curry's paradox*, see Chapter 4.

[3] Much more about revenge to come. See the introduction of [Beall, 2007].

[4] The canonical presentation of the argument I've just sketched is in [Priest, 1979].

[5] Mates similarly cautions the reader not to be distracted while "the rabbit of paradox is being brought out of the hat" [Mates, 1981, p. 5]. He says the paradoxes are "both intelligible and insoluble."

[6] My favorite example is the "magic trick" of being impaled by a sword, or stabbed with an ice pick, without any pain or blood. This is done by the magician having a special kind of scar tissue in the relevant place – a *fistula*, like an earring hole. So the magician really does just stick a sharp object into themself; that's it. Cf. "Miracle Man," *Time*, June 23, 1947.

[7] *Mereological nihilists* say that tables and trees don't exist, though they disagree about living people. See [Unger, 1979; van Inwagen, 1990].

as "is a baby" and assumptions about the nature of time come with enough doubt that you think this oddity can be safely ignored (at least until Section 1.2). After all, a little reflection will show that our days are filled will vagueness, from the dawn (when exactly is it?) to rainstorms (how may drops does it take to be raining?) onward. And how could something so ubiquitous be paradoxical?

So here is the trick again, starting with something that must be true, could not be more certain – Aristotle's definition of truth (enjambment added):[8]

To say of what is that it is not, or of what is not that it is,
is false,
while to say of what is that it is, and of what is not that it is not,
is true.

A *truth predicate* $T(x)$ takes a name for any sentence and is satisfied depending on whether or not that sentence holds, whether the state of affairs described by that sentence is the case. It says of what is, that it is, and of what is not, that it is not. In 1936, Tarski put a precise schematic form around this: with arrows for implication, for any sentence φ,

Truth schema $T(\ulcorner\varphi\urcorner) \leftrightarrow \varphi$,

where $\ulcorner\cdot\urcorner$ is a name-forming operator on sentences. It is true that φ if and only if φ: this is the schema for *naive truth theory*.

Why "naive"? Because there is a sentence ℓ called "the liar" that says of itself that is it false; with $\neg$ representing negation:

$$\ell \leftrightarrow \neg T\ulcorner\ell\urcorner. \tag{0.1}$$

Sentence ℓ says that sentence ℓ is not true. Putting (0.1) together with the instance of the truth schema $T\ulcorner\ell\urcorner \leftrightarrow \ell$, we have

$$T\ulcorner\ell\urcorner \leftrightarrow \neg T\ulcorner\ell\urcorner \tag{0.2}$$

using the transitivity of biconditionals (if $p \leftrightarrow q$, and $q \leftrightarrow r$, then $p \leftrightarrow r$). Then we reason as follows. Either ℓ is true, or it is not. If ℓ is true, $T\ulcorner\ell\urcorner$, then $\neg T\ulcorner\ell\urcorner$, by (0.2) and modus ponens (if p and $p \rightarrow q$, then q). Thus by reductio (if $p \rightarrow \neg p$ then $\neg p$), ℓ is not true,

$$\neg T\ulcorner\ell\urcorner.$$

But if ℓ is not true, then that is just what ℓ says (!); going through (0.2) again,

$$T\ulcorner\ell\urcorner$$

after all. So we have established

$$T\ulcorner\ell\urcorner \,\&\, \neg T\ulcorner\ell\urcorner. \tag{0.3}$$

Contradiction. And then, assuming that the truth schema is contraposable (contraposition is if $(p \rightarrow q)$ then $(\neg q \rightarrow \neg p)$), then

$$\ell \,\&\, \neg\ell. \tag{0.4}$$

---

[8] From *Metaphysics* 4.1011b25, echoed in Plato's *Cratylus* 385b2 and *Sophist* 263b.

Thus we appear to have proved both a sentence and its negation using an obviously true axiom (scheme) and elementary propositional logic.[9]

The paradoxes are no trick.[10] Listen to Tarski:

> In my judgment, it would be quite wrong and dangerous from the standpoint of scientific progress to depreciate the importance of this and other antinomies, and to treat them as jokes or sophistries. It is a fact that *we are here in the presence of an absurdity*, that we have been compelled to assert a false statement ... *[Tarski, 1944, emphasis added]*

The paradoxes are there in the most basic places: sets; truth; raindrops. If there were no genuine paradoxes, only apparent ones, then there would still be a psychosociological project of explaining why so many people have found the paradoxes compelling. But after a while, there being a sufficient number of lucid expert witnesses should start to suggest that maybe they have all witnessed something real. Consider, then, one very gentle and elegant move. The paradoxes look unavoidable, cannot be eliminated, because they are "what they always seemed to be, proofs" [Routley, 1979, p. 302]. The conclusions of the paradoxical arguments are *true*.

0.1.2 *There Are Contradictions in Plain Sight in Space*

Paradoxes are not confined to abstruse contemplation of self-referring sentences or vague predicates. They have a simple visual presentation. Here is a circle:

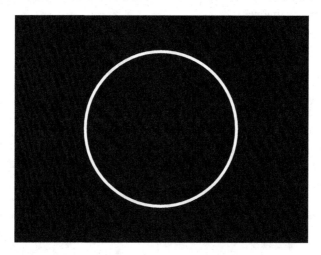

It divides the plane in which it sits into an interior and an exterior. There is a continuous path between any two points in the exterior, and any two points in the interior. But a path from a point in the exterior will not be able to reach a point in the interior without crossing the circle, which forms the *boundary* of the two parts. It looks like the plane is divided

---

[9] See Section 1.1.2, [Priest, 2006b, ch. 1], and book-length treatment in [Beall, 2009].
[10] "There are scarcely any philosophical problems of greater urgency than the liar paradox, for there are scarcely any concepts more central to our philosophical understanding than the concept of truth. ... Quite unmistakably, our present way of thinking about truth and reference is inconsistent" [McGee, 1990, p. vii].

exclusively and *exhaustively*: assuming continuity of the plane, every point is either in the interior or the exterior, and not both. Except – what about the points on the boundary itself?

An even simpler version of this problem is in one dimension. Given a continuous line segment,

it is obvious that the line can be perfectly divided into, or is composed of, two perfectly symmetrical halves.[11] But, as your idealized knife comes down to make the cut, it touches the *center point* of the line. There must be a point there, because the line has no gaps. But points themselves have no extension and so cannot be divided; so your knife must slip either to the left or to the right of the center point. Then the resulting two pieces are not perfectly even: one piece has the center point, and the other does not.

Dividing anything in two – e.g. the left side of a line and the not-left side, the truths and the falsities, the babies and the non-babies – calls attention to the logical assumption of bivalence: that every sentence is either true or false. The implications of this assumption have been well appreciated at the "cosmic" level. According to Gödel, paradoxes are in fact due to the (purportedly mistaken) notion of

dividing the totality of all existing things into two categories. *[Gödel, 1964, p. 519]*

But grand talk about dividing the totality of all existing things,[12] the *universe*, distracts from the fact that we have the very same sort of problem, not at the level of the universe, but at *any* medium-sized object. We do not need special properties of connectedness or closed curves to appreciate a problem here; we just need to look at the sun in the sky (or the moon, since you shouldn't stare at the sun) and wonder how it appears to be a distinct object that is nevertheless smoothly embedded in phenomenal space.

The problem localizes in asking, which portion of reality are your hands, and which portion not? The microscopic particles of matter grazing the surface between your skin and the air are a *question*. There are few things more certain than holding up a hand, and saying "here is a hand," as Moore observed.[13] At the same time, "there are always outlying particles, questionably parts of the thing, not definitely included and not definitely not included" [Lewis, 1999, p. 165]. What looks like a special problem for the universe is a commonplace problem occurring on the end of your arm.

It *seems* like a circle can sit in the plane. It *seems* like there is a universe, in which some things are hands and everything else is not. And – here's the pointy end of the stick – it seems like these things must be able to be true without plunging us into abyssal and vexatious mysteries.

Things are not always as they seem.

---

[11] Gödel once remarked (in an unpublished note; see [Putnam, 1994]) that if a geometric line segment is divided evenly at a point, it would be natural to expect the two halves of the line to be perfectly symmetric mirror images. See Section 9.2.1.
[12] Petersen echoes Gödel's point [Petersen, 2000, p. 384]: "The point is a highly metaphysical one: is it possible, in principle, to divide the world (or the universe, if you prefer) into two disjunct parts, the union of which is the world?"
[13] "Proof of an External World" [1939] in [Moore, 1993, pp. 147–170].

0.1.3 Paradoxes Are Resolved by Reversing the Order of Explanation

Sometimes things are as they seem. Usually, even: accepting things as they are, mostly, is a necessary assumption for getting around in the world. We trust perception, testimony, and the conclusions of informal reasoning as overwhelmingly veridical. Exceptions are exceptional, surprising, unsettling. *Exceptio probat regulam in casibus non exceptis*, my mother always said.

If the first phase of philosophy is to make what is unproblematic into a problem (Cartesian doubt!), then the next phase is to make what is problematic into what is not, to "show the fly the way out of the bottle,"[14] so to speak. There is a place for scientific surprises, but there is also a place for theories that make it possible to understand the world as we find it, theories that do not propound a long story about why everything we seem to see and think is wrong. An account of reality should be available of the reality we live in.[15] Dealing with the paradoxes, it has always seemed to me, should not require an elaborate apology for why the world is not really as it seems.

An impressive practitioner of this method is Richard Dedekind. In 1872, Dedekind famously advanced on the problem of characterizing continuity. In thinking about a geometric line, Dedekind took it as an adequacy condition that the line have no gaps. That is the linear continuum as we find it. And so he proposed, in essence, that *any gap in the line be thought of as itself a point* – indeed that the line is entirely made of "cuts," pairs of sets comprised of everything to the left and everything to the right. Any possible *counter*example is reconsidered as a natural *example*.

Or again, in a magnificent 1888 treatise, Dedekind faced a problem known since Proculus in the third century, that given two concentric circles, a radius cuts each circumference exactly once:

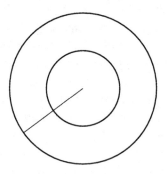

---

14 [Wittgenstein, 1953, §309].

15 "…the world as I found it" [Wittgenstein, 1922, §5.631]. This methodology is also found in Husserl and the phenomenological tradition – placing focus on the world of appearance and experience. See Tieszen, 2005, ch. 2, 3; [van Atten, 2007].

Apparently, then, there are exactly as many points on both circumferences, since the radius establishes a one-to-one pairing of the points on the inner circle and the points on the outer circle. But, even if the set of points on both circles is infinite – *obviously* the outer circle is bigger, so its circumference has more points!? Dedekind's move was to treat this anomaly as a *definition*: a set is infinite if and only if "it is similar to a proper part of itself" [Dedekind, 1901, def 64]. He didn't deny the data, or try to "solve" the problem, or fall back on quietism or eliminativism about circles. The anomaly that infinity is "bigger than itself" (captured in the simple $\infty + 1 = \infty$) is what makes infinity infinite. The puzzle is the answer. He took his task to be to describe the world as we find it – to explain, rather than explain away.

"What we see of the things are the things," writes Pessoa. "Why would seeing and hearing be to delude ourselves / when seeing and hearing are seeing and hearing?" The world I live in has days and nights and forests and cities, flocks of birds and bundles of recycling. There are words for these things, names and predicates, but not *only* words; the world has the things the words name, too. So there are sets, and there are properties, and there are boundaries, and some are vague and some are inconsistent. The *ordinary* leads us to an impasse. Dedekind shows how an impasse can be turned in to a way out.

0.2 The Choices

Given a paradox, there are only three options:[16]

- Reject the reasoning of the argument as invalid
- Show one of the premises is false
- Accept that the conclusion is true

A brief and biased review of how these options play out, say in the case of the liar (Section 0.1.1), is in order. (It would mostly carry over for the paradoxes of set theory, as per the next chapter, too.) This is well-worn ground, and I only intend a thumbnail sketch of some vast and important literature; various distinctions, objections, and replies are omitted.[17]

On some very familiar assumptions, truth is what is the case, every proposition is either true or false, and no proposition is both true and false. There must be a set of *all and only* the true propositions, divided out perfectly from the falsities. This is to be an *exclusive* and *exhaustive* division. Except, again – what about propositions on the "boundary"? This is the liar.

---

[16] Of course, not only three. You could excuse yourself and go for a walk. Or in a more philosophical vein, you could try to undercut the trilemma as somehow misconceived or based on some bad presupposition, like "the paradoxical sentences are meaningful." I will assume that the paradoxes are meaningful problems demanding a direct response; and so in that sense, these other options are not options.

[17] More details are in [Priest, 2006b, ch. 1]; see also opening chapters in [Field, 2008; Scharp, 2013], and many other sources, e.g., [Beall et al., 2018].

0.2.1 Incompleteness

0.2.1.1 The "Classical" Solution

Many different sorts of proposals fall under the heading of "classical" solutions to the liar paradox, but most of them share some key features, around the strategy of imposing (or discovering) that truth is somehow indexed, or stratified, structured in a *hierarchy*.

Classical theories deny, or restrict, the completely unrestricted truth schema $T(\ulcorner\varphi\urcorner) \leftrightarrow \varphi$. This is closely related to Tarski's 1936 theorem, that no (consistent) language can contain its own truth predicate, or be *semantically closed*, on pain of the liar. To spell this out, Tarski gives a construction with an infinite hierarchy of *metalanguages*,[18]

$$\mathcal{L}_0, \mathcal{L}_1, \mathcal{L}_2, \ldots$$

Each language $\mathcal{L}_{n+1}$ at level $n + 1$ can look back at (all) the previous one(s) $\mathcal{L}_n$, but not itself; truth claims are only made about "earlier" sentences, so if φ is a sentence of language $\mathcal{L}_n$, then $T\ulcorner\varphi\urcorner$ is a sentence of $\mathcal{L}_{n+1}$. Self-referential sentences involving the truth predicate are impossible, and so the liar sentence is never formed. The truth schema can hold, but only over a restricted language.[19]

The problems for this approach are basic.[20] First, natural languages *do* contain their own truth predicate. Not only is there no evidence for the existence of any "metalanguages" in natural language, a priori, there cannot be any language that is "beyond" [$\mu\epsilon\tau\alpha$] language. To his credit, Tarski is explicit on this point: natural language *is* semantically closed, and therefore, in his view, mathematically intractable; he abandons trying to analyze the full concept of truth. Later approaches, such as [Kripke, 1975], would purport to be otherwise, but Tarski's approach is not intended to solve the problem. It is to provide a replacement that does not have problems.[21] Second, the hierarchical solution appears to misdiagnose the problem. There are unproblematic cases of self-reference involving truth – such as this (true) one right now – so banning it outright is overkill. More generally, there is nothing particularly ungrammatical or categorically wrong with the liar sentence as such: it seems well formed; it has the right type of subject for its predicate, as opposed to something like "the Pope's chair is a prime number." If one wants to try to block the paradox by restricting the expressive power of the language, it should at least be by a more surgical incision.[22]

The most important problem for any hierarchy, though, is the question: how can a hierarchical theory be *true*, according to itself? True claims must be indexed to some level of the never-ending hierarchy, but the claim "all true claims must be indexed to some level of the hierarchy" cannot be so indexed. In a straightforward sense, then, a hierarchical

---

[18] [Tarski, 1956a].
[19] Increasingly sophisticated formal theories of truth have followed. For the state of the art, see [Halbach, 2014].
[20] For a polemical account, see [Routley, 1979].
[21] Cf. the *inconsistency theory* of truth [Azzouni, 2006; Scharp, 2013].
[22] As Russell wrote of similar such solutions in set theory, they "seem to be created *ad hoc* and not to be such as even the cleverest logician would have thought of if he had not known of the contradictions" [Russell, 1959, p. 61].

theory of truth *is not true according to its own account*.[23] This is an extremely bad feature for a (putatively true) theory of truth to have. Cutting to the chase,

The paradoxical aspect of Tarski's theory, indeed of any hierarchical theory, is that one has to stand outside the whole hierarchy even to formulate the statement that the hierarchy exists. ... The paradoxes themselves are hardly less paradoxical than the solutions to which the logical community has been driven. *[Putnam, 1990, pp. 14, 17]*

The same problem arises with any solution to the paradoxes, via stratification or otherwise, that attempts to block quantification over all truths, all sets, all propositions, etc. The truth schema appears to talk about all propositions at once; but to say, à la Russell, that "No proposition may quantify over all propositions," is self-undermining. The prohibition prohibits itself [Priest, 2002a, ch. 9].

This problem – that the solution to a self-referential paradox is itself self-referentially undermining – is an instance of *revenge*, the observably repeated pattern in which the basic notion that a solution requires is not available by the lights of that very solution.[24] A neat sum of the history:

It slowly dawned on us that it is unbelievably difficult to say anything at all about the aletheic paradoxes without contradicting oneself. *[Scharp, 2013, p. 2]*

Revenge is what makes the paradoxes paradoxes, strikingly different from other types of hard problems: when you try to solve the problem, *you get the same problem back!* You try to block self-reference, and find that you must self-refer to do so. You try to ban universal quantification, and find that you must universally quantify to do so. Some problems are impossible to solve simply due to their complexity. But the liar is not complex. It is the very simplicity of the problem that makes it so troubling; it is the endless resilience of the paradox that makes it so compelling.

Classical solutions would appear to maintain the law of excluded middle (LEM), p or $\neg p$. In this way, they maintain, with respect to Gödel's diagnosis, that for any property φ, the universe can be divided in to two categories, the φs and the $\neg\varphi$s. But this preservation of bivalence is largely an illusion, because the classical position is a *conditional*: *if* there were a universe, then it could be divided in two. But hierarchies can never be finished, never take full stock of themselves, on pain of a liar contradiction; since the classical position eschews all inconsistency, there cannot be a level from which to talk (truly) about all levels. On strictly classical grounds, there is no such thing as universal quantification of an absolutely general sort, because on pain of contradiction there is no completed hierarchy of truths or domain over which to quantify.[25] On pain of contradiction, the classical position reduces to this arresting claim: *There is no universe* [Halmos, 1974, p. 7].[26]

---

[23] Fitch decried the "dim outlook" for philosophical metatheory since Tarski, because "there seems to be no final formal language adequate for dealing with its own semantical concept of truth" [Fitch, 1964, p. 397].

[24] On revenge in general (but in the context of one proposal to deal with the paradoxes), see Priest [2006c, 2007].

[25] See [Rayo and Uzquiano, 2006]. We return to this issue when we get to set theory Section 1.1.

[26] At the close of his charming autobiography *I Want to Be a Mathematician*, Halmos credits to himself the abbreviation "iff" and the little square "tombstone" marker at the end of proofs.

A standard response to the problems I've just indicated, then, is to adopt some kind of "inexpressibility" or *ineffability* thesis. A famous example of this is Wittgenstein at the end of the *Tractatus* distinguishing what can be "said" from what can be "shown." Suggestions along these lines attempt to come right up to, but not cross, the line of what is consistently allowable; they claim we can *intimate*, somehow, the existence of hierarchies or the universe, but without quite *indicating* them, somehow. We try to *show* but not *say*.[27] Apologetics of this sort remain a relatively popular means of navigating around the paradoxes. An ungenerous reconstruction of these doctrines is that some things that we appear to be able to express, and need to be able to express if various kinds of discourse (especially discourse about paradox solution!) are legitimate, are nevertheless inexpressible. This is to deny semantic closure (see Section 0.2.2.2), saving classical logic de jure, but de facto speaking of sentences that cannot be spoken of, or else speaking into an openended universe that does not in toto exist.

This brings us to a second approach, which is more explicit about possible gaps.

0.2.1.2 The "Gappy" Solution (Paracompleteness)

A venerable alternative to the classical solution is a nonclassical one: it is to reject (or not accept) the law of excluded middle. Some sentences are neither true nor false-they are *gaps*. In particular, this is the case with the liar; so the T-schema can be maintained, and while the liar circle $\ell \leftrightarrow \neg\ell$ holds too, it is no contradiction, because in the derivation at the end of Section 0.1.1, the step from the biconditional $\ell \leftrightarrow \neg\ell$ (0.2) to the contradiction $\ell \,\&\, \neg\ell$ (0.3) is blocked.[28] The classical approach, in giving up on a total theory, is in some ways already gappy. The full gap approach says that the liar has no truth value (or demures from asserting that it has a truth value). Logics without the law of excluded middle go back to the 1920s at least. Logics with truth value gaps are called, not entirely happily, *paracomplete*.[29]

The problems with this approach are a bit more subtle, but similar. As in the classical case, the gappy solution is itself inexpressible on gappy terms. The negation of the LEM ("for some p, neither p nor not p") is not, cannot be on pain of inconsistency, a thesis of standard paracomplete logics. The status of the liar sentence for a gap approach is that ℓ is neither true nor false–but "ℓ is neither true nor false" cannot be stated *in* the official gap theory; a gap theory cannot affirm that gaps *are* gaps [Priest, 2006b, 1§3].[30] The basic notion that a solution requires is itself not available by the lights of that very solution. As with Tarski, gap theory appears to need a separate, more expressive *metalanguage* in which

---

[27] It is not an accident that Wittgenstein concludes with a self-contradiction: "Whereof we cannot speak must be passed over in silence" [Wittgenstein, 1922, §7] speaks whereof we cannot; cf. Priest [2002a, ch. 12]. Related to this are attempts to claim that we *can* express the (apparently) inexpressible, but without dire consequences, due to some device such as Russell's 1908 "systematic ambiguity in his theory of types" (cf. Whitehead and Russell 1910).

[28] Although not in *intuitionistic* logic, which maintains reductio, $(p \rightarrow \neg p) \rightarrow \neg p$. See Priest [2008, ch. 6, 7].

[29] For specific use in truth theory (and a different paracomplete logic), a paradigm is [Field, 2008]. The gap approach is different from, but related to, a "meaninglessness"-type diagnosis of the liar. For early work by Routley treating such sentences as "nonsense" with a third truth value, see [Goddard and Routley, 1973].

[30] See also [Beall et al., 2018]. Details change depending on the logic of the implication connective underlying the truth schema, and in particular whether it contraposes; cf. Chapter 10.

to describe gap theory. Again an open-ended "topless" hierarchy appears overhead.

The simple paracomplete approach I've just caricatured, then, faces revenge – as various "gap" theorists are all well aware, going back to Kripke's remark that "the ghost of the Tarski hierarchy is still with us" [Kripke, 1975, p. 714]; contemporary proponents of paracomplete theories will have various detailed answers for the revenge problems. So as remarked previously, I don't intend for the sketch of the debate given here to be complete or decisive. But the outlines of the problem are stark. The gap approach is to separate the space of propositions from two categories to three, so that some propositions are true, some are false, and some are *neither* (*tertium datur* after all). If the original liar was on the border between truth and falsity, then a three-valued approach just has *two* new boundaries at which to strike. This is illustrated by the "revenge liar" sentence, "this sentence is false or has no truth value" – which, if it has no truth value, is true. Alternatively, a single-sorted gap theory does not resort to any metalanguage, and then is simply "quiet" on the paradoxes, which seems closer to the intent, but which is terminal in terms of explaining paradoxes.

One reading of the classical approach, in instituting an open-ended hierarchy of metalanguages, is to see it as urging us to be more careful about distinguishing truth from theories thereof. While *truth* is exclusive and exhaustive, there is no *theory* – no complete and tractable (recursive) description – of truth. The best we can do is approximate. Such is our human frailty. So the classicist and the gap theorist are in some agreement. Our account of truth cannot be complete, or maybe more radically, there is no complete thing called truth to give an account of. These approaches differ mainly on the degree or quality of the limitations. Both then face a similar sort of revenge.

0.2.1.3 A Dilemma

In summary, the project of describing the world in ways where a truth predicate neatly separates everything exclusively and exhaustively appears to be impossible. There is an apparently intractable dilemma:[31]

- A theory of *only* truths will leave some truths out. *(incompleteness)*
- A theory of *all* truths will let some falsities in. *(inconsistency)*

In inquiry, we may avoid all falsity, or seek every truth, but we cannot do both. There is *no* way out of the paradoxes that leaves untouched everything one may have initially wanted, and yet removes everything one may have initially not wanted. The paradoxes are *paradoxes*.

---

[31] Priest sets up a trilemma, by distinguishing *incompleteness* approaches (something is not assigned a truth value by the theory) versus *inexistence* approaches (there's nothing there to assign a truth value to) [Priest, 2007], and elsewhere deals with the related ineffability approach. For purposes here, any "ineffability"-type solutions (that something apparently sayable cannot be said) would, for these purposes, fall in the "incompleteness" category.

0.2.2 Paraconsistency and Dialetheism

The *paraconsistent* approach to the liar paradox and similar problems in set theory (Chapter 1) is to let go of the first horn of the preceding dilemma, and take the second: to let some falsity in, by not assuming that all contradictions are absurd. We may be here in the presence of a contradiction, but not necessarily an absurdity. The T-schema is true; the derivation of the liar is valid; a contradiction is true; life goes on. In particular, paraconsistency rejects that if one contradiction is true, then *everything* is true. The step from some contradiction to any arbitrary conclusion is called *ex contradictione quodlibet* or *explosion*:

$$\varphi, \neg\varphi \therefore \psi,$$

with "$\therefore$" for the moment read as "therefore." This is valid according to classical (and other) logic(s), and says that any arbitrary contradiction entails any arbitrary conclusion whatsoever. Minimally, in paraconsistent logics explosion is generally *invalid*.[32]

In the derivation of the liar in Section 0.1.1, nonparaconsistent logics will go from (0.4), $\ell \ \& \ \neg\ell$, to a next line,

$$\bot. \tag{0.5}$$

The symbol "$\bot$" in logic represents the worst thing that can happen; put more optimistically, it is something that *can't* happen. Everything whatsoever follows from $\bot$,

$$\bot \therefore \psi.$$

A theory is *trivial* if it contains every sentence; a trivial theory is absolute absurdity; a theory with $\bot$ is trivial.[33] Paraconsistent logics deny that every contradiction leads to $\bot$. Paraconsistency says that not everything is true, even if some contradictions are.

Classically, any inconsistent theory is trivial – absurd. Paraconsistently, a theory may be inconsistent but nontrivial. Nontriviality, or *coherence*, then, often replaces the notion that would have been played by consistency for nonparaconsistent reasoning, e.g., asking whether a proposal makes a minimum amount of sense.[34] Classically, a contradiction is absurd, so any contradiction is the worst thing that can happen. Paraconsistency can be taken as the doctrine that a contradiction is not always the worst thing that can happen. Since contradiction does seem to be the sort of thing that happens, it is unhelpful to panic or accept rational chaos when it does; just consider the damage of an explosive inference, like "War is wrong. Therefore, if we go to war, it is okay to bomb civilians."[35]

Dialetheism goes one step further. According to this thesis, some propositions of the form $\varphi \ \& \ \neg\varphi$ are true; some true propositions have true negations. There are *gluts*. And

---

[32] For a good critical exposition and discussion of "paraconsistent dialetheism," see Field [2008, chs. 24–27].

[33] Assuming the transitivity of logical consequence, which I do, though see [Ripley, 2013].

[34] The distinction is sometimes glossed as *simple consistency* (consistency) versus *absolute consistency* (nontriviality) [Routley et al., 1982, ch. 1].

[35] The example is due to Alan Weir [Weir, 2004, p. 406].

this is not to be read in any sort of fanciful or speculative sense; dialetheism holds that some contradictions are true *in the actual world*.[36] Nonparaconsistent logics in turn have consequence relations that are not truth preserving; the argument from φ, $\neg\varphi$ to ψ may have true (as well as false) premises, but a false (and not true) conclusion. A true contradiction is a counterexample to the validity of *ex falso*. If dialetheism is right, a paraconsistent logic is required, as appropriate to the nature of truth.

A *theory* is a set of sentences, including all logical consequences. Some theories are attempts to describe the world; I'll say that a theory that succeeds in this attempt is a *true* theory. Paraconsistency is about theories that are inconsistent, containing both some sentence and its negation, but not incoherent. A dialetheic or *glut-theoretic* approach takes it that some paraconsistent theories are true – that there are entities in the world that are best described, or can only be described, by inconsistent sets of sentences. In this sense, such an approach holds that the world is contradictory, that ours is an inconsistent world.

In the following sequence of subsections, I will highlight some of the main features of the kind of dialetheic paraconsistency I am endorsing. This is mainly to set the scene, not to argue points.

0.2.2.1 Revenge

Dialetheic paraconsistency, prima facie, does not face the same revenge problems that other theories do. (Whether there is some other revenge lurking, after all, we will return to in Section 3.3.2 and Chapters 4, and 10.) The reason incompleteness approaches (classical, paracomplete) suffer from revenge is that they are trying to avoid inconsistency; then, in expounding the means by which inconsistency is to be avoided, they stray into the inconsistent. The dialetheic program is not aiming for consistency. It explicitly accepts some contradictions. When the inevitable contradictions come, then, there is no blowback. We "do not need to keep running through richer and richer meta-languages in order to chase our semantic tails. ... We embrace some contradictions in the semantics, and get it all from the start" [Shapiro, 2002, p. 818] (noting that Shapiro is not himself a dialetheist).

Try to avoid contradicting yourself, and you contradict yourself.[37] First, then, this is itself an argument for dialetheism – a kind of reductio, that nondialetheic approaches lead to dialetheism anyway.[38] Second, apparent immunity to revenge puts dialetheism in a singular position to reach for universal theories, as I now emphasize.

---

[36] A related position is that there are *some* inconsistent worlds out there in modal space, but maybe not ours; see [Martin, 2014]. I intend this book to be about the (actual) world; but if you prefer to read this as a story about some other world, just to see what it is like, then you can.

[37] Whitman: "Do I contradict myself? / Very well then I contradict myself / (I am large, I contain multitudes)."

[38] Methodologically, the form of the argument could be construed as a disjunctive syllogism (see Section 0.2.2.3): either the classical, gappy, or glutty position is correct; but the classical and gappy theories are not, so .... Disjunctive syllogism is not valid according to paraconsistent logic, as we will see. But this is not the form of argument I am using. Rather, it is an argument by cases: either the classical, gappy, or glutty position is correct; but the classical and gappy approaches both lead to gluts anyway, so the glut position is correct no matter what.

0.2.2.2 Closure

A theory with an unrestricted T-schema is semantically closed (Section 0.2.1.1) or *universal*.[39] More generally, I will call a theory *closed* if it is in a language that expresses unrestricted notions (such as truth, validity, membership, provability, etc.) that are crucial for setting up the theory itself, and can validate what is true of those concepts. A closed theory of truth expresses (truly) everything that is (truly) expressible. As Fitch stated:

> The way now appears open to construct formal languages that can adequately deal with their own concepts, and we are no longer forced to try to climb either an unending topless ladder of formal metalanguages, or a ladder of formal metalanguages that ends with natural language at the top. *[Fitch, 1964, p. 397]*

A theme of logic in the twentieth century was to deny closure, in the form of the incompleteness theorems.[40] For we all know what happens when mathematical practice is reconstructed using classical logic: Hilbert's program fails.[41] In thumbnail, Hilbert's program conjectured a purely *formal*, mechanical proof system that can *completely* account for all of mathematical truth, in the sense that for any sentence in the language, either it or its negation will be a proven theorem, and there is a procedure for checking which of the two it is.[42] And yet, we revisit a familiar site:

Gödel's Theorem [1931] Let φ be a sentence in the language of arithmetic. The following conditions are jointly inconsistent:

(1) Either φ is provable, or $\neg\varphi$ is provable.
(2) Never are both φ and $\neg\varphi$ provable.
(3) The proof relation is computable.

The proof: the existence of an arithmetic version of the liar sentence, "This sentence is not provable in arithmetic." The standard gloss is to say that no computable theory will ever deliver all the truths of, e.g., arithmetic – that axiomatic theories always fall short of the world.

But this is a nonsequitur. We have *no reason* to think that there is a mismatch between these theories and the world. What we know is this: Gödel proved that any *consistent*, tractable account of the world (assuming that arithmetic is in the world) falls short. Received wisdom ignores the possibility that the Gödelian facts simply leave us with a failed hypothesis [Routley and Meyer, 1976] *that the complete theory of the world is consistent*. The only possibility for a complete and tractable theory is that it be inconsistent.[43] A dialetheic paraconsistent approach does not dispute the Gödelian facts, but by adopting a different logic, it can draw a different lesson from them. To put it a bit

---

[39] Tarski and later [Fitch, 1964] call it "universality." See Routley [1979, p. 322]; Mares [2019]. The term *semantic* closure is specifically about capturing truth [Priest, 2006b, ch. 2]; closure (simpliciter) is a natural generalization on the same move.

[40] A careful tour of the standard material is [Smith, 2007].

[41] Although see Franks [Franks, 2009].

[42] In echo of Hilbert's call, Myhill: "It will never be possible to acquire rational knowledge of the truth of any mathematical proposition without at the same time acquiring rational knowledge of its provability" [Myhill, 1960, p. 467].

[43] This interpretation of Gödel's theorem is implicit in [Priest, 1979] and explicit in [Beall, 1999]; cf. [Meyer, 1996].

glibly, if it is impossible that there be a theory of everything, then the theory of everything will be an impossible theory.

Not to say that inconsistency will guarantee a successful "theory of everything," but it is a prerequisite – a necessary precondition for the possibility, to put it in Kantian transcendental terms. If the theory is closed, it is inconsistent. The incompleteness solutions face revenge because they require more expressive power than they themselves allow. Attempts to truly place limits on a truth predicate require appeals to truth that outstrip those limits. A "naive" truth theory rejects the Tarskian infinite hierarchy of metatheories. Instead, closure is made an adequacy condition. If the task is to chart out the metaphysical universe of truths, and to do so in a mathematical way, then those charts are themselves mathematical objects and what they say should be true; so the charts themselves should be on the chart. The spurious twentieth century object-language/metalanguage distinction dissolves, and we are free to try again.[44]

Routley extols the benefits of pursuing closure via dialetheism:

The liberating effect of giving up the classical faith ... is immense: ... one is free to return to something like the grand simplicity of naive set theory, to semantically closed natural languages (having abandoned the towering but ill-constructed and mostly unfinished hierarchies of formal languages), and to intuitive accounts of truth, of proof, and of many other intensional notions. *[Routley, 1979, p. 302]*

Nietzsche said that for Socrates, "to be beautiful everything must be intelligible."[45] For everything to be intelligible, some things must be inconsistent.

0.2.2.3 Logic

A full presentation of the logic I plan to use for mathematical proofs in this book will come later (Chapter 4), but we need to have some details in view now. With t, f standing for truth values, conditions on extensional logical connectives (conjunction &, disjunction $\vee$, negation $\neg$) and quantifiers ($\forall, \exists$), following the logic "logic of paradox" LP [Priest, 1979], are as given in Box 0. These conditions say how to compute truth values from some given base set of atomic values.

Read aloud, these are completely familiar semantics: a conjunction is true iff both conjuncts are, and so forth. What may be unfamiliar is that the conditions are being stated in a single sorted language, where we do not presume that "and," etc., are classical. Indeed, if one endorses dialetheic paraconsistency, the locution "p is true" is compatible with "p is false" and "p is not true." Orthodox logical training might incline you to read the conditions as exclusive, where sentences take only one value, but nothing on the page makes it so.[46] The copula – the "is" of predication – is not univocal in general, and it is not here.

---

[44] Meyer: "One aspect of a familiar myth must go; namely, that Gödel has consigned us to a winding staircase of increasing epistemological uncertainty about mathematics, each level of which, though it may appear sound from above, seems shaky when on it; and where one may always dread that, climbing as we do through the dark, the next level will lead directly into the void" [Meyer, 1976, p. 11].

[45] In *The Birth of Tragedy*. Cf. Plato in *The Republic* 507b–509b, 509e–511d.

[46] See [Weber et al., 2016] and [Sylvan, 1992].

Box 0 **Truth conditions for logical connectives.**

$$\neg\varphi \text{ is t} \quad \text{iff} \quad \varphi \text{ is f}$$
$$\neg\varphi \text{ is f} \quad \text{iff} \quad \varphi \text{ is t}$$

$(\varphi \,\&\, \psi)$ is t iff $(\varphi$ is t) and $(\psi$ is t)	$(\varphi \vee \psi)$ is t iff $(\varphi$ is t) or $(\psi$ is t)
$(\varphi \,\&\, \psi)$ is f iff $(\varphi$ is f) or $(\psi$ is f)	$(\varphi \vee \psi)$ is f iff $(\varphi$ is f) and $(\psi$ is f)

$$\varphi \text{ is t} \quad \text{or} \quad \varphi \text{ is f}$$

$\forall x\varphi(x)$ is t iff for all $x(\varphi(\ulcorner x\urcorner)$ is t)	$\exists x\varphi(x)$ is t iff for some $x(\varphi(\ulcorner x\urcorner)$ is t)
$\forall x\varphi(x)$ is f iff for some $x(\varphi(\ulcorner x\urcorner)$ is f)	$\exists x\varphi(x)$ is f iff for all $x(\varphi(\ulcorner x\urcorner)$ is f)

(where $\ulcorner x\urcorner$ is a name for each x)

The semantics permit

$$(p \text{ is t}) \,\&\, (p \text{ is f})$$

for some p. And notably, the clauses for universal quantification $\forall$ are restored to their natural state – it is true that everything is φ, just in case, for every thing, it is true that thing is φ – without anxiety over the existence of a universal domain.[47] We will return to these issues in Chapter 3, and I will defend specifics of the semantics in Chapter 10.

The final propositional clause, bivalence, codifies that every proposition has at least one of the two truth values, and with it the law of excluded middle (LEM). Unlike the other clauses, this is not a (bi)conditional assertion; it is a bare descriptive statement of fact.[48] I do flag immediately that the LEM is not the same as assuming that for every proposition p, either p is true or else p is *absurd*. Assuming that would be wrong because falsity can be distinct from absurdity. ("Pluto is a planet" is false [since 2006]), "Pluto is a small family of factory workers made of broccoli" is absurd.) Negation according to these semantics does validate, along with the LEM, *double negation* laws,

$$\neg\neg p \qquad \text{iff} \qquad p$$

and *de Morgan laws*,

$$\neg(p \,\&\, q) \quad \text{iff} \quad \neg p \vee \neg q$$
$$\neg(p \vee q) \quad \text{iff} \quad \neg p \,\&\, \neg q,$$

so it is sometimes called de Morgan negation.

For a set of atomic propositions, an *assignment* or *valuation* is a way of relating every member of the set to at least one of the two truth values t, f. The atomic values are extended

---

[47] We do need to assume that each thing has a name, but this can be done by either taking everything to be its own name, or using a freely available naming device from (naive) set theory, as in Section 1.1.2.1.

[48] The LEM will be used without restriction or even much comment. (Most works in which the law of noncontradiction is assumed would not usually offer any justification for the assumption.) In terms of arguing for the LEM, I doubt it can be done directly in a non-question-begging way. Priest offers a "teleological" argument for LEM [Priest, 2006b, §4.7], but later accepts criticisms that it is circular [Priest, 2006b, p. 267].

to all formulas via Box 0. Then the understanding of *validity* is a minor tweak to the usual. Where an argument is a collection of premises and one conclusion,

Valid: An argument is valid iff on any assignment of truth values, when all the premises are *at least* true, the conclusion is *at least* true.

Derivatively, a *theorem* is the conclusion of a valid argument with no premises: a proposition that is always at least true. In the negative direction,

Invalid: An argument is *invalid* if there is an assignment of values where all the premises have some truth, but the conclusion has no truth at all.

In nonparaconsistent logic, any argument with inconsistent premises is valid *(ex falso quodlibet)*, because inconsistent premises cannot be true. That is not the case here: if φ is both true and false, but ψ is not true,

$$(\varphi \text{ is t}) \,\&\, (\varphi \text{ is f}) \,\&\, \neg(\psi \text{ is t}),$$

then $\varphi, \neg\varphi \therefore \psi$ is invalid (again for now using $\therefore$ as "therefore").

What about a conditional, some connective "$\rightarrow$" for implication? Pinning down a decent conditional will take all of Chapter 4. For now, at least this much is taken for granted. Modus ponens is *valid*:[49]

$$p, p \rightarrow q \therefore q.$$

Now, the standard conditional of classical logic is $p \supset q := \neg p \vee q$, the *material conditional*. But *disjunctive syllogism*

$$p, \neg p \vee q \therefore q$$

is *invalid*, because arguments that can have true premises and a false conclusion are not valid. (If $p, \neg p$ are both true, but q isn't true, that is a counterexample to disjunctive syllogism.) Disjunctive syllogism would render the logic nonparaconsistent, restoring explosion $p, \neg p \therefore q$, so it is invalid on pain of triviality. The proof is due to C. I. Lewis:

1	p	(premise)
2	$\neg p$	(premise)
3	$\neg p \vee q$	(from (2), $\vee$-introduction)
4	q	(from (1) and (3), disjunctive syllogism).

Line (3) follows from (2) just by the preceding truth conditions for a disjunction; so if this derivation is to be halted, it must be at the last step. While disjunctive syllogism was one of the Stoics' "Five Indemonstrables," denying it is the prerequisite of paraconsistent research; and "we suppose that it is better to deny an Indemonstrable than a Demonstrable" [Anderson and Belnap, 1975, p. 488].

---

[49] "Modus ponens is a sine qua non of any implication connective" [Priest, 2006b, p. 86].

Putting the validity of modus ponens together with the invalidity of disjunctive syllogism, a conditional $\rightarrow$ cannot be the *material* conditional, $\supset$. What $\rightarrow$ *is*, again, remains to be said later. For now, we note some further basic constraints, assuming a conditional $\rightarrow$ does satisfy modus ponens:

- All the conditionals we will consider obey reflexivity, $p \rightarrow p$, and transitivity: if $p \rightarrow q$ and $q \rightarrow r$, then $p \rightarrow r$. (The material "conditional" does not obey transitivity.) (Counterexample: letting p be true, q be both true and false, and r be false.)
- To maintain paraconsistency, a conditional may contrapose, or weaken, but not both. Suppose both. Weakening is $p \rightarrow (q \rightarrow p)$. From that, by contraposition (and transitivity), $p \rightarrow (\neg p \rightarrow \neg q)$. Then $p, \neg p \therefore \neg q$ is valid for all q, by modus ponens. With double negation elimination, pick $\neg q$ for q, and that's full explosion.
- The notion "p implies absurdity," or $p \rightarrow \bot$, which does not obey LEM as before, does deliver a form of explosion,

$$p, p \rightarrow \bot \therefore q,$$

and so a form of disjunctive syllogism,

$$p \vee q, p \rightarrow \bot \therefore q,$$

which can be useful. So various "classical" behaviors of negation – namely, exclusion – are still available, just not all packed together in one logical connective.

More about conditionals to come.

Of course, these are all in some sense *choices*, and for every option (drop disjunction introduction instead of disjunctive syllogism; drop modus ponens) there is at least one reasonable logician who has endorsed or explored that option. Walking through all those options would be a book in itself (cf. [Humberstone, 2011]). Instead, I will show in detail what happens when one follows *one* such suite of choices where it leads.

0.2.2.4 Truth and Counterexamples

An immediate consequence of dialetheism is arrestingly simple. Dialetheism, the claim that some contradictions are true, is itself false.[50] In light of its own logic, dialetheism itself is a dialetheia [Priest, 1979, p. 238]. From the law of bivalence and de Morgan laws, all contradictions, even the true ones, are false:

$$(\varphi \text{ is t } \& \varphi \text{ is f}) \text{ is f.}$$

We agree, then, with Leibniz: "We judge to be false anything that involves contradiction, and as true whatever is opposed or contradictory to what is false" [Leibniz, 1714, p. 31]. The law of non-contradiction, $\neg(\varphi \& \neg\varphi)$, is a theorem of our logic. All its instances are true. It will also have false instances.

---

[50] "The subject of paradoxical assertions is one full of surprises. However, that it should be so is not particularly surprising" [Priest, 1979, p. 240].

It is important to see that dialetheism *cannot* be made into a consistent theory. Contradictions are always false (at least); true contradictions are falsities that are *also* true. Any theory which contains contradictions will contain falsities:[51]

Truth and falsity come inextricably intermingled, like a constant boiling mixture. One cannot, therefore, accept all truths and reject all falsehoods ... *[Priest, 2006b, p. 100]*

The question then is which way we incline – toward accepting all truths (and so some falsity), completeness, or rejecting all falsehoods (and so some truths), consistency.

As may be becoming evident, dialetheic paraconsistency recasts the force of counterexamples. The existence of a counterexample does not always rule out a law. The law of noncontradiction holds, always; but it does have some exceptions, too. This on its own may seem to rob counterexamples of all their (important) force in our reasoning. That would be an overreaction, though; for dialetheists, counterexamples continue to play a crucial, just more nuanced, role in developing theories. (See [Priest, 2006a] and Chapter 3.) Nevertheless, to ensure that we *retain* enough force from counterexamples, the notion of *in*validity encodes the venerable idea that if the premises of an argument are (or can be) true, but the conclusion is (or can be) false, then argument is invalid. For the conditional, this will trickle down as a *counterexample* law,

$$p \mathbin{\&} \neg q \mathbin{\therefore} \neg(p \rightarrow q),$$

which we will endorse. Similarly, as with negation, the usual quantifier dualities hold – some x is φ iff not all xs are not; all xs are φ iff no x is not. So we have that $\forall = \neg\exists\neg$ and $\exists = \neg\forall\neg$. But, both

$$\forall x \varphi x \qquad \text{and} \qquad \exists x \neg \varphi x$$

can hold at the same time, in the same way (and then by the usual rules, $\neg\forall x \varphi(x)$ and $\neg\exists x \varphi x$ would both hold, too). The Cartesian method is relaxed: some certainties are dubitable, just as some truths are also false.

There will be more to say about counterexamples and invalidities at the end of the book (Chapter 10), when we have more concrete experience working with the notions to base the discussion on. For now, we have the scaffolding of dialetheic paraconsistency (as I understand it) in place, and some common language is established. To close off this section's survey of various choices of logical approaches to the liar, I will say something about how I see the role of *logic* in any "solutions" to the paradoxes.

0.2.3 The Role of Logic and the Role of a Solution to the Paradoxes

According to Frege,[52] logic is about the *laws of truth*:

Being true is different from being taken to be true, be it by one, be it by many, be it by all, and is in no way reducible to it. It is no contradiction that something is true that is universally held to be false.

---

[51] Cf. Tarski [1944, p. 368].
[52] And see the introduction to [Smith, 2012].

By logical laws I do not understand psychological laws of taking to be true, but laws of being true. If [something] is true ... then it remains true even if all humans should later hold it to be false. *[Frege, 1903a, p. xvi]*

On this view,[53] logic is *not*, primarily, about *reasoning*, nor about the vehicles of reasoning, thought and language. The laws of truth of course have something to do with describing good reasoning (in thought or language), but only in the way that pure mathematics has something to do with international finance. Prime numbers apply to cryptography, but are not *about* or beholden to cryptography. Logic is about the rules for how truth and falsity of different propositions relate.

That is important, because if the *only* desiderata on, say, a naive truth theory is that it saves the T-schema, then it is very easy to produce a naive truth theory.[54] There are at least two elegant approaches: embed in the trivial logic (every sentence is a theorem, every argument valid); or the null logic (no theorems, no validities), which has the extra feature of consistency. Neither of these will do, it goes without saying, because there is another desiderata on a naive truth theory. Not only must it be naive (having a full truth predicate) and be a theory (some but not all arguments are valid), but it must be about *truth*. A theory of truth must give the best description or model we can manage of truth, including logical truth. This is why approaches that cannot, by their own lights, be true (e.g., the hierarchical approach Section 0.2.1.1) seem inadequate.

How, then, does logic relate to dealing with the paradoxes? We know that changing our theory of logic changes the impact of the paradoxes. But getting clear on a suitable logic is only the first step. It simply puts out the fire, stops the bleeding. Determining which steps in paradoxical arguments are wrong is necessary, but the solution to the paradoxes – or really, their resolution – is found in *applying* an appropriate logic to describe the space paradoxes inhabit. If we think that a lesson of (some) paradoxes is that there are true contradictions, for example, then negation is dialetheic, but dialetheic negation is not the "solution" to the liar, any more than a ruler and compass are the solution to a geometry problem. Finding the surrounding picture that makes sense of the liar is the solution. The paradoxes are not ultimately about reasoning, or even the laws behind good reasoning, the laws of truth. The paradoxes are *about* things like the foothills of mountain ranges, and bunches of grapes, and twilight – they are about the *world* – and in the project I am setting out, their resolution is to be found in *using* logic to describe and understand the world.

0.3 Prospectus: Fixed Points

Let's look ahead now.

Logicians for over a century have thought that the danger posed by inconsistency is *explosion, ex falso quodlibet*. Paraconsistent logic neutralizes that threat. But we've all had

---

[53] I leave aside that a few lines after the preceding quote, Frege considers what to make of people who reason differently than classical logic: "I would say: here we have a hitherto unknown kind of madness."

[54] "It is not at issue that we can devise formal theories that are consistent, or even provably consistent ... It is disturbing to see how many logicians think that the problem has been solved once some formal construction, which is (putatively) consistent, has been given" [Priest, 2006b, p. 9].

it wrong. The real trouble, the threat that will eventually preoccupy us, is that contradictions lead to *implosion*. Let me briefly explain this gnomic threat.

A good first observation about many of the paradoxes is that they involve self-reference. As an attempt at a diagnosis, this does not cut at the joints of the problem, as we've already noted, but it can be made more precise. The paradoxes involve *fixed points*. For example, the liar is a fixed point on negation, $\neg\ell \leftrightarrow \ell$. Negation, over the space of propositions, divides the true from the false but has a self-intersection, a fixed point. A fixed point on an operation is an object that is the same input as output. For any function f, some x is a fixed point on f if and only if

$$x = f(x).$$

The identity operator, for example, has *everything* as a fixed point: everything is identical to itself. It is not too much of a stretch[55] to read Descartes as using a doubt operator, "I doubt(x)," and his cogito as fixed point for it:

I doubt($^\ulcorner$ I am $^\urcorner$) iff I am.

As Archimedes said, "Give me a fixed point and I can move the world."

Fixed points are double-edged. They are stable, and remarkably useful, appearing everywhere from thermodynamics to economics. Fixed point constructions are the standard method for "solving" the paradoxes, starting from [Kripke, 1975]. But sometimes fixed points are *too* stable. A fixed point is when an operation coincides with identity, and some operations seemingly *never* coincide with identity. Negation, for example, *seems* to be such an operation. Paradoxes arise when fixed points appear that, prima facie, shouldn't be there, where none can be.

We ask, why are there paradoxes? The question is now restated, why are there fixed points? Here is the start of an answer. In the sort of closed system I've begun clearing the ground for, there is no metalanguage or "typing," because the system is intended to be universal (Sections 0.2.2.2 and 3.1.3.1), about *everything*, things qua things. At this level of generality, there is only one sort of object: objects. Relations and mappings are objects; so some objects are also relations. A clear and early example of this type of system is the *untyped λ calculus*, due to Church, which abstracts on the idea of functions. In this calculus, there is no distinction between operations and things they operate on; every object can be applied to any other, without restriction.[56] And in the untyped lambda calculus, one can prove (entirely orthodoxly) that *every function has a fixed point*. In a slogan: "there is no term that is never the identity" [Odifreddi, 1989, p. 82].

To capture that terms $F, G, \ldots$ can be both operations and inputs of operations, with no type distinctions, I will simply call them *objects*. Here is the (abstract) theorem.[57]

---

[55] Barwise and Moss similarly gloss the cogito as a *circular* phenomenon [Barwise and Moss, 1996]. Treatment of the cogito as a full diagonal, not just left to right, is found in [Boos, 1983, p. 285].

[56] Notation: Whitehead and Russell [Whitehead and Russell, 1910] wrote "$\hat{x}.\varphi(x)$" to mean "the set of all x such that $\varphi(x)$"; Church moved the hat over to the left and wrote "$\lambda x.\varphi(x)$" for something like "the operation on x such that $\varphi(x)$." For an excellent resource on the λ calculus, see [Hindley and Seldin, 2008].

[57] Going back to Turing and other early papers in recursion theory – Kleene's "second recursion theorem" in 1938, from Kleene and Rosser's 1935 inconsistency proof. See Kleene [1952, ch. 12]; Barendregt [1984, p. 24]; [Moschovakis, 2010].

Fixed Point Theorem (1936):   *For all F, there exists an X such that*

$$FX = X.$$

Proof There is an object Δ that applies any object t to itself,

$$\Delta t = tt.$$

When applied to itself, Δ self-reproduces:

$$\Delta\Delta = \Delta\Delta.$$

Then take any F and apply it to this self-reproductive operation to form the new operation Δ_F,

$$\Delta_F t = F(tt).$$

Letting $X = \Delta_F \Delta_F$, then

$$X = F\Delta_F\Delta_F$$
$$= FX,$$

showing that X is a fixed point for F. That's it.   □

This theorem has some elegant generalizations:

1. For all F and G, there are X and Y such that $X = F(XY)$ and $Y = G(XY)$.
2. For all $F_0, \ldots, F_n$, there exists $X_0, \ldots, X_n$ such that $X_0 = F_0(X_0, \ldots, X_n)$ and ... and $X_n = F_n(X_0, \ldots, X_n)$
3. (Turing 1937) There is a $\mathcal{Y}$ for all F, the fixed point combinator, such that

$$\mathcal{Y}F = F(\mathcal{Y}F).$$

This theorem also has some dire consequences. It resulted, in 1942, in *Curry's paradox* [Curry, 1942], the most diabolical of all the paradoxes, which renders Church's original foundational system incoherent, and makes the naive closure project almost completely intractable, about which more later (Chapter 4).

Generally, any space has the *fixed point property* (Section 9.3.2) iff *every* operation (of a certain kind) f on that space has a fixed point. What the preceding theorem foreshadows is that, in a completely general closed framework, the universe $\mathcal{V}$ has the fixed point property (according to Theorem 13 to be proved in Chapter 5) – *every single map* from $\mathcal{V}$ to $\mathcal{V}$ has a fixed point, without restriction. The proof is more or less the one we've just seen. A completely untyped universe – as ours would seem to be – will, according to this theorem, have the fixed point property.

That's all so much syntax in the void, though, until we think about operations that really don't seem to have fixed points. Take something like the *blip function* (cf. Bell, 2008, p. 5). This is an operation that takes every number x to 0, except when x is 0:

$$f(x) = \begin{cases} 1 & \text{if } x = 0; \\ 0 & \text{otherwise} \end{cases}$$

The function "blips" up to 1 at 0. Can the blip function have a fixed point? Suppose some $z = f(z)$. If $z = 0$, then $0 = f(0) = 1$ by definition of f and the assumption that z is a fixed point; but if $z \neq 0$, then $z = f(z) = 0$ so $z = 0$ after all. Either way, apparently,

$$0 = 1.$$

No one, dialetheist or otherwise, is going to accept that $0 = 1$. It's an absurdity, assuming that our universe is such that there are at least two objects that are in no way identical. To make this very stark, just take a space with at least two occupants, call them t and f, and the function $\neg$ that swaps them, $\neg(t) = f$ and $\neg(f) = t$. If *every* mapping on that space has a fixed point, then so does $\neg$, so either $t = \neg(t) = f$ or $f = \neg(f) = t$. In either case, by transitivity of identity, $t = f$. If *every* mapping on a space has a fixed point, then the only thing that space could be, apparently, is a single point: implosion.

This leaves us with a decision to make. The "classical" reaction is that, if $\mathcal{V}$ is a structure for which *every* operation has a fixed point, then that structure can *only* be a point; and since our world is not a point, our world is not $\mathcal{V}$; there is no $\mathcal{V}$. A different reaction is that $\mathcal{V}$ is *our world*, and our world is not (it seems) a point; then our task is to come to grips with the existence of fixed points. A start, but only a start, is to accept that our world can be both inconsistent and yet still viable. Even then, it needs to be shown how the global fixed point property does not devastate our life projects. For make no mistake – even in inconsistency-tolerant contexts, not just any old nonsense is allowed. That is why, after all, we bother with "paraconsistent" logic at all. Unless Parmenides and other monists are right, our universe is not a structureless point. And yet, it appears that on a sufficiently comprehensive accounting of our universe, under closure, it does sustain fixed points, paradoxes. Explaining why there are paradoxes boils down to explaining how this could be.

Explaining how this could be will involve, eventually, working through the ideas that support perhaps the loveliest fixed point result of all: *Brouwer's fixed point theorem*. The theorem states that no matter how you spin, twist, or rearrange the points on a disc (including the points on its boundary), as long as the rearrangement "makes no jumps," then some point does not move. If you are holding a paper map of a region where you are, and you crumple up the map and drop it, then by Brouwer's theorem some spot on the map is exactly above the location it is a representation of. And in a way, that is why, to put it picturesquely, the universe $\mathcal{V}$ has the fixed point property: if we try to devise a map, a system that mathematically represents $\mathcal{V}$, and do so from within $\mathcal{V}$ (as we must), then some "spot" on the representation is of itself.[58]

A special case of Brouwer's fixed point theorem is visible in the *intermediate value theorem*, which says that you can't get continuously from one side of the room to the other

---

[58] A protracted meditation on this theorem is found in [Lawvere and Schanuel, 2009]. For more on fixed points, see [Smullyan, 1991] and [Humberstone, 2006]. For proof of the fixed point theorem using diagonalization via a list of all terms, see Odifreddi [1989, p. 152].

without crossing the middle (no teleportation!); for a fixed point, consider the diagonal *identity line*:

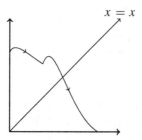

To get from the y-axis to the x-axis, you have to cross the diagonal; the curve describing your path must have a fixed point.

This simple fact is what is driving the paradoxes: you can't be a baby and then later not a baby without the change occurring somewhere. No continuous curve across the unit square is never the identity. Everything has a fixed point–and yet our world is not a point. Proving Brouwer's fixed point theorem (or suitable modifications thereof) using paraconsistent mathematics all the way is the touchstone for trying to explain how the universe may support paradoxes without collapse, and is the final technical target of the book.[59]

0.3.1 *The Plan from Here*

This book is about paradoxes. We have known about them for a very long time. The purpose ahead is to interpret the paradoxes as describing legitimate (mathematical) objects, to read the metaphysics right off the formalism, to glimpse how it is that the world can sustain these objects.

A paradox, etymologically, is a *surprise* [*para/doxos*, παραδοξος, "beyond belief"]. In the case of genuine paradoxes, the surprise never goes away. "Revenge" is when, after an apparent solution to a paradox, the surprise sneaks up on you from a different direction. An explanation of the paradoxes, an answer to why there are paradoxes, then would tell us why the paradoxes are surprising. And if it is a faithful explanation, one that does not deny the data, then even after learning why there are paradoxes, the surprise – the *awe* – will remain. "Wisdom begins in wonder," says Socrates; and maybe wisdom remains in wonder, too.

A different sort of project is trying to "solve" the paradoxes by removing the surprise. This is what a lot of people seem to want: "See, we never should have *expected* such and such, so the fact that it leads to contradiction is in retrospect no surprise." If the phenomenon of revenge is real, then those projects are hopeless. They always lead to new but really the same surprises. I don't say that we never should have expected a solution to the paradoxes; it would have been nice, like it would have been nice (but impossible) to be able to tune

---

[59] According to Terui [Terui, 2014], Brouwer's fixed point theorem is *equiconsistent* with naive set comprehension in various nonclassical logics – showing it is true is exactly to show that comprehension is not absurd.

a piano perfectly[60]; but that's not the world we live in. The fact that purported solutions always lead to revenge is an ongoing source of surprise.

This book does not offer a solution to the paradoxes; there is no such thing. It seeks resolution via explanation. The paradoxes are surprising because they point out impossibility in the everyday. They are found in the utterly ordinary, and so nothing is ordinary, and that itself is an endless surprise. It is unfathomable that the world is so unfathomable. The paradoxes are eternally surprising; they are surprising *because* they are surprising. The paradoxes are paradoxes.

* * *

Chapter 1 describes the paradoxes at issue in detail. In Chapters 2 and 3, we will spend some time thinking about how methodologically to respond, in particular, how we might go about trying to classify the paradoxes (e.g., as the points on truth's outer circle?), and to what extent any solution to the paradoxes can avail itself of any part of classical logic/mathematics. These will suggest a top-to-bottom paraconsistent approach, and that the resulting view of the paradoxes will be "local," rather than captured by big global schematics. I will set as a target to develop enough mathematics to (re)prove the motivating paradoxes. After a chapter settling on an appropriate paraconsistent logic for inconsistent mathematics, Chapter 5 establishes some antinomies of naive class theory. Then chapters on arithmetic, real analysis, and topology establish respectively discrete, continuous, and topological versions of the sorites paradox, as well as more general results to do with boundaries, connected space, and fixed points. The final Chapter 10 deals with some serious objections to this project, and concludes by facing directly the question of why there are paradoxes.

---

[60] You can't do this, because the ratio 2:1 (an octave) is not a power of 3:2 (a fifth). Pianos cannot be tuned so that all octaves and fifths are perfect [Lamb, 2014]. On the unnaturalness of world, see [Dietrich, 2015].

1

Paradoxes; *or*, "Here in the Presence of an Absurdity"

Wherein we officially meet some paradoxes: of sets, vagueness, and spatial boundaries. The chapter is expository, laying out intuitive arguments for thinking that some of these paradoxes are genuine; it will be the task of later chapters to see how much of this reasoning can be brought up to logical code.

1.1 Sets

A starting point for taking the paradoxes arch-seriously comes from naive set theory.[1] Set theory provides a very natural and intuitive language and basic toolkit for the rest of mathematics. It also provides an ontology for mathematics, insofar as (it is generally thought) any mathematical object can be reduced to, or at least modeled by, sets. Set theory is a *foundation*. Paradoxes there are paradoxes at the source.

1.1.1 An Analytic Definition

The concept of a set is simple to state. A *set* is any collection of objects that is itself an object, with its identity completely determined by its members. A set is the unique extension of a predicate or property.

Many textbooks open by claiming that "set" cannot be formally defined,[2] but this isn't so; we've just had a fine definition. This is the *naive set concept*, and it can be completely characterized by the following principles:

Abstraction: $x \in \{z : \varphi(z)\} \leftrightarrow \varphi(x)$ and
Extensionality: $x = y \leftrightarrow \forall z(z \in x \leftrightarrow z \in y)$.

---

[1] For standard presentations of set theory, some good "classic" sources are [Fraenkel, 1953; Levy, 1979], the more advanced [Kunen, 1980], and of course the compendious [Jech, 1974]. For more attention to philosophical issues, see [Potter, 2004]. Parts of this section go back to [Weber, 2009].
[2] E.g., [Quine, 1969] among others. "We cannot say with any kind of conviction what sort of things sets are, so we attempt a type of ostensive definition of them through axiomatization or 'listings'" [Hallett, 1984, p. 303].

Abstraction says that something is in the set of φs if and only if it is a φ, without exception or qualification.[3] The axiom of extensionality vindicates the definite descriptor "the." These clauses fix the meanings of $\in$ and $=$, the only nonlogical parts of the vocabulary of set theory, giving *existence* and *uniqueness* conditions over the universe of sets. Existential generalization on abstraction gives the further quantified principle,

Comprehension: $\exists y \forall x (x \in y \leftrightarrow \varphi(x))$,

that for any property there is some set of all and only the things with that property.

Frege construed sets as the ontology of predication. In his *Grundgesetze* [Frege, 1903b], he stated the set concept in a single axiom, the infamous equivalence

Basic Law V: $\{x : \varphi(x)\} = \{x : \psi(x)\} \leftrightarrow \forall x (\varphi(x) \leftrightarrow \psi(x))$.

Frege's axiom looks obvious to the point of banality. The φs are the ψs exactly when all and only φs are ψs. Indeed. These clauses look very much like analytic definitions of predication. Peano's choice of the "$\in$" symbol, from the Greek verb $\epsilon\sigma\tau\iota\nu$, "to be," suggests that set membership is intended to capture the "is" of predication, or more metaphysically, property instantiation.[4] Sets are predicates in extension.[5]

These are *definitions*, in the old Socratic sense. For Socrates proposes to describe the world in terms of collections of things that share precise necessary and sufficient conditions, forms, or models that provide a standard by which we may be able to say that such and such an x is φ, such another not φ (e.g., *Euthyphro* 6e [Cohen et al., 2000, p. 95]). The naive set concept captures, in slogan form, sets as the metaphysics of definitions.

The reason textbooks claim that there can be no definition of set, then, is not that the concept is somehow opaque or ambiguous; and it is not because the concept is so familiar or primitive that it admits no definition. The reason is that the set concept is inconsistent. The concept is not indeterminate, or underdetermined; it is *over*determined, famously and paradoxically inconsistent.[6] After reviewing why, I will argue that naive set concept is

---

[3] Priest and Routley: "The naive notion of set is that of the extension of an arbitrary predicate .... This is as tight an account as can be expected from any fundamental notion. It was thought to be problematical only because it was assumed (under the ideology of consistency) that 'arbitrary' could not mean arbitrary. However, it does" [Priest et al., 1989, p. 499]. Or Priest again: "[A] set just is the extension of an arbitrary condition, and that's that" [Priest, 2006b, p. 29]. (Cf. Forster, in defense of a (consistent) universal set [Forster, 1995, ch. 1].) These statements motivate the *absolutely unrestricted* or *generalized comprehension scheme*, explicitly introduced by Routley and then Brady, which allows the set being defined to appear in its own defining property, "impredicative" instances of the form

$$x \in y \leftrightarrow \varphi(x, y);$$

[Routley, 1977, p. 915; Brady and Routley, 1989, p. 419; Brady, 2006, p. 177]. Others who endorse the naive set concept require a *restricted* form, where the set being defined *cannot* appear in the description defining it [Priest, 2006b, ch. 2]. But the distinction makes little difference, as circular sets can be produced by the sedate version as the unrestricted (Theorem 13 of Chapter 5); nothing much is gained or lost either way. The redundancy is pointed out at [Petersen, 2000, p. 383, footnote 14].

[4] In Peano's 1889 *Principles of Arithmetic*, at [van Heijenoort, 1967, p. 89].

[5] "By the law of excluded middle, ... for any predicate there is a set of all and only those things to which it applies (as well as a set of just those things to which it does not apply). ... Our thought might therefore be put: 'Any predicate has an extension'" [Boolos, 1971, p. 216].

[6] "Naive set theory is simple to state, elegant, initially quite credible, and natural in that it articulates a view about sets that might occur to one quite naturally. ... Alas, it is inconsistent" [Boolos, 1971, p. 217].

correct, and not in spite of but *because* of its paradoxicality. Its inconsistency cannot be removed without doing fatal damage to the concept.

1.1.2 The Antinomies

Naive set theory is full of paradoxes, or "antinomies." Consequently, the general consensus at the start of the twentieth century was of a *crisis in the foundations of mathematics*.[7] This is a story that has been told many, many times – even in a 2008 graphic novel, *Logicomix* – so I presume familiarity. Here then, in brief, are the famous paradoxes of naive set theory. It has seemed to many, from [Russell, 1905b] to [Priest, 2002a, ch. 8, 9], that they are all related (see Chapter 2).

1.1.2.1 The Liar [500 BCE]

Naive set theory includes naive truth theory. (One reasonable, but wrong, hypothesis about this is that naive set theory is inconsistent *because* it includes naive truth theory.) For any sentence (a closed formula, with no free variables) φ, just consider the set

$$\ulcorner\varphi\urcorner := \{x : \varphi\},$$

the set of all x such that φ, which exists by naive comprehension. If φ is true, then $\forall x(x \in \ulcorner\varphi\urcorner)$. For every φ there is such a set, and if $\ulcorner\varphi\urcorner = \ulcorner\psi\urcorner$, then by Basic Law V, $\varphi \leftrightarrow \psi$, so the naming is unique. So we can define a truth predicate, where for some arbitrary but fixed t,

$$T(x) := t \in x,$$

and, in particular, $T(\ulcorner\varphi\urcorner)$ is $t \in \ulcorner\varphi\urcorner$. Then $t \in \ulcorner\varphi\urcorner \leftrightarrow \varphi$ and naive set theory has vindicated the truth schema,

$$T(\ulcorner\varphi\urcorner) \leftrightarrow \varphi.$$

By making the naive set assumption, we have already assumed naive truth theory [Priest, 2002b, p. 363; Beall, 2009, p. 114].

To get a liar, we prove a special case of a general fact about naive comprehension, the fixed point theorem (Theorem 13 of Chapter 5). Consider the open formula $\neg T(x)$. By naive comprehension, there is a set L such that

$$x \in L \leftrightarrow \neg T(\{z : t \in L\}).$$

So by instantiation, $t \in L \leftrightarrow \neg T(\{z : t \in L\})$. Letting ℓ be the sentence $t \in L$, then $\ulcorner\ell\urcorner = \{z : t \in L\}$, and ergo

$$\ell \leftrightarrow \neg T(\ulcorner\ell\urcorner)$$

is a liar sentence. The liar contradiction follows as in the Introduction.

---

[7] Of that crisis, Fraenkel et al. say that "a treatment of the logico-mathematical antinomies is a task that cannot be dodged" [Fraenkel et al., 1958, p. 5]. Though for an alternative view of the history, see Lavine [1994, p. 3].

1.1.2.2 Cantor's Paradox [c. 1895]

Set theory became an independent discipline when Cantor proved in 1874 that the set of all subsets of X, the *powerset*

$$\mathscr{P}(X) = \{x : x \text{ is a subset of } X\},$$

must be bigger in size than X itself. For Cantor's great insight is that the sizes of infinite sets can be tracked, via the notion of a one-to-one correspondence. Even if we cannot *count* all the members of a set, we can say whether or not it is the same size as another set, by trying to pair off their members exactly. If any attempted pairing off between two sets fails, then they are not the same size, or *cardinality*.

Take then a function f from X to $\mathscr{P}(X)$. Consider the *diagonal subset*

$$r_X = \{x \in X : x \notin f(x)\}$$

of all the members of X that are not in the subset they map to. Noting that r_X is a subset of X, $r_X \in \mathscr{P}(X)$, Cantor proved that f cannot map anything from X to r_X. For if some $x \in X$ had the ill fortune to pair off with this diagonal subset, $f(x) = r_X$, then $x \in r_X$ if and only if (iff) $x \notin r_X$, which is (or classically entails) a contradiction. So, by reductio, nothing in X maps to this subset, and so the powerset $\mathscr{P}(X)$ has more members than X.

That is *Cantor's theorem*. But what about the universe of *all* sets, V? Surely V is the biggest size there is: any set is *in* the universe (containment), and any set is a *subset* of the universe (inclusion). It would seem that

$$\mathscr{P}(V) = V.$$

If sets are *identical*, then of course they are the same size. So then by Cantor's theorem, the powerset of the universe is greater than, but not greater than, the universe itself. This is also known as Frege's paradox for Basic Law V.

1.1.2.3 Russell's Paradox [1902]

Russell found his eponymous paradox based on his own study of Cantor's proof. He transmitted it to Frege in 1902, and it is only a small exaggeration to say that this destroyed the latter's life's work.

Focus on a special case of Cantor's diagonal process, where X is instantiated by the universe of sets, and f is just the identity $f(x) = x$. Then Cantor's $r_V = \{x \in V : x \in f(x)\}$ is just

$$r = \{x : x \notin x\},$$

which is called the *Russell set*: the set of all sets that are not members of themselves (or "nonselfmembered," in pseudo-German). Then $r \in r$ iff $r \notin r$. Hence, by the law of excluded middle, $r \in r$ and $r \notin r$.

Russell's antinomy is just the tip of the iceberg.[8]

1.1.2.4 Mirimanoff [1917]

A set M is well-founded (by $\in$) iff from M there is no infinitely descending membership chain

$$\cdots M_2 \in M_1 \in M_0 \in M.$$

A well-founded set M cannot be a member of itself, because if it were then $\cdots \in M \in M \in M$ would be an infinite chain. But the set of all well-founded sets

$$\mathfrak{M} = \{M : M \text{ is well-founded}\}$$

is itself well-founded, since for every $M \in \mathfrak{M}$ there is no infinite descent from M by definition. Therefore, $\mathfrak{M} \in \mathfrak{M}$. But then, by well-foundedness, $\mathfrak{M} \notin \mathfrak{M}$.[9]

1.1.2.5 Burali-Forti [1897]

Ordinals are a generalization on the natural numbers. The study of ordinals was one of the cornerstones of Cantorian set theory [Cantor, 1895], carried on by Hausdorff [Hausdorff, 2005]. Upon later reductions of mathematics to set theory, the ordinals came to form the central load-bearing column of the mathematical universe. The first ordinals are $0, 1, 2, \ldots$ followed by the first *transfinite* number greater than all of these, ω, followed by further transfinite successors and limits, with an order relation. The ordering has its members in a perfectly straight line, and any part of the line has a first (but not necessarily last) member; the ordinals are not only well-founded (as before) but *well-ordered*. And ordinals are *transitive*: anything that precedes an ordinal in the well-order is itself an ordinal.

Ordinals are the order types of well-ordered sets. Von Neumann found that it works very nicely to think of an ordinal number α recursively, as the *set* of all ordinals β that precede it, so that an ordinal is a well-ordered, transitive set of ordinals. But the set of all ordinals, On, is itself a well-ordered, transitive set of ordinals – so On is an ordinal, and, being the set of all ordinals, is also the *greatest* ordinal. But every ordinal has a *successor*, which is strictly greater. Therefore, the successor of On is strictly greater than, and also not strictly greater than, On.

Indeed, with the ordering relation on the ordinals represented by $\in$, and an ordinal taken to be the set of all the ordinals that come before it, $\alpha = \{\beta : \beta \in \alpha\}$, then

$$On = \{\alpha : \alpha \in On\}$$

is clearly an ordinal, and self-membered at that. But since the ordering relation $\in$ on ordinals is well-founded, too, there are no self-members; so the contradiction is just

$$On \in On \quad \text{and} \quad On \notin On,$$

which is what you should expect from the biggest number.

---

[8] "[A]lthough logic forces us to accept that there isn't any such [Russell] set, it's highly paradoxical that there isn't. ... [I]sn't a collection or totality just the same thing as a set? How COULD there NOT be a set containing all and only the sets that don't contain themselves?" [Boolos, 1998, p. 148].

[9] See Hallett [1984, § 4.4]; Barwise and Moss [1996]. A variant of this paradox is Smullyan's hypergame.

This is my favorite paradox. The ordinals are *designed* to be recursive, to be a foundation for arithmetic and transfinite induction. For the ordinals to do their job, they need to be self-reproducing in an orderly, automatic way. But once you start a flower forever blooming from within itself, it cannot be stopped.[10]

Several other more "semantically" flavored paradoxes appeared in a rush.[11]

1.1.2.6 König [1905]

A set is *countable* iff it is the same cardinality as the natural numbers. A number is *definable* iff there is a finite sequence of (English) words that refers to it. There are only finitely many words; so there are only countably many finite sequences of words; so there are only countably many definable numbers. So take the least indefinable number. The previous sentence just defined it.

1.1.2.7 Richard [1905]

Similarly to König's, consider the set of definable real numbers on the interval $[0, 1]$. By a diagonal construction, there is a definable real number not in this set.[12]

1.1.2.8 Berry [1906]

Consider the set of natural numbers definable by less than 19 syllables. There is a least such number; this number is defined by less than 19 syllables.

1.1.2.9 Grelling [1908]

A term is homological iff it exemplifies what it describes, e.g., the word "word" is a word. A word is heterological if not homological. Then "heterological" is heterological iff it is not.

There are many others, but you get the point. These are apparently sound arguments with apparently false conclusions. We must deny the reasoning, deny a premise, or accept the result. In all cases, the premise is naive set comprehension.

1.1.3 Naive Sets and Iterative Sets

1.1.3.1 Why Are There Set Theory Paradoxes?

The antinomies of naive set theory are overwhelming. Obviously true assumptions (in each case, that there is some set of things with some clearly defined property, such as "is an ordinal") lead by simple reasoning to contradictions. By 1925, Hilbert is apocalyptic, calling for a solution to these problems "not merely for the interests of the individual sciences, but rather for the *honour of the human understanding itself*"; for the paradoxes of

---

[10] Burali-Forti himself thought this was a counterexample to trichotomy [van Heijenoort, 1967, p. 105]. Sometimes the contradiction is expressed as $On = On + 1$. See [Moore, 1982].

[11] On dividing these paradoxes into different categories (set theoretic vs. semantic) as Ramsey suggested, and whether the division is good or not (it is not), see Priest [2002a, ch. 10].

[12] See van Heijenoort [1967, p. 143]; and Priest [2002a, ch. 9].

set theory, appearing "ever more severely and ominously," had "a downright catastrophic effect in the world of mathematics":

> Let us admit that the situation in which we find ourselves with respect to the paradoxes is in the long run intolerable. Just think: this paragon of reliability and truth, the very notions and inferences, as everyone learns, teaches, and uses them, lead to absurdity. [Hilbert, 1925, p. 375; emphasis in the original text]

How could it be? I think the answer lies in thinking about what a set is.

An otherwise disparate collection is encircled by a predicate, a property, a definition. Birds become a flock, stairs a flight, people a crowd (a *Menge*), counting numbers the naturals $\mathbb{N}$. Objects become a set. A set is a multiplicity that forms a unity, a many that is also a one. Metaphysically speaking, a set is a *composition*. But this is not a definition; this is a mystery. Here *are* birds. Here *is* a flock. Even at the level of grammar, the former are plural, while the latter is singular.[13] How is a many also a one?

The beauty of naive set theory is that it provides just the right way to say that a flock is a *set* of some birds, not unlike Plato's forms in the *Republic* (596a6–7). The way this is possible is that sets play two roles at once, roles that are in tension with each other. A set is no more and no less than its members *together*. This is the ultimate reason why set theory is inconsistent. Let us return to the explication of the nature of sets; Bolzano takes arbitrary assemblages to be cohesive entities [Bolzano, 1973, p. 128]:

> I permit myself, then, to call any group you please, in which the nature of the connection among the parts is to be regarded as an indifferent matter, a *set* [*Inbegriff*].

And here, more famously, is Cantor echoing the same thought: in the 1895 *Beiträge* [Cantor, 1915, p. 85],

> A set is any collection into a whole of definite, well-distinguished objects of our intuition or thought.

Or earlier, in the 1883 *Grundlagen*,[14]

> By a "manifold" or "set" I understand generally any multiplicity which can be thought of as one [*jades Viele, welches sich als Eines denken lasst*], that is to say, any totality of definite elements which can be bound up into a whole by means of a law.... By this I believe I have defined something related to the Platonic ειδος.

Sets are extensional collections, yes, but they are also grasped (the etymology of "inbegriff"), "by means of a law." This is our ingress:

On the one hand, sets are *extensional*, in that they are determined by their members. To pick out a set, there is nothing more one needs to know about it than its members. Set theory is often called the theory of extensions par excellence.[15] On the other hand, sets are *intensional*, too, in that they are each determined by a property. Within naive set theory is

---

[13] Cf. Lewis [Lewis, 1991, p. 81, emphasis in the original text] on compositions and the things that compose them: "It just *is* them. They just *are* it."

[14] From [Hallett, 1984, p. 33; Jané, 1995, p. 391].

[15] Forster calls "the most tough-minded expression" of extensionality, that "the only thing a set theorist can know about = is that it is a congruence relation with respect to ∈" [Forster, 1982, p. 2].

naive property theory, obtained just by dropping the extensionality axiom, and reading "∈" as property instantiation: $\varphi(x)$ just in case x instantiates the property (or form, or *eidos*) of φ. Sets *must* be more than extensional, if they are to be useful for any more than some (small) finite combinatorics. As Weyl aptly observes,

No one can describe an infinite set other than by indicating properties which are characteristic of the elements of the set. *[Weyl, 1919, p. 23]*

A collection requires a predicate.[16] But sets *cannot* be more than extensional, since they are properties in extension. Governing naive set theory is a membership-based notion of identity, and a property-based notion of membership. It is *both* "top-down" and "bottom-up." And this means a collision of extensionality with intensionality.

Here then is the nexus of the paradox. Sets are themselves objects, intensions over and above their members, *more* than the sum of their parts.[17] Meanwhile, the extensionality principle governs all sets, forcing them to be *no more* than the sum of their parts. Since these together characterize sets, sets are both intensional and extensional. Frege's Basic Law V expresses, correctly, that there is a one-to-one correspondence between extensions and properties; Cantor's theorem expresses, correctly, that there are more properties than extensions.[18] This is unstable. This is Cantor's paradox: there are more objects than fit in the universe, because there are properties with the same extension; but because (by Basic Law V) there are exactly as many properties as extensions, the universe is bigger than itself.

My view is that this contradiction inherent in the notion of naive sets is a *good thing*. Naive sets provide a foundation that has both these qualities, intensional comprehension and extensional tractability. Theories that try to do without one or the other will inevitably be incomplete. The dual nature of naive sets is a feature, not a bug; they offer Platonic forms with precise identity conditions. And, most tendentiously, if the comprehension axiom turns out to be *true*, then this is an even more attractive reason to use it as a foundation.

1.1.3.2 From Twentieth-Century Set Theory to Twenty-First

The naive comprehension principle is an axiom in the old sense: simple, self-evident, inalienable. I suggest the mathematics of collections *cannot* abandon the naive view without ceasing to be a theory of all collections. To see this, let us review the past century's attempt to evade the paradoxes. This was done, first, by replacing naive set theory with *axiomatic* set theory (from 1908 to the 1920s); and then later (from the 1930s to the 1960s) by clarifying

---

[16] Cf. Beall's diagnosis of the semantic paradoxes of truth theory: God would have no need of a truth predicate [Beall, 2009, p. 1], because God in his infinitude does not need to save time by generalizing. By these lights, God would have no need of a comprehension principle, either.

[17] Unlike *pluralities*, e.g., some birds, which advocates of *plural logic* say do not always give a further entity, "a plurality," e.g., a flock of birds, over which to quantify. A "plurality" is not itself an entity; referring to "it" in the singular is only a *façon de parler* (cf. proper classes in set theory; Section 1.1.3.2). The most prominent argument that not all pluralities are sets is that, if they were, one could derive Russell's contradiction [Boolos, 1998, p. 67]. A naive set is also different from a *mereological sum/fusion* – the smallest portion of reality that has all the φs as *parts*. (*Mereology* is the theory of parthood.) A fusion exists, as a numerically distinct entity from its members, unlike plurals; but the parthood relation is more like subset than membership. See [Lewis, 1991]; Potter, 2004, ch. 2; [Cotnoir and Baxter, 2014].

[18] Or just think of the difference between the properties "equilateral triangle" and "equiangular triangle" even though their extensions in Euclidean geometry are identical. This basis for the paradoxes of naive set theory is spelled out in [Zalta, 2007].

the "intended model" of the new axiomatic set theory, and urging that this was the real (consistent) set concept all along. How did this strategy fare?

Zermelo's 1908 axioms met with "intense criticism," not only over the axiom of choice but because of his decision to molest the comprehension principle; Schönflies, Bernstein, and Poincaré all rejected this possibility out of hand.[19] Zermelo's axioms name some key properties sets have, without explaining what a set *is*.[20] The aporia is apparent in the opening pages of most set theory books. Devlin opens as many do with exposition of basic Cantorian theory.

> In set theory, there is really only one fundamental notion: the ability to regard any collection of objects as a single entity (i.e. a set). *[Devlin, 1979, p. 1]*

Or, as a very rigorous textbook puts it,

> the idea of the collection of all objects having a specified property is so basic that we could hardly abandon it. *[Takeuti and Zaring, 1971, p. 9]*

Reformed set theory since the paradoxes retains "as many as possible of the naive set theoretic arguments which we remember with nostalgia from our days in Cantor's paradise" [Potter, 2004, p. 34]. And since the Gödel/Cohen independence results,[21] it has been known that the axioms of ZFC – Zermelo–Fraenkel set theory with the axiom of choice – are highly *incomplete*, leaving key questions about sets permanently unanswered.[22]

Now, the reader may wish to remind me, gently but firmly, that I appear to be trying to re-litigate a closed case. During the twentieth century, much work went in to a replacement idea, *iterative sets*, and their main formal theory, ZFC.[23] An iterative set is formed from *already* existing objects. Therefore, iterative sets cannot be, are not candidates to be, members of themselves. The set theoretic universe is a cumulative hierarchy, in which sets may only be formed from preexisting members, starting with the empty set – formed from "all" the sets that exist at the start.[24] Then the idea is that none of the antinomy-inducing collections are (iterative) sets. The solution is similar to Tarski's truth hierarchy, with the

---

[19] [Moore, 1982, pp. 111, 117].

[20] "Axiomatization went hand in hand with the divorce from any attempt to understand what sets are or what conceptual role they play" [Hallett, 1984, p. 303]. The point is made explicitly and overtly not least by Zermelo himself, at the outset of his 1908 axiomatization: "At present, the very existence of the discipline [of set theory] seems to be threatened by the existence of certain contradictions or 'antinomies' that can be derived from its principles – principles necessarily governing our thinking, it seems – and to which no entirely satisfactory solution has yet been found" [Zermelo, 1967, p. 200]. See Woods [2003, p. 334]. In 1914, Hausdorff expressed doubt about Zermelo's system: "At present, these extremely ingenious investigations cannot be regarded as completed, and introducing a beginner by this [axiomatic] approach would cause great difficulties. Thus we wish to permit the use of naive set theory here" [Hausdorff, 1957, p. 2]. Zermelo's axioms are supposed not even intelligible without naive acquaintance with sets [Devlin, 1979, p. 49]. "A survey of the the axioms does not suffice to reveal the source of their attraction" [van Aken, 1986, p. 992]. Von Neumann, at the close of his own sophisticated 1925 axiomatization, sighs that despite much work, still he must "entertain certain reservations" because "for the time being no way of rehabilitating this theory is known" [van Heijenoort, 1967, p. 413].

[21] See [Cohen, 1966; Smullyan and Fitting, 1996].

[22] As witness to the ongoing dissatisfaction with rejecting naive comprehension, there remains a steady stream of research attempting either to approximate it with ever-stronger new "large cardinal" axioms [Kanamori, 1994] or to restore some semblance of the principle, via modalities or other devices.

[23] There are many other alternative set theories, before even considering nonclassical logic; [Holmes et al., 2012]. Most land on saying that some collection is not a set.

[24] Note that this – *if there are no sets, then all sets are in* $\varnothing$ – is a form of explosion: $p \therefore \neg p \supset q$. So the *ex nihilo* construction in this form is very classical.

added weight of the legitimacy and importance of actual mathematics (say as it is needed for science) and therefore the pragmatic obligation to get some kind of working fix on the table so that we may get on with life.

Let us ask, though, whether iterative sets can be sufficient for *explaining* what sets are. Without impugning the impressive work of the last century, there is an obvious problem in explaining what sets are in terms of iteration: iterative sets are formed from *some collection* of preexisting members. It presupposes collections. And not only that, but it supposes those collections arise in an orderly, indexed process. For example, in [Potter, 2004], following an idea from Scott, it is shown how one can justify the axioms of ZF from some "stage axioms"; these axioms postulate (a) the existence of cumulative stages, (b) an ordering relation such as "earlier than," and (c) indices to keep track of the process. But collections, order relations, and a (transfinite) index set are *exactly the sorts of things developed from within set theory*. This is "assuming a considerable amount of 'set theory' in order to define our *set theory*" [Devlin, 1979, p. 49, emphasis in the original text]. The iterative notion is therefore not the primary intuition because it *presupposes set theory*, including the notion of ordinality. One of the central purposes of set theory is to deliver a theory of ordinals, not the other way around.[25] So the iterative notion is not the last word on sets [Weir, 1998, p. 780]. Like a Tarskian hierarchy, the iterative universe is simply unable to account for itself.

There are also problems with making the iterative conception more than an attractive metaphor. The iterative idea does appeal to our physical intuition about collecting up objects into a bag; you need to have objects *before* you can put them in a bag! Beyond small finite sets, though, the intuition is exhausted – "naturally we are not thinking of *actually building* sets in any sense" [Devlin, 1979, p. 43]. Talk of "collecting" must be taken metaphorically. And only metaphorically:

The notion that an infinite set is a "gathering" brought together by infinitely many individual arbitrary acts of selection . . . is nonsensical. *[Weyl, 1919, p. 23]*

To replace the metaphor with mathematics, we can follow Hilbert's strategy in geometry [Hilbert, 1902a] of replacing Euclid's imperative phrasing ("now you construct a triangle like so") with a declarative ("triangles exist"). Instead of "forming" the ωth stage, we simply profess that it is there: *there is an ωth stage*. This is now mathematically cogent, but it leaves the constructive intuition behind and with it any explanatory power of the iterative view. It calls into question whether there is any nonmetaphorical content to the iterative view. Unless we are some version of strict constructive finitists, or believe some kind of "idealized" constructing angels,[26] then sets simply exist or they do not.

---

[25] Priest and Routley: "We do not deny that once one has a notion of set one can non-circularly produce . . . the cumulative hierarchy. But to suppose one can use the notion of an ordinal to produce a non-question-begging definition of 'set' is moonshine" [Priest et al., 1989, p. 500].

[26] Agents human enough to make the "gathering together" talk more than a pretty locution, but superhuman enough to perform transfinite tasks up to arbitrary places in the ordinals. See Potter [Potter, 2004, pp. 36–40] for discussion of construction metaphors, idealized constructors, etc.

To see that the classical solution to the paradoxes is not entirely happy, I would gesture at two problematic ideas that underwrite ZFC. One is the tortured doctrine of *proper classes*, which introduces entities that are like sets (classes) in every way, except they have a special property – if they were sets, they would be inconsistent![27] The related doctrine of *limitation on size* says that some collections are "too big" to be sets – namely, those collections that are the same size as a proper class.[28] Suffice to say that both of these ideas are rather euphemistic – as Boolos says, "you can't get out of this paradox merely by substituting one word for another" – and could be criticized at some length if one were so inclined (e.g., the very enjoyable [Weir, 1998]). As Woods puts it, these problems are "central, deep, and disabling" [Woods, 2003, p. 162]. These doctrines suggest, on my reading, that practitioners who are committed to doing some mathematics mainly want to get past paradoxes for the sake of getting on to some results. And indeed *this* seems like one of the better arguments in favor of "classical" set theory. Fair enough. But the argument is irrelevant in the present context. The natural notion of set is inconsistent, and attempts to replace it either acknowledge the inconsistency or rest content with mathematical storytelling.

Rather than relitigating the past, or peevishly throwing stones at the towers of modern ZFC, I can find much to agree on in statements such as this one:

It should not be forgotten that the paradoxes never applied to any type structure, and in this sense they are *not* paradoxes of set theory .... If we call such things *properties* then it is clear that the paradoxes are a real problem to be dealt with before a thoroughgoing theory of properties can be developed. *[Drake, 1974, p. 14]*

Or, to a lesser extent, this one:

It has been pointed out that the paradoxes of Russell, Burali-Forti etc, never really caused a crisis in mathematics (where one deals only with unproblematic examples of sets) but rather in logic (and general set theory), where one tries to provide a general and universal frame for mathematics and in particular for arbitrary sets. *[Reinhardt, 1974, p. 190]*

These statements acknowledge unsolved problems in what is being called "property theory" and "logic" or "general set theory." So let "iterative set theory" carry on unperturbed, as long as it is admitted that ZFC sets are a subconcept of what collections are, as long as it is clear that iterative sets are not *all* the sets or have somehow *solved*, rather than deferred, the paradoxes.[29] After all, "simply saying that we ought never to have expected any property whatever to be collectivizing, even if true, leaves us well short of an account which will settle which properties are" [Potter, 2004, p. 27]. As Dummett puts it,

---

[27] As such, proper classes cannot, by definition, be members of anything. See [Maddy, 1983]. Here is Halmos: "[I]t seems a little harsh to be told that certain sets are not really sets and even their names must never be mentioned. Some approaches to set theory try to soften the blow by making systematic use of such illegal sets but just not calling them sets; the customary word is 'class'" [Halmos, 1974, p. 11]. Shapiro and Wright sum up neatly: "Invoking proper classes is an attempt to do the very thing we are intuitively barred from doing. ... *Set* is supposed to encompass the *maximally general category* of entities of the relevant kind" [Shapiro and Wright, 2006, p. 272].

[28] "A great diffculty of the theory is that it does not tell us how far up the series of ordinals it is legitimate to go" [Russell, 1905b, p. 44]. See [Hallett, 1984].

[29] Reinhardt goes on to say, "We now consider such a frame to have been provided for set theory by the clarification of the intuitive idea of the cumulative hierarchy" – a kind of disciplinary "monster barring" and "withdrawal to a safe domain" [Lakatos, 1976] by drawing the boundaries of "set theory" to exclude problems in "logic" and "general set theory."

A mere prohibition leaves the matter a mystery. ... To say, 'If you persist in talking about the number of all cardinal numbers, you will run into a contradiction', is to wield the big stick, but not to offer an explanation. *[Dummett, 1991, p. 315]*

There remains a more general theory, naive set theory, with a lot of work still to do. If there is something true about collections that is not captured by a theory of collections, then that theory is eo ipso inadequate as a theory of collections – sets. If the naive set concept were not ineluctable, then its inconsistent consequences would be reason to reject it. But it "seem[s] forced upon us in such a way," writes Slaney, "that we should in all intellectual honesty take [it] seriously" [Slaney, 1989, p. 472].[30]

1.1.3.3 Prospects

Classical logic makes set comprehension absurd. It pushes us to say that there is no universal set, and thus that there is no such thing as universal quantification ("for all" claims) in absolute generality, because there is no universe to quantify over. This has been cause of much surprise and consternation. The consternation is due to the deep sense that comprehension is not absurd, because there is a collection – a set – of all sets, and we do indeed universally quantify. That, indeed, saying "all universal quantification is impossible" is, in the long run, intolerable.

At the end of an exhaustive survey of possible approaches to the paradoxes of set theory (the Burali-Forti paradox in particular), Shaprio and Wright say that "frankly, we do not see a satisfying position here," but they do mention one other option:

Allow the quantification and the predicates, allow the associated order-types, allow that they are ordinals as originally understood, ... and just accept that there are ordinals that come later than all the ordinals. Cost: none – unless one demurs from the acceptance of contradiction. *[Shapiro and Wright, 2006, p. 293]*

As with the solutions to the liar, there is no option for resolving the paradoxes that gives us everything and costs us nothing. But the paraconsistent dialetheic approach dangles the possibility of an absolutely comprehensive theory of sets, where the paradoxes can finally be accepted for what they are, proofs.

Cantor knew that there exist what he called "inconsistent multiplicities"; he even used them in a proof in a letter to Dedekind, proving that every cardinality is an aleph.[31] Potter warns us to be wary of the Panglossian view that "the paradoxes are not really so paradoxical if we only think about them in the right way" [Potter, 2004, pp. 26, 37]. The paradoxes are paradoxes. Forster advises:

In the ZF world, ... the paradoxes are viewed as large holes in the ground that one might fall into. ... However, it is *always* a mistake to think of *anything* in mathematics as a *mere* pathology, for there are no such things in mathematics. ... One should think of the paradoxes as supernatural creatures,

---

[30] For more pugilistic motivations for naive set theory, and attacks on ZFC and its associates, see [Priest and Routley, 1983]; cf. Priest [2006b, ch. 2].

[31] "As we can readily see, the 'totality of everything thinkable,' for example, is such a multiplicity" [Cantor, 1967, p. 114]. Cantor had been thinking about absolute paradoxes at least since 1895 [Lavine, 1994, p. 55].

oracles, minor demons – on whom one should keep a weather eye in case they make prophecies or by some other means divulge information from another world not normally obtainable otherwise. One should approach them as closely as is safe, and from as many different angles as possible. *[Forster, 1995, p. 11, emphasis in the original text]*

The question turns on just how close is still a safe distance. The idea of maintaining comprehension in a nonclassical logic goes back at least to Skolem [Skolem, 1963]. The idea of mathematics founded on self-evident axioms goes back even further. Naive set theory in paraconsistent logic is presented in Chapter 5.

1.2 Vagueness

Let's leave set theory for a while and go outside. It was raining very heavily earlier, but now we can go out. It is not raining, although it is still raining a bit. It is what in Ireland is sometimes called a "soft day."

Like most predicates, "is raining" is *vague*.[32] And vagueness gives rise to the sorites paradox. The sorites paradox, like the liar, is attributed to Eubulides in the fourth century BCE. The sorites paradox may seem different from the inconsistencies of naive set theory or truth theory. But the sorites exhibits some of the same and most important features, especially in the way that most solutions to it fall prey to revenge, and in the way it connects with the spatial/boundary paradoxes of Section 1.2.1.[33]

1.2.1 The Sorites Paradox

Olympus Mons is the tallest mountain in the known universe.[34] It is an extinct shield volcano on Mars, and its summit stands 26 kilometers over the surrounding plains – three times the height of Mt. Everest. Unlike Everest, though, Olympus Mons is much wider than it is tall, about 600 km wide, so its slope is generally extremely gentle, about 2.5 degrees around the caldera at the top and 5 degrees on the wider foothill; atmosphere notwithstanding, one could easily walk most of the path to the top, and down again, without need of climbing gear or any special athletic skill. A path down the gentle slope of the mountain (which, when viewed from the summit, extends beyond the horizon) can be sketched out by a discrete linear order,

$$a_0 < a_1 < \ldots < a_n,$$

with the indices natural numbers and a_n a point at surface level.[35] The top, point a_0, is very high up. The bottom of the mountain is not high up any more.

---

[32] The discussion here is mostly neutral as to the locus of vagueness – whether is it linguistic (vague predicates), metaphysical (vague properties), or ontic (vague objects). (The idea of *ontic vagueness* is unpopular, but can be made at least an intelligible thesis [Barnes, 2010].) Throughout, as before, "predicate" can be substituted mutatis mutandis for "property," depending on your preferences.

[33] Thanks to Mark Colyvan for collaboration on topics here [Weber and Colyvan, 2010].

[34] Gary Hardegree (in conversation) objects that surely there are taller mountains out there in the big, big universe. Yes, probably. But they are not (yet) *known*.

[35] Or "martian geodetic datum." Since Mars has no sea, it has no sea level, so measuring the elevation of Olympus Mons is more complicated than I've suggested. See Carr [2007, p. 51]; Frankel [2005, ch. 6].

Let's have our points spaced one centimeter apart. Then any two consecutive points are too similar in all relevant respects, too close to each other, for one to be high up but not the other. This is an instance of the *tolerance* principle;[36] any φ is tolerant when it obeys the following:

Tolerance: It is not the case that two things are very, very φ-similar, and yet one is φ, but the other is not.

Formalizing a bit, then, we have the (material) *conditional* version of the sorites paradox, following Hyde's very useful classification of the sorites paradoxes [Hyde and Raffman, 2018]: TOLERANCE is that, for all i, it is not that case that a_i is high up but a_{i+1} is not, and the soritical argument is spelled out:

a_0 is high up
a_0 is high up $\supset$ a_1 is high up
a_1 is high up $\supset$ a_2 is high up
$\vdots$
a_{n-1} is high up $\supset$ a_n is high up
$\therefore$ a_n is high up,

with $\supset$ the material conditional (Section 0.2.2.3), "a_i is not high up or else a_{i+1} is." Then the conclusion follows by a sufficient number of applications of disjunctive syllogism, or "material modus ponens":

high(a_0)
not high(a_0),　or high(a_1)
　　　　not high(a_1),　or high(a_2)
　　　　　　　not high(a_2),　or high(a_3)
　　　　　　　　　　　　　$\ddots$
　　　　　　　　　　　　　high(a_n).

But a_n is at the foot of the mountain. It is *not* high up. Generalizing, we get the *inductive* version of the paradox, which uses mathematical induction rather than material modus ponens:

a_0 is high up
For each i in the sequence, (a_i is high up $\supset$ a_{i+1} is high up)
$\therefore$ For every i in the sequence, (a_i is high up).

This argument has led from truth to falsity.

This is a *paradox* in the sense defined in the Introduction. It is an apparently sound argument with an apparently false conclusion. We must deny the reasoning, deny a premise,

---

[36] The name "tolerance" comes from [Wright, 1976, p. 334]: a predicate φ is tolerant "if there is also some positive degree of change in respect of φ insufficient ever to affect the justice with which [the predicate] applies to a particular case."

or accept the result. In this case, we cannot accept the result: not all points in the path are high up. Accepting "all points are high up" would make a nonsense out of the predicate. And all the more so since this is just an example. If sound, the conditional inductive sorites would render any vague predicate completely vacuous. Everything is red, everyone is bald, all jokes are funny. In fact, if "vague" is itself a vague predicate (as some have urged [Sorensen, 1985; Hyde, 1994]), and we accept soritical reasoning such as the preceding, then it would follow that *all* predicates are vague, and in particular *truth*, and then, like all grains of sand are heaps, so too would everything be true. Absurd.

So, as Sorensen is careful to explain [Sorensen, 2001, ch. 1], the obvious solution to the paradox now is to deny the conclusion, and instead derive the falsity of TOLERANCE, via a *line-drawing* form of the argument. Informally, on Olympus Mons, the soritical situation is thus described with three true sentences:

> Some point is high up.
> Not all points are high up.
> ∴ Some point is high up while the very next one is not.

Here the conclusion is the negation of TOLERANCE. This is a sound argument, stemming from a truism about finite sets of natural numbers: if some number is such and so, then there is a *least* such number. Not only is the argument valid, but the premises are true. Since the vague predicate eventually ceases to apply, there *must* be a line – a sharp cutoff – between what is high up and what is not.

Except TOLERANCE is not so easily dismissed. Our new conclusion, that "high up" comes abruptly to an end at some exact centimeter-length patch of martian hillside, smacks just as false, just as unacceptable as the initial conclusion that "high up" never ends. This is especially pressing when we observe that the centimeter-length patches could be made as small as we like. However it is that a baby becomes a non-baby, it is not in the duration of a millionth of a picosecond, right? It is unavoidable – there *must* be such a line, a counterexample to an otherwise disastrous proof. But all the same, TOLERANCE seems basic: vague predicates do not, almost by definition, have sharp cutoff points. That's a truth of the world we live in. Here then is the appearance of *revenge*, and the point at which the paradox becomes genuine.

The original problem from the conditional/inductive sorites was that it would be absurd if all points were equally "high up" (bracketing any mystical epiphanies). The only (apparent) option was to affirm that not all points are high up, and rejig the argument accordingly, from the conditional version with a false conclusion (of the form $p, q \supset r \therefore s$) to its contrapositive, the line drawing version ($p, \neg s \therefore q \& \neg r$). But now there is a new problem, which is that the line-drawing version has a false conclusion, too: it seems absurd to think there is some exact spot – right there, and nowhere else – at which a predicate like "high up" cuts off. The proposed solution is every bit as counterintuitive as the original, since the original was driven by the conviction that TOLERANCE is true. This is how we know we are dealing with a genuinely paradoxical problem. Like the derivation of a Kantian antinomy, there are parallel reasons that

TOLERANCE is true and TOLERANCE is false

– a contradiction.

As with all serious paradoxes, our intuitions combine with the facts to box us into a corner, between impossibility and inevitability. Some mountains are tall; some are not; and the notion is tolerant: there aren't two mountains that are extremely close in height, but one of them is tall and the other is not. The notion is, nevertheless, as prone to having a "cutoff" as any other: since tallness gives out, it must give out *somewhere*. The existence of a cutoff point is inevitable. It is also incredible. It is a paradox.[37]

Going forward, it would be good to have a (provisional) definition of what it means to be vague. Susceptible to the sorites? Satisfying tolerance? Many accounts of vagueness begin by describing the phenomenon as a kind of *indeterminacy* – a predicate is vague when there are cases where it is too hard to say whether, e.g., a mountain is tall or not. But this is already a theory-laden diagnosis,[38] casting the whole situation in an epistemological light, and one pushing toward a gappy solution. I suggest defining a predicate φ as *vague* in a more theory neutral way, if and only if the following are true:[39]

- φ is *nontrivial*: something is φ, and something is not φ.
- φ satisfies TOLERANCE.

A nice consequence of the definition is duality: if φ is vague, then $\neg\varphi$ is vague. A more basic consequence is that, given some simple mathematics, a contradiction follows, as sketched previously; the specific mathematical principles involved will be laid out in Section 3.3.3.2.

1.2.1.1 A Continuous Sorites

Before moving to the options, let's look at a generalization of the paradox, a *continuous sorites*, which concerns a smooth (rather than incremental) transition. Increasing the level of abstraction in this way forces attention on cutoff points of vague predicates. This generalization shows that nothing about the paradox relies on the discrete nature of the presentation. Indeed, in presenting the previous paradox for "high up on Olympus Mons," I needed to "digitize" it in order to express the paradox – taking the continuous slope of the mountain and breaking it into centimeter-sized pieces. Vague properties as they are found

---

[37] It is worth noticing just how many apparently independent philosophical problems and positions have the sorites paradox as a key part: *the problem of the many* [Unger, 1980]; *universalism* and *nihilism* as answers to the special composition question [van Inwagen, 1990, ch. 12], based on the *argument from vagueness* [Lewis, 1986, p. 212]; and four-dimensionalism about time as a solution to the Ship of Theseus [Sider, 2001], to name a few.

[38] See [Bueno and Colyvan, 2012]. Thanks to Mark Colyvan here.

[39] Thanks to Lloyd Humberstone for useful suggestions here. And also, once upon a time, for teaching me the correct way to use the abbreviation "cf.," among other things.

in nature are often not broken up into units; but there still seems to be the same "slippery slope" problem about continuous properties – maybe even more so. We ought to be able to formulate the sorites paradox in terms of continuous transitions and not merely discretise continuous cases; otherwise, it would look like the problem is to do with $\mathbb{N}$, not vagueness. But the smaller the increments, the more compelling the sorites argument.

The sorites may be generalized to the continuous case.[40] The argument, which is entirely classical, draws out consequences of two properties of the real numbers $\mathbb{R}$, which is that (i) any set of reals bounded from above has a least upper bound, its supremum or sup; and (ii) the reals are dense, in the sense that if $x < y$, then there is a real z such that $x < z < y$. From (i), every set of reals bounded from below has a greatest lower bound, its infimum, or inf. Now consider a vague predicate φ mapped onto a real-number interval $[0, 1]$, exhaustively partitioned into two nonempty sets,

$$A = \{x \in [0, 1] : \varphi(x)\}$$
$$B = \{x \in [0, 1] : \neg\varphi(x)\},$$

with $a < b$ for all $a \in A, b \in B$. We assume that $\varphi(0)$ and $\neg\varphi(1)$. The nonempty set A has a least upper bound, call it sup A. Now, φ is vague, hence TOLERANT; therefore, since points vanishingly close to sup A are φ, then also $\varphi(\sup A)$. By a symmetrical argument, $\neg\varphi(\inf B)$. By the linear order on $\mathbb{R}$, one of the following must be true:

$$\sup A < \inf B$$
$$or \quad \inf B < \sup A$$
$$or \quad \sup A = \inf B.$$

Since the reals are dense (between any two reals is another one), we have the following contradiction. If sup A and inf B are different numbers, then there is some z between them, sup $A < z < \inf B$ or inf $B < z < \sup A$. But then φz and $\neg\varphi z$, by examining both cases: if sup $A < z < \inf B$, then anything less than inf B is φ but anything greater than sup A is not; if inf $B < z < \sup A$, then anything less than sup A is φ but anything greater than inf B is not. On the other hand, if sup $A = \inf B$, then again φ sup A and $\neg\varphi$ sup A. This exhausts all the cases. Therefore, there is a point both φ and $\neg\varphi$, a contradiction.[41]

What can we learn from this version of the paradox? We see how the sorites can be constructed so that it relies upon a property of the real line – the property of being *connected* (see Section 1.3.1.2). Because the reals are connected, a continuous path must cross over from A to B *at some distinct point*. If A and B are partitioned by a vague property, then that point of crossing is inconsistent. A very common response to the discrete forms of the sorites paradox is to see a problem with exclusively and exhaustively separating objects into two closed categories, φ and not. This problem is well expressed in terms of connectedness,

---

[40] Due to James Chase [typescript], via Mark Colyvan.
[41] The argument can be represented in analogy to the discrete inductive form of the sorites, via Cauchy sequences, appealing to what Priest calls the Leibniz continuity condition [Priest, 2006b, ch. 11]: whatever is going on arbitrarily close to some limiting point is also going on at the limiting point; *natura non facit saltus*. See Weber and Colyvan [2010, p. 316].

and is the key in generalizing from the discrete to the continuous. We can use this property to generalize again, for a metric-free topological version, but this will be better to return to at the end of the chapter.

1.2.2 The Options

In the Introduction we looked at dialectics of dealing with paradox – the options breaking into classical, paracomplete, and paraconsistent directions. The patten repeats here. There are already many impressive surveys on vagueness and analyses of the problem, so as with truth, I don't attempt that scholarship here.[42] The aim of the discussion is to suggest that, like the liar, all standard solutions to sorites (except, perhaps, nihilism) face revenge, either at the first level or higher orders, because they are committed to sharp cutoffs in some way. That means they deny TOLERANCE, which I think is getting it half-right.[43]

1.2.2.1 Classical Solutions

According to a venerable tradition – Frege, Russell, and Quine – vague predicates are not amenable to logic. The sorites paradox is a reductio against the existence, as far as logic is concerned, of most ordinary properties. Another tradition extends this nihilism to ontology, as mentioned in the Introduction: mereological nihilists take the sorites paradox as reductio against the existence of most ordinary objects. Any "object" that gives rise to sorites arguments does not even exist. Nihilists give some sophisticated explanations as to why our ordinary beliefs are in such massive error. But it would take us off-topic to dwell on this option. Nihilism is its own sort of revenge. The world I live in includes tables and chairs and tallish mountains and rainy days, and these things can be reasoned about using logic. Our job is to explain, not explain away. Stories about why we are wrong about almost everything are about some other world.

Another classical solution to the sorites paradox is to accept the conclusion of the line-drawing argument – there is indeed a sharp cutoff point for vague predicates – but to explain our incredulity about this by positing that the cutoff is *unknowable*. This is called epistemicism. Vagueness is hence a knowledge problem [Williamson, 1994; Sorensen, 2001]. Since I agree on much of the epistemicist's setup of the problem – that our job is not to get rid of cutoffs but rather to explain why they are "embarrassing" – all that matters for the purposes of this chapter is that epistemicism by design includes the existence of sharp cutoffs for vague predicates.

1.2.2.2 Gappy Solutions

Vagueness is, perhaps, the most widely accepted reason for dipping into nonclassical logic. And the most widely accepted way to do that is to deny bivalence, or the law of the

---

[42] See, e.g., [Keefe, 2000], [Hyde, 2008], and [Smith, 2008].

[43] Priest urges that a solution to the sorites paradox can *only* be in form of explaining why the existence of a cutoff is counterintuitive [Priest, 2003] in [Beall, 2003]. A different sort of option I don't discuss is *contextualism* (e.g., [Shapiro, 2006]).

excluded middle. The basic idea is that some people are bald, some are not, and some are *indeterminate*, neither bald nor not bald. The aim here is to accept TOLERANCE in some sense, but the vague property does not spread everywhere, because somewhere along the way the vague property falls into a gap. For some points, the answer to whether they are high up or not is "neither."

There are many ways to go with this idea, but as with truth, the main problem for gappy solutions is revenge. For these approaches all still end up with a sharp cutoff point for the vague predicate somewhere.

- According to the popular *supervaluationism* of van Frassen and others, being, e.g., bald admits of different evaluations. Someone is "super"-bald iff they are bald on every admissible valuation. Then there is a cut between bald and "super"-bald – the first number of head-hairs that come out as bald on every valuation.
- In *subvalutationism*, someone is supra-bald iff they are bald on *some* admissible valuation.[44] This position is symmetrical to supervaluationism, as all involved attest, and so comes with sharp cutoffs, too.
- Using the *fuzzy* logic of Hajek, or the fuzzy set theory of Zadeh, baldness comes in degrees. Then there is a cut between those who are bald to degree 1, and those who are not, because "is bald to degree 1" is itself determinate, not fuzzy.[45]

This is called the problem of *higher-order* vagueness. It is a clear case of revenge: eliminating one cutoff gives rise to two new ones. The very notion invoked to solve a problem (that some things are neither φ nor not φ) then gives rise to a problem of exactly the same sort. See [Colyvan, 2008b].

Gap solutions to the sorites attempt to preserve TOLERANCE, at least in spirit, but at some level must (like everyone) deny it, by allowing sharp cuts.[46] I think if that's what is going to happen anyway, then as with the liar paradox we should be upfront about it. The gap theory takes only one horn of the dilemma, that there is something wrong with saying 'this is *exactly* what it takes to be a tall mountain'. But this is a dilemma with two horns. Let's look at the glutty solution.

1.2.2.3 Glutty Solutions

Because there is a difference between being high up and not, the difference must begin somewhere, and so there must be cutoffs. TOLERANCE is true, but it is also false. The idea of the glutty solution is just to accept the existence of the cutoff and explain it.[47] Because TOLERANCE is both true and false, borderline cases of φ are both φ and not φ. The guiding

---

[44] See [Hyde, 1997; Beall and Colyvan, 2001a].

[45] For a formidable elaboration of a fuzzy approach, see [Smith, 2008].

[46] Although maybe not so for gap theorists who are so thorough as to adopt a different view of mathematics. In a Brouwerian setting, the intermediate value theorem fails [Bishop and Bridges, 1985] and with it the argument that a continuous shading off from red to not red must cross a sharp line somewhere; similarly for Heyting arithmetic, where the least number principle is weakened [Heyting, 1956].

[47] See [Priest, 2019a]. From their start, paraconsistent logics were intended for application to vagueness [Jaśkowski, 1969]; cf. Priest and Routley [1989a, p. 389]. See also [Hyde and Sylvan, 1993] and [Beall and Colyvan, 2001b] on ontic (inconsistent) vagueness and seeing contradictions.

picture in the background is of the extension of a vague predicate φ overlapping with its complement, which looks like this:

$$\overbrace{a_0 < \ldots < a_{j-1} < a_j}^{\varphi} < \underbrace{\ldots < a_k < a_{k+1} < \ldots < a_n}_{\neg\varphi}$$

Vagueness is not indeterminacy, but *overdeterminacy*. The point of the approach is to admit the joint truth of two claims: TOLERANCE and *nontriviality* – the empirical fact that some points on Olympus Mons are high up, but others are not. Put these together with a mathematical principle asserting that if a change occurs, it must occur somewhere, and a contradiction follows.

How can this be coherently maintained? TOLERANCE is extensional: it just says that *either n cm is not high up, or else $n + 1$ is*. And since in a paraconsistent logic, disjunctive syllogism is invalid (the material conditional does not obey modus ponens; see Section 0.2.2.3), the apocalyptic conclusion that all points are high up does not follow.[48] Reconsider the classical conditional sorites premises: they present pairs, a_i, a_{i+1}, such that either a_i is not high up or a_{i+1} is high up. At each pair, to conclude that a_{i+1} is high up would be to take as valid the argument

$$\varphi(a_i), \neg\varphi(a_i) \vee \varphi(a_{i+1}) \therefore \varphi(a_{i+1}),$$

implicitly assuming that it couldn't be that a_i both is high up and is not. If there can be, or is, an inconsistency, then the conclusion that everything is high up does not follow, and the sorites comes to an abrupt but natural end. Rather than the soritical reasoning running unchecked to an absurdity, on this account, the sorites argument halts because it reaches an inconsistency. The inconsistency is *revealed* by the soritical reasoning: TOLERANCE on vagueness, which we must respect if we are trying to understand the phenomenon, insists that being high up does not abruptly come to a halt; since it does, one of the sorites premises is false as well as true, and nontriviality is preserved.

I will put more detail on this in a moment. For now, the crucial element of "saving the appearances" while escaping revenge is again a major virtue of the dialetheic approach. Revenge for the other approaches is an attempt to deny cutoffs, only to have them recur at higher levels. The dialetheic approach simply accepts the cutoff, and makes this possible via inconsistency. A baby becomes a non-baby by being, for a while, both a baby and not a baby. If you've ever lived with a baby, this should not seem so implausible.

1.2.3 Paraconsistency, Definite Descriptions, and Uniqueness

In the previous section, I suggested that the phenomenon of vagueness is that tolerance is both true and false; as such, vague predicates like "is high up" have *inconsistent cutoff*

---

[48] What about phrasing tolerance with a conditional that does obey modus ponens? See [Beall and Colyvan, 2001a]. Depending on the conditional, that would deliver a principle that is simply false. See Section 2.2.3.5 and [Weber et al., 2014].

points. Some altitude is both high up and not high up. I will now argue that, correlatively, there can be *more than one* "first" high up point, and that the abundance of legitimate cutoffs is *why* it seems vague: there are so many "the first" to choose from, we can't believe any of them are really the first. But they are.

How can there be more than one "the first" cutoff point? *Definite descriptions* such as "the first high-up point" can, on a paraconsistent analysis, be satisfied by multiple objects. Russell's 1905 analysis [Russell, 1905a] suggests that "a_k is the unique φ" should be glossed as

a_k is φ, and for all j, if a_j is φ then $a_j = a_k$.

Russell uses a material conditional. Unpacking the description, then:

a_k is φ, and for all j, either a_j is not φ, or $a_j = a_k$.

Now establishing that $a_j = a_k$, given that a_j is φ, would require an application of disjunctive syllogism. If, however, it is possible that a_j is both φ and not, then such an inference would be incorrect. What does this portend?

Using the idea of definite descriptions, we will say that some z is a *cutoff* for φ on the condition that it is the first non-φ, e.g., z is not φ but nothing before z is not φ, or

$$\neg\varphi z \mathbin{\&} \forall y(y < z \supset \varphi y).$$

Now, consider what counts as a cutoff in the following stripped down scenario:[49]

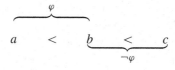

It may appear that b is the obvious cutoff point, and that c is not a cutoff. But this would not be the whole story. Because b is overdetermined, it will turn out that both b and c are cutoffs for φ, even while they are numerically distinct. For everything before b, namely a, is high up, but b is not high up. So b is (at least) a cutoff. On the other hand, everything before c is high up (because b also *is* high up), but c is not, so c is a cutoff. So both b and c are cutoffs. On assumption, b was distinct from c. Therefore, being a cutoff does not imply being unique. Uniqueness of the cutoff is not necessary.

If we read the Russellian analysis as suggested – that a_k is *the* φ – then there can be *more than one least* such and so. A definite description, "the first φ," may be satisfied by more than one object. According to the paraconsistent picture of sorites, a Russellian definite description holds, as it should, without the further consequences that are only drawn if inconsistency is discounted. If vagueness behaves anything like the picture presented here, drawing these further consequences about uniqueness would be as disastrous as inferring that everyone is bald. The notion of "least" is not univocal in inconsistent contexts.[50]

---

[49] A four-element model was first suggested by Sam Butchardt. Is there a one-element model that would do the same job? Stay tuned.

[50] What about rephrasing Russell's definite description scheme with a conditional $\rightarrow$ that *does* obey modus ponens, e.g., "$\varphi(x) \mathbin{\&} \forall y(\varphi(y) \rightarrow x = y)$"? As with the issue of rephrasing the sorites argument with a detachable conditional, this

One more feature of the model should be pointed out. Both b and c are cutoffs, and not. That is, b is high up, so b is also not a cutoff:

$$\neg\varphi b \mathbin{\&} \forall y(y < b \supset \varphi y) \quad \text{and} \quad \neg(\neg\varphi b \mathbin{\&} \forall y(y < b \supset \varphi y))).$$

Meanwhile, b is not high up, so some things before c are not high up and therefore c is (at least) not a cutoff:

$$\neg\varphi c \mathbin{\&} \forall y(y < c \supset \varphi y) \quad \text{and} \quad \neg(\neg\varphi c \mathbin{\&} \forall y(y < c \supset \varphi y)).$$

Thus it is false, as well as true, that they are cutoffs; but they are the only cutoffs. While it is therefore true that in the model there is a cutoff for φ,

$$\exists z(z \text{ is a cutoff for } \varphi),$$

it is, consequently, also false, and hence has a true negation:

$$\neg\exists z(z \text{ is a cutoff for } \varphi).$$

Read literally, this says that *there is no cutoff at all*. This accounts for our incredulity about the existence of a cutoff point: no n is the cutoff point, because $\exists k(\varphi k \mathbin{\&} \neg\varphi(k-1))$ comes out as both true and false. There is more than one cutoff point, and none.

How does this *explain* the sorites? In part, I think the pull of TOLERANCE is due to an implicit, utterly reasonable question. If TOLERANCE is false and there really is is a cutoff point, *where* could it be? Intuitions cry out that it is *impossible* to believe that vague predicates have sharp boundaries. It is incredible – there *cannot* be such a sharp boundary; otherwise, why can't we identify it? One has the feeling that if there were a sharp boundary, we could say where it is. And while I suspect this sort of worry places too much emphasis on epistemic accessibility,[51] I do think part of any approach to the sorites paradox needs to offer some sort of answer to this reasonable question. The model here goes some way in doing so. Why can't we *identify* cutoffs? Because we expect any such identification to be *unique*, fixing the one-and-only first φ, and such cutoffs are not unique. In the three-element model, just by dint of cutoff b being overdetermined, c turns out to be a cutoff. Despite appearances, one cutoff point cannot be said to be a more natural or obvious cutoff point than the other, and this is how to vindicate our competing intuitions: *the first φ is not the only φ*. The progression is like descending from Olympus Mons itself. A single inconsistent cutoff appears; but then, by inconsistency, there cannot be just one cutoff; and so finally, since any one cutoff is a defeater for all the others, *there is no cutoff*, which is what we expected all along.

depends on the conditional (e.g., for using relevant implication in related applications, see [Dunn, 1987]) but the upshot is that anything stronger than the material hook is implausible as an expression of "the." We will return to this in the context of mathematical functions versus relations in Chapter 5.

[51] Williamson writes: "When we conceive that something is so, we tend to imagine finding out that it is so. We are uneasy with a fact on which we cannot attain such a first-personal perspective. We have no idea how we ever could have found out that the vague statement is true, or that it is false, in an unclear case; we are unable to imagine finding out that it is true, or that it is false; we fallaciously conclude that it is inconceivable that is it true, and inconceivable that it is false" [Williamson, 1994, p. 3].

"How did you go bankrupt?" asks a character in Hemingway. "Two ways," is the answer: "Gradually, then suddenly." The sorites is a *paradox*, because the situation it describes is unbelievable, even while the principles generating that description make it unavoidable. The sorites paradox is that vague predicates cannot have sharp boundaries, even as they must. The dialetheic paraconsistent solution to the paradox is that it is true. Every TOLER-ANCE pair $\varphi_n \supset \varphi_{n+1}$ is true, but some things are φ and some things are not. There is some cutoff point, but there is also no such cutoff, since both tolerance and its negation are true. The intuitive data are (putatively) preserved, without absurdity, and the sorites is explained as a feature of more fundamental inconsistency. The paraconsistent mathematics to back this up is in Part III. Connections between sorites and the other paradoxes are discussed at the end of Section 1.4.

1.3 Boundaries

It is getting dark. Let's go back inside. It is twilight now, both day and night, but soon it will be only night.

Night is when we are in the the shadow of the Earth. But we do not think of the darkness of the night as one big shadow. Why not? Casati suggests it is because the shadow of night has no *edges*, or at least not ones we usually see,[52] and a shadow is ordinarily thought of as something with edges, or boundaries. Shadows are strange in themselves;[53] and yet, even for entities that minimally qualify as entities, there is a clear intuition that shadows must have an edge in order to exist. Boundaries are integral in some way to there being objects. Euclid [Book 1, Def 13–14, emphasis added] [Euclid, 1956]; emphasis added:

A boundary [terma] is the extremity of any thing. A boundary is the limit within which anything is contained. *A figure is given by its boundaries.*

Without the limit of a figure, there is no figure at all.

Even more so than with sets, truth, or vagueness, objects with boundaries are so much a part of our everyday experience that it is at first highly implausible that they be paradoxical. We navigate boundaries *all the time*, from not bumping into people when we walk down the street, to turning off the water tap before the sink overflows. The boundaries of human bodies are basic to our phenomenal experience.[54] We are inclined to assume, incorrectly, that if something is a commonplace, then it could not be problematic; but material objects in space are every bit as puzzling – and paradoxical in *the same way* – as sets, semantics, and sorites.[55]

---

[52] In [Casati, 2003]. A planet's *terminator* is the edge of the night.
[53] Cf. the *Yale shadow paradox*, 1975 (see [Sorensen, 2008]), used to motivate dialetheism in [Mares, 2004b].
[54] As Merleau-Ponty argues; see [Dillon, 1997].
[55] Thanks to Aaron Cotnoir for collaboration on this topic in [Weber and Cotnoir, 2015].

1.3.1 The Problem: Symmetry and Connected Space

There is a Great Red Spot on the planet Jupiter.[56] Estimates vary as to how old it is (in the vicinity of 350 years), but all agree it is an anticyclonic vortex, a gigantic and remarkably persistent storm at least twice the size of the Earth.[57] Although the Spot is dynamic, fluctuating in size and shape, it is stable and stark, visible through Earth-based telescopes. What is it that makes the Great Red Spot a distinctive part of the Jovian sky?

The Great Red Spot is a very large and red example of a much more general phenomenon – the existence of material, ordinary objects. What is it that makes the Great Red Spot an *object* at all? This is like the question that we asked, and answered, around set composition, about how the *many* particles become *one* spot (Section 1.1.3.1). For the world is unquestionably full of *stuff*;[58] some of this stuff *composes* so as to make other stuff, called *ordinary objects*;[59] and a natural observation about ordinary objects is that each of them has a *boundary*. Aristotle defined the boundary of an object as "the first thing outside of which no part is to be found, and the first thing inside of which every part is to be found" (*Metaphysics* V, 17, 1022a4–5). As Varzi puts it, "whether sharp or blurry, natural or artificial, for every object there appears to be a boundary that marks it off from the rest of the world" [Varzi, 2004]. A boundary contributes crucially to there being a figure at all, a *thing*, and not merely a disenfranchised mess of unrelated *stuff*.

Boundaries are puzzling, too. The sorites paradox has already shown this: the existence of twilight threatens to, but cannot, mean there is no difference between day and night.[60] Now focus on the dual problem: could there be a difference between day and night *without* the intermediary of twilight? Take the following puzzle: consider the Great Red Spot on Jupiter, and trace a path from a red-colored point inside the Spot to a non-red colored point outside the Spot. What happens as we pass through the boundary of the Spot?

The most obvious answer is that we pass from the last red point to the first non-red point (passing over any question of whether *points* have a color). But on the assumption that space is *dense* (recalling the continuous sorites in Section 1.2.1), there would be an infinite number of points in between. What color are they? Do they belong to the Great Red Spot, or to its complement in the Jovian atmosphere? Either some point both is and is not part of the Great Red Spot, or else neither, leaving a "gap" on the face of Jupiter. It is *arbitrary* simply to assign the boundary to the Spot, say, and leave its complement in the atmosphere unbounded [Varzi, 2004], or vice versa. That would violate the *principle of sufficient reason* (PSR), our Premise 0 from the preface. It is incomplete not to assign the boundary at all; but the remaining alternative is inconsistent, as I will now try to bring out.[61]

---

[56] There is also a giant hexagon on Saturn. Look it up.
[57] [Rogers, 1995, pp. 191–196].
[58] "Stuff" is a technical term; as in [Miller, 2009], it is contrasted with "thing": there is *a* thing but *some* stuff.
[59] Another technical term; see [Thomasson, 2007], also [Koslicki, 2008].
[60] Example attributed to Dr. Samuel Johnson.
[61] For discussions of similar material, see Varzi [1997]; Casati and Varzi [1999, p. 74]; Dainton [2010, ch. 17]; Hudson [2005, chs. 2, 3]; Priest [2006b, chs. 11, 12, 15]; Arntzenius [2012, ch. 4]; Hellman and Shapiro [2018, 7§4].

1.3.1.1 Symmetry

As with the sorites paradox, the boundary puzzle does not go away once we bring it mathematical rigor. It becomes more puzzling.

As mentioned in the Introduction, the modern theory of continuity was developed in part by Dedekind [Dedekind, 1901], as a reverse engineering feat around the intermediate value theorem. He wanted to ensure that, given a starting point inside a set of points, e.g., the Great Red Spot, and a finishing point outside it, that a continuous path must *cross* the boundary between them. As for which side the boundary line belongs to, Dedekind dismissed the problem: one or the other, but whichever we like, he said.

If now any separation of the system R into two classes A_1, A_2 is given which possesses only *this* characteristic property that every number a_1 in A_1 is less than every number a_2 in A_2, then for brevity we shall call such a separation a *cut* [Schnitt] and designate it by (A_1, A_2). We can then say that every rational number a produces one cut or, strictly speaking, two cuts, which, however, we shall not look upon as essentially different [Dedekind, 1901, § IV, emphasis in the original text.]

By definition, a *closed* object includes its boundary. The boundary of an *open* object belongs to its complement. So Dedekind's solution is that a boundary splits a (continuous) space into two parts, where one part will be open and the other closed. The same goes for dividing a one-dimensional line, as we saw (in the Introduction): the center of the line is an indivisible geometric point, so any division of it leaves one side open and the other side closed.

Dedekind's solution is ingenious, but it solves a mathematical problem by creating a metaphysical one. Given no other facts about the space, *why* do the open/closed properties fall where they do? Neither macroscopic analysis nor mathematical argument settle the matter, since physically the object loses its definition at very small scales, and logically, the properties of being open and closed may be interchanged without any rational difference.[62] The principle of sufficient reason[63] motivates the following constraint:

Symmetry: Objects that otherwise have no relevant difference between them have no relevant difference between their boundaries.

From SYMMETRY follows a corollary: if there is an object with a closed proper part, then, absent any further compelling reason otherwise, the relative complement of that part is closed too, if no reason can be given why one side should get the boundary rather than the other. Note carefully, however, that SYMMETRY does *not* say that *every* division of an object is symmetric. That would be silly. Nor does it say that all objects are closed. SYMMETRY says that there is *some* way to divide an object without creating an imbalance. SYMMETRY is violated when *every* way of dividing objects forces an arbitrary difference between its otherwise indifferent parts.

---

[62] [Pratt-Harmon, 2007, p. 14], although see Chapter 9.
[63] "The principle of sufficient reason, on the strength of which we hold that no fact can ever be true or existent, no statement correct, unless there is a sufficient reason why things are as they are and not otherwise – even if in most cases we can't know what the reason is" [Leibniz, 1714, p. 32]. For more on the PSR, and its role in Spinoza, see [Della Rocca, 2008].

To put SYMMETRY the other way around, then, if there *is* a difference between the boundaries of two objects a, b, (which we will write $\partial(a), \partial(b)$ (a precise definition of ∂ is given in Section 9.2.1)), a principled and explicable difference that persists across all cases, then there is an explicable difference between the objects after all. This is itself a continuity intuition (cf. Section 8.3.2), and closely related to tolerance – roughly, if a is very similar to b, then $\partial(a)$ is very similar to $\partial(b)$. Where one object is open and another is closed, there should be some reason as to why; to ask which is which about Jupiter and its Spot is to ask "an embarrassing question" [Casati and Varzi, 1999, p. 87].

The puzzle is not merely physical or mathematical. An object may be *open and closed* (even classically, where this is sometimes called being "clopen"), but two open-and-closed objects *cannot*, on the classical account, touch (Section 1.3.1.2). What metaphysical sense can we make of the classical account of bounded objects in connected space? Are there mysterious metaphysical forces such that some objects (ones that are both closed) simply must repel one another?[64] Or, if not mysterious forces, then primitive mysterious *differences*, a "preordained asymmetry"? Nothing at this level of generality accounts for some objects being open and others closed. And the level of generality is important: we are not trying to understand why some objects are round, and others square, or why some objects are flowers and others are trees; we want to know why for the continuum, and so for any continuous object, symmetric division always fails. The reason must be something about objects qua objects – not a case-by-case inspection of each situation, but something about the pure geometry of how things sit together in the world, just by virtue of their being things. No matter how serendipitous one thinks the conferring of structural properties has been, differences at this level of generality call out for an explanation. The SYMMETRY problem is the postulation of a distinction without a difference.

1.3.1.2 Connectivity

The bite of the boundary puzzle comes from the assumption that space is connected. The key fact is just this:

Separation: A space is *separated* iff there is an exclusive and exhaustive division of the space into two closed parts.

Spaces that are not separable are *connected*:

Connection: A space is *connected* iff every exhaustive division of the space into two closed parts is not exclusive (has a nonempty overlap).

Since the basics of topology tell us that the complement of a closed set is open,[65] both parts of the partition in a separation are open, too. Connectivity is enmeshed with the concepts of open and closed sets: if a space is connected, then according to classical topology following

---

[64] "A certain class of objects are unaccountably differential to one another – always just managing to step out of one another's way – while they bang heedlessly into the members of another class of objects. Surely repulsive forces would *have* to be posited to explain such behavior" Zimmerman [1996, p. 12]. See Sider [2000, p. 587] for a reply.

[65] Kelley [1955, p. 40].

Dedekind, it cannot be that it divides symmetrically. That would just be a separation.[66] What this shows is that a space that satisfies the SYMMETRY intuition cannot, apparently, be connected. And this is itself a problem, because the space of naive experience *is* connected. There are no rips or tears.

In classical topology, the definition given in SEPARATION is equivalent to another: that *the only sets in connected space that are both open and closed are the entirety of the space, and the empty set.*[67] Why? How does this rather recondite restatement follow from intuitive claims about space? Well, consider an apparently connected space like Jupiter, and some nonempty proper part of it, like the Great Red Spot, G. The boundary of G, intuitively, is made up of all the points that are "extremely close" to both G and its complement $\overline{G}$. Suppose G is both open and closed. Then:

(1) If G includes its boundary $\partial(G)$, if it is closed, then all the points in $\partial(G)$ are points in G.
(2) If G is open, then all the points in $\partial(G)$ are points in $\overline{G}$.
(3) Ergo if G is both open and closed, then for any $x \in \partial(G)$,

$$x \in G \qquad \text{and} \qquad x \in \overline{G}$$

which is a contradiction, $x \in G$ and $x \notin G$.
(4) By classical reductio, then, there is no $x \in \partial(G)$, and the boundary is *empty*, $\partial(G) = \varnothing$.
(5) But an empty boundary between G and its host space means that G is *separated* from the host space, and that the overall surface is *disconnected*.

Therefore, if the Great Red Spot is really a nonempty proper subspace of Jupiter's atmospheric surface, *and* it is both open and closed, then there is a rip in the fabric of the space, and so the space is not connected. Thus a connected space has no nonempty subspaces that are both open and closed.

The existence of bounded, discrete entities grates against the connectivity of space (cf. Section 9.2.1). The problem aligns closely with the sorites, as discussed in Section 1.2, with SYMMETRY for TOLERANCE and SEPARATION for the line-drawing negation of TOLERANCE. If objects differ, they must inevitably begin to differ somewhere, but any exact point where the difference begins looks to "cut" our lived-world impossibly.

Reasoning has led us into a corner once again. On the one hand, objects that are touching share a boundary; space is connected. On the other hand, objects do not arbitrarily differ; boundaries are not *random*; and so *both* objects in the connection can still have their boundary as a part. This breaks the space into two subobjects that are both open and closed. If a subobject in space is both open and closed, then the space is not connected – but there *are* subobjects in space that are both open and closed, and yet space *is* connected! The underlying principles here – intuitions about the topology of our lived experience – are

---

[66] Cf. [Paul, 2006].
[67] [Kelley, 1955, p. 53; Hocking and Young, 1961, p. 15].

perhaps not as axiomatically compelling as naive set comprehension, but they are robust, as we're about to check by considering if there are other options.

1.3.2 Options

The previous section has already detailed the orthodox cause and solution to the problem, for example defended in [Bolzano, 1851]: a boundary may be part of an object *or* its complement, but not both. Other approaches all endeavor to explain how a boundary never arbitrarily belongs to an object over its complement, always assuming the connectivity of space. The increasingly familiar options of incompleteness and inconsistency return; but now, unlike in the case of abstracta such as propositions or sets, the implications are literally palpable.

1.3.2.1 Underlap

Analogous to the nihilist view of vagueness, or the nihilist view of mereology, is the so-called *eliminativist* response to this problem: boundaries do not exist. At least, not as parts of objects: boundaries are either abstractions (equivalence classes) of convergent series of nested bodies, the boundary of the space-time receptacle of that object, or some other ersatz replacement concept.[68]

Eliminativism automatically avoids assigning boundaries arbitrarily, but only because it does not assign boundaries at all. Since we are interested in explaining boundaries, not explaining them away, this way of (vacuously) accommodating SYMMETRY is a nonstarter. It may be a *problem* (paradoxical, even) to face liar sentences, Russell sets, vague predicates, and boundaries, but that does not excuse us from doing so.

Analogous to the "indeterminateness" intuition from vagueness, it is natural enough to think that a boundary may be neither part of an object nor part of its complement.[69] And in analogy to the paracomplete/gappy approaches to the sorites paradox, proposing an "underlap" between objects faces basic revenge problems. For what separates the Great Red Spot from its complement? A boundary exists between them, neither part of the Spot nor its complement – which leads to a new question: since the boundary and the Spot are disjoint, what separates the boundary from the Spot itself? What separates the boundary from the Spot's complement? Presumably, by parity of reasoning, there must be a *new* boundary between the Spot and its boundary. If we reapply the underlap solution, *this* boundary is part of neither the Spot, nor the boundary of the Spot, and we're off on a regress of infinitely many distinct gaps between any two objects in contact. To the extent that one boundary needs explaining, multiplying new boundaries only multiplies the problem. This is very similar to the chase of cutoffs in higher-order vagueness (Section 1.2.2.2);[70] this is paradigmatic revenge.

---

[68] The most notable eliminativist is Whitehead; see [Clarke, 1985; Hellman and Shapiro, 2018].
[69] Varzi [Varzi, 1997, p. 27] attributes this view to Leonardo.
[70] Thanks to Marcus Rossberg here.

1.3.2.2 Overlap

In what is known as the *coincidence* view, there are two coinciding boundaries: one that is part of an object, and one that is part of its complement.[71] These two boundaries occupy the same space. The theory claims that all entities whatsoever have boundaries as parts as a matter of metaphysical necessity [Smith, 1997, p. 18]:

... the possibility of a coincidence of boundaries is essential to the concept of what is continuous *[Brentano, 1988, p. 5].*

On this view, all entities are *closed.* In this way, there is no arbitrary choice between two adjacent objects as to which is to be open and which closed. Both must be closed, each including their own boundary.[72]

The coincidence view approaches, but does not embrace, acceptance of contradiction. Even if the boundaries coincide, they are distinct. The view then suffers from basic revenge. For, on this view, there are many otherwise indistinguishable boundaries, all colocated and doing the same work. While SYMMETRY is nominally respected in that there is no arbitrary choice between bounded and unbounded objects, there still is a choice as to which object a given boundary belongs. The Great Red Spot has a boundary b_1 and so does its complement b_2. But given the coincidence of b_1 and b_2, what makes it the case that it is b_1 and not b_2 that is the Spot's boundary? The two boundaries occupy exactly the same set of points; to mark any difference at all beyond two names, it must be that boundaries are essentially "directional," intensional. A given boundary is always directed toward the (only) entity that it bounds. (A point in the interior of a solid sphere is a boundary in *every* direction.) But there is no explanation for a given boundary's direction. Boundaries are attached to their objects in an essentially arbitrary way. So the violation of SYMMETRY is very basic: there is a distinction between two boundaries, recorded by directionality, but no *difference* between them. This is little better than the original problem, which was explaining why a boundary went one way rather than the other. This is revenge.

1.3.3 Toward a Paraconsistent Topology

The coincidence view is approaching a simpler *glutty* view: a nonempty boundary may simply be part of *both* an object and its complement. Points very near both the Great Red Spot and its complement are both parts of the Spot and not. It is perhaps Brentano's coincidence view that best shows how our competing intuitions are irreconcilable – if consistency must be maintained. Objects in contact overlap, if they are to touch symmetrically. Overlap means that the objects share some points, either their entire boundaries or some relevant portions thereof; when an object touches its complement, there will be points both on and

---

[71] See [Chisholm, 1984; Brentano, 1988]; cf. [Smith, 1997].

[72] Concomitantly, there are a great deal of coincident objects: "Each point within the interior of a two- or three-dimensional continuum is in fact an infinite (and as it were maximally compressed) collection of distinct but coincident points ... (Not for nothing were the scholastic philosophers exercised by the question as to how many zero-dimensional beings might be fitted onto the head of a pin)" [Smith, 1997, p. 10].

not on an object, calling for a paraconsistent treatment. Being disjoint (no overlap) may hold simultaneously with overlap. Thereby, SYMMETRY can be satisfied simultaneously with CONNECTION.

That's the thought, in any case. The project of understanding boundaries (at this level of generality) comes down to a need for an account of proper, nonempty subspaces of connected space that may themselves be both open and closed. And, it should be added, the theory that vindicates this still must bear enough resemblance to ordinary topology to be modeling the same concepts. It is left to Chapter 9 to see whether enough precision can be given to this naive idea. For the moment, let's circle back to make the connections between this paradox and the others clearer, by looking at how to express the sorites paradox in purely topological terms.

1.3.3.1 A Topological Sorites

As with the continuous sorites, the sorites paradox does not require a digital or discrete presentation; in fact, it does not even require an ordering relation. Or so a *toplogical sorites* purports to show.[73]

A function f is *locally constant* iff for each $x \in X$ there is a "neighborhood" of nearby points such that the restriction of f to that neighborhood is constant, all taking the same value. A *globally constant* function always takes the same value, without restriction to a neighborhood. It is a classical lemma that *if f is locally constant in a connected space, then f is globally constant.*

Let X be connected, and f a function from X to some set Y. Consider some region A of X. Now the range of f, Y, can be thought of as the pair of truth values $\{0, 1\}$, in which case the *characteristic function* σ of the set A is defined thus:

$$\sigma_A(x) = \begin{cases} 1 & \text{if} \quad x \in A, \\ 0 & \text{if} \quad x \notin A, \end{cases}$$

where $A = \{x : \varphi(x)\}$, the analogous $\sigma_\varphi(x) = 1$ if $\varphi(x)$ and 0 otherwise. But if f is locally constant on A, so that $\sigma_\varphi(x) = 1$ for every $x \in A$, then by the preceding lemma, f is globally constant: $\sigma_\varphi(x) = 1$ for every $x \in X$.

In keeping with the proposed definition of vagueness offered at the end of Section 1.2.1, say that a predicate is *vague* iff something satisfies it, and its characteristic function is locally constant – that's TOLERANCE – but not globally constant. As ever, in connected space, vagueness threatens to trigger a slippery slope avalanche: either nothing satisfies a vague predicate, or everything does.

Let X be connected and A be a subset of X, with A the extension of a vague φ. Then by the previous reasoning about locally constant functions, either $A = X$ or $A = \varnothing$. This gives us a *topological inductive* version of the sorites:

---

[73] This is a completely classical argument. As such, I only sketch the argument; see [Weber and Colyvan, 2010] for unpacking, or just [Jänich, 1984, p. 14], where I got it from. For criticism of this sorites construction, see [Rizza, 2013].

Some member of X is φ.

X is a connected space, and φ is vague.

∴ Every member of X is φ.

Contrapositively, the result is a *topological line-drawing* sorites: for some vague φ,

Some things in X are φ.

Some are not.

∴ X is separated.

The boundary of the φs in X is empty, on pain of contradiction. The usual sorites paradox is a special case of the boundary problem in space. The glutty interpretation of the line-drawing form, following from the preceding, is that X is only "separated" in the sense that φ has an *inconsistent boundary*.

* * *

With the topological sorites in view, let's see how it helps with the more usual sorites, and paradoxes in general. Here is *not* how the boundary puzzle was presented: given this object,

then is its boundary black or white? This form of the question is static, third-personal. It has all the needed components for the puzzle, but does not feel puzzling. No, rather the question is posed: *as you cross the threshold, what do you see?* It takes longer acquaintance with the situation to generate the relevant phenomenology of paradox (unbelievable!) with a static object, as the topological version of the sorites attempts to do. The sorites paradox feels most acute when it is dynamic, for example in the form of a "forced march" (... and is this man bald? What about this man? And this ... ?).[74] But topologically, no paths are required. The underlying structure of the space that makes a problem for the march is already there. Our first personal experience is how we come to learn about facts, but we (perhaps to our disappointment) eventually begin to learn that the facts don't care whether we interact with them or not. Marching has nothing to do with it.

I do not think that, if only we could speak or think about these problems clearly enough, they would be resolved. I think the history shows that the more clearly we think about these problems, the more irascible they become. Absurdity arises whether or not we the

[74] See [Horgan, 1994].

humans bring our demands to bear on the world; we are not so powerful that we can make the otherwise-sanguine world absurd. We notice the paradox at the penumbra because it is there to be noticed – we are here *in the presence* of an absurdity. So, while our participation in the process makes the problem vivid, whatever is driving revenge *would already be there even if no one ever tried to solve the paradoxes.*

If there were only paradoxes to do with sets or semantics, that would be quite a lot – but the effect would remain intangible, abstruse.[75] The paradoxes could be compartmentalized away, as a quirk about human cognition, representation, language, or the like. Similarly, if there were only paradoxes to do with our responses to vague stimuli, that would be quite a lot, but not the job of pure metaphysics or mathematics to answer. The paradoxes go beyond subjective and mental experiences. I think the topological case indicates that the problem can be generated with no more than some points in space.

Just as logic is not a solution to the paradoxes (Section 0.2.3), the paradoxes are not about *us*. To put it colorfully, the paradoxes would be paradoxical even to God, and "God has no need of any arguments, even good ones" [Meyer, 1976, p. 94]. If that is right, our attention can focus at the right level to be trying to understand the paradoxes: what it is about the *world* as such, about objects and their boundaries, that makes there be paradoxes?

1.4 Conclusion

This chapter has presented several interconnected paradoxes. They are a problem for anyone with basic convictions about *comprehension*, *continuity*, and *composition*. Start with something obviously true (set comprehension; tolerance; the existence of boundaries in connected space), and you find a contradiction – but any attempt to evade the contradiction leads to revenge, because what you started with was obviously *true*.

One thing these paradoxes all have in common is that they are *resilient*. This is the revenge phenomenon, and a tell-tale sign of a genuine paradox. Revenge is what drives, I think, an acceptance-based approach of dialetheic paraconsistency (though not without its own costs; see Section 3.3.2, Chapter 4, and Chapter 10). The paradoxes are not solvable. Simple immunity to revenge is not a virtue just on the basis of scoring points against competing views; immunity to revenge shows that a position has got to grips with the real problem, while those that evade and are stung, evade and are stung, again and again, have not. Any solution does not merely fail, but recapitulates the original paradox.

Tappenden points out the striking fact that to state what is *wrong* with these paradoxes is just to *state* the paradoxes. In the case of the liar, we have a sentence ℓ for which

(*): ℓ is true iff ℓ is not true.

Sentence (*) "seems to capture one feature of the sentence ℓ in virtue of which ℓ is paradoxical," says Tappenden (he uses "k" instead of "ℓ"); he explains [Tappenden, 2002, p. 552]:

---

[75] Semantic dialetheism is discussed in [Mares, 2004b] and defended by Beall [Beall, 2009]. From another direction, *fictionalism* about dialetheism is discussed in [Kroon, 2004].

Imagine that you are explaining the liar to someone who does not catch on right away. You might well say: "Here is what is funny about ℓ. If ℓ is true then it is not true, and if ℓ is not true then it is true." You have apparently just contradicted yourself by uttering a variation on (*). But (*) seems to be the right thing to say in the situation. And does it not get across precisely what is odd about the liar?

This is an intuitive way of explaining revenge. An attempt to solve the problem – or even articulate it – triggers the problem itself! As with the liar, so the sorites.[76] Tappenden describes a parallel scenario [Tappenden, 2002, p. 553]: you explain that "if any two samples are observationally indistinguishable to you, then one looks red to you if and only if the other looks red to you." Just like (*), this "seems like just the thing to say [to] correct the other person's mistakes ... But if we are attempting to say something true, we are failing." And spatial boundary paradoxes suggest a common structure underlying these puzzles: that there is inconsistency at the edges of things.

In the last two chapters, I have tried, in effect, to suggest that this is the case for all the relevant paradoxes: simply to describe them (ordinals are sets of proceeding ordinals; some people are babies and some are not; closed objects in connected space can touch) is to find ourselves led into contradiction. And then, I've suggested ([Priest, 2002a] argues the point at length) that if we try to find our way out of the contraction, by denying the data in one fashion or another, that leads into contradiction *again*. In abstract, the structure of revenge is as follows:

Step 1 – there is a paradox, at the boundary between two (exclusive/exhaustive) categories.
Step 2 – solve the paradox by introducing a new, third category.
Step 3 – a new (is it?) paradox arises at the two new boundaries.

The pattern of the paradoxes is a particular sort of valid argument that ends in a particular sort of contradiction. The argument is one in which *denial of a premise* leads to a regress, or more euphemistically, a "hierarchical" picture (iterative sets, higher-order vagueness, gappy boundaries) where the hierarchy in some direct sense then affirms the very premise being denied; cf. [Priest, 2007].

In emphasizing revenge, I am *not* endorsing a principle like the following: "If two paradoxes respond the same way to treatment, they are the same type of paradox." That is not correct.[77] Rather, the problem in both the liar and the sorites (and by extension, the Russell paradox) is that there is something on a boundary where it shouldn't be – between truth and falsity, high and not high, the Russell set and its complement. We are compelled to admit that it really is there, because attempts to remove the bad overlapper lead to more boundaries with more overlappers, and not as a one-off accident but *no matter what*, a

---

[76] The thought that the sorites and the paradoxes of self-reference have something important to do with each other is also found in [Dummett, 1978, ch. 12; McGee, 1990; Sorensen, 1994, p. 53; Field, 2008, p. 106]. More recently, prominent approaches to paradox using substructural logic ([Zardini, 2011; Ripley, 2013]) are developed to treat both the liar and the sorites.
[77] See [Colyvan, 2008b].

structural necessity.[78] The standard consistent/incomplete responses lead to the same end. They lead me to a paraconsistent approach – not as a last resort, but as first step toward something new.

* * *

In this chapter, I have set the scene by making the following case. Naive set theory is true. Therefore, there are true contradictions – paradigmatically, the liar and Russell paradox, among more sophisticated others. Naive sets provide a simple account of multitudes grasped together as a singular entitity. In the case of material, ordinary objects, this cohesion is particularly vivid when looking at the boundaries of things, and points on the boundary. The sorites paradox is the sharp edge of the boundary paradox, occurring on spaces where we find it particularly noticeable that there are boundary points that seem both part and not part of an object, multiple "first" cutoff points. Any response to these paradoxes leads, because of revenge, either explicitly or implicitly to a contradiction.

I've made this case at full speed, and don't pretend to have given proper consideration to all the arguments. We've seen some paradoxes, and I have broadly recommended dialetheic paraconsistency as a promising response. We could keep weighing up the options in the abstract, but to understand the glutty option properly, I think we need to see what it can actually do, in terms of making sense of the paradoxes. That's what I plan to pursue here. The question now, then, is: how is the dialethetic approach to proceed?

---

[78] Compare, for example, your plan to live a life of unfettered adventure, and your plan to settle down securely with a long-term partner; see [Slote, 2011], echoing Kierkegaard's injunction in *Either/Or* that whether you marry or do not marry, "you will regret it either way." : "Marry and you will be unhappy; do not marry, and you will be unhappy."

Part II

How to Face the Paradoxes?

How wonderful that we have met with a paradox.
Now we have some hope of making progress.
– Niels Bohr

2

In Search of a Uniform Solution

I find myself in complete agreement with you on all essentials. ...
There is just one point where I have encountered a difficulty. ...
—*Russell to Frege, 1902 [van Heijenoort, 1967, p. 124].*

2.1 In Search of an Explanation

After more than a century since Russell wrote to Frege, we have moved from the shock of observing that there *are* paradoxes, to seeking a deeper theory *of* the paradoxes. For a long time, our depth of understanding of the paradoxes stood at roughly the level that high school students understand clouds in the sky. Clouds, they can tell you, have different *shapes*. This one is a stratus, those there are cirrus, and that big one, of course, is the cumulonimbus. These names tell you, roughly, how high up the cloud is (officially, its *étage*), how it is shaped, and perhaps whether rain is likely or not. But these names *explain* almost nothing. They merely describe and classify the cloud as it externally presents itself. So with paradoxes. We have many names, and we know the places these paradoxes occur. This one uses negation, that one implication, this one sets, that one vagueness, etc. A better understanding of the paradoxes would come, we have begun to hope, through some kind of "grand unified theory" of the paradoxes, or at least an abstract characterization of a wide class of paradoxes.

One can see Russell already attempting such a "unified theory" of paradox.[1] The value of such a characterization is attached to what Priest calls the

Principle of uniform solution: If two paradoxes are of the same kind, they should be solved in the same way.

Unifying the paradoxes would make it possible to deal with them all at once, as a family. Whatever works for one will work for all. And if a proposed solution only works for some paradoxes in a unified family, then that solution is not adequate, "bound to appear somewhat one-eyed, and as not having come to grips with the fundamental issue" [Priest, 2002a,

---

[1] In [Russell, 1905b].

p. 166]. A unified theory would not only have predictive power, in the sense of isolating the essential conditions for a paradox to arise. Unification would provide a satisfactory *explanation*.[2]

In this chapter, we will look at two proposals, each providing clues to answering this question: they aim to provide a general structure of the paradoxes. Lawvere's 1969 fixed point theorem is a direct generalization of Cantor's diagonal argument, presented in the language of category theory. The other is also a generalization of the diagonal argument, given first by Russell in 1905 and then modified by Priest in 1991 as the *inclosure schema*. We will look at both these insightful proposals, to see what they accomplish, what they do not, and where to go from here.[3]

2.2 Two Schemas

2.2.1 Lawvere's Schema

We begin from far away, with a very general abstract schematic that captures much about what is happening in paradoxical situations.

Set theory developed in part as an abstract language suitable for founding real analysis.[4] Similarly, category theory was developed for the needs of (abstract) algebra. Since much of the point of category theory is to provide a new language, many people are not conversant in it and find it obscure. It is not my purpose here to be obscure. It is possible, following Yanofsky [Yanofsky, 2003], to get a sense of Lawvere's idea [Lawvere, 1969] in the language of set theory (Chapter 5).

First let's have Lawvere's theorem as classically stated, and then unpack it. It is a version of the 1936 fixed point theorem we met in the Introduction and will meet again at Theorem 13, all of which are related to Cantor's diagonal argument. Let X, Y be sets. A mapping from X to Y

$$g : X \longrightarrow Y$$

is *representable* by a mapping from pairs in X to Y

$$f : X \times X \longrightarrow Y$$

iff there is a $t \in X$ such that

$$g(x) = f(x,t).$$

Then we have *Lawvere's Diagonal Theorem*:[5]

---

[2] Hajek points out the the principle of uniform solution is a continuity principle: two paradoxes cannot be very, very "close" to one another while their solutions are "far apart" [Hájek, 2016, §3]. Uniform solution fits with Kitcher's *unificationist account* of explanation, which places a premium on making connections between apparently disparate phenomena [Kitcher, 1989, p. 423, 437]; cf. Mancosu [2011, §6.2].

[3] Priest provides detailed replies to many of my arguments in this chapter in [Priest, 2019b].

[4] See [Dauben, 1979].

[5] Calling this the "diagonal theorem" comes from the (accessible) [Lawvere and Schanuel, 2009, p. 304], where representation is described as X having "enough points to parameterize all the maps [from X to Y] by means of some single map" from $X \times X$ to Y.

If there is an $f : X \times X \longrightarrow Y$ such that any $g : X \longrightarrow Y$ is representable by f, then there is a fixed point

$$e(y) = y$$

for every $e : Y \longrightarrow Y$.

This is category theory, so there is a commutative diagram, with Δ a map that takes $x \in X$ to $\langle x, x \rangle$ in the Cartesian product $X \times X$ (see Section 5.2.2.3):

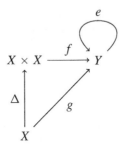

The proof is direct. Assume that any mapping g is representable. Fix some $e : Y \longrightarrow Y$, and then define g as

$$g(x) = e(f(x,x)).$$

Since g is representable by f on assumption, $g(x) = f(x,t)$ for some t and for all x, so in particular

$$g(t) = f(t,t),$$

which by substitution means that

$$g(t) = e(f(t,t))$$

$$= e(g(t)).$$

Then $g(t)$ is a fixed point for e, as required.

We already saw in Section 0.3 that the existence of a fixed point for every mapping over a space seems to put severe constraints on what that space could be. For the preceding example, think of Y as a set of truth values, like $\{0, 1\}$. Think of a truth predicate as a mapping that takes truths to 1 and falsehoods to 0. Think of the map e as "negation," with $\neg(0) = 1$ and $\neg(1) = 0$. Then the theorem shows that every map on Y has a fixed point, which for classical negation means $0 = 1$. So the only (classical) solution for Y is a single point, $Y = \{\bullet\}$ [Yanofsky, 2003, p. 363]. This is an algebraic way to say what Tarski showed: a classical system that is able to express its own notions – e.g., semantically closed – has no structure.

Many paradoxes, including the ones we have been mulling over and beyond, fit this schema. Here is how the Russell paradox fits Lawvere's schema. Keep Y as $\{0, 1\}$ and let X be the collection of all sets. Consider the mappings

$$f(x, y) = \begin{cases} 1 & \text{iff } x \in y; \\ 0 & \text{iff } x \notin y \end{cases}$$

$$g(x) = \begin{cases} 1 & \text{iff } x \notin x; \\ 0 & \text{iff } x \in x \end{cases}$$

The Russell set is $r = \{x : x \notin x\}$. Then

$$g(x) = f(x, r)$$

for all x, showing that f represents g. This can be checked:

- $g(x) = 1$ iff $x \notin x$ iff $x \in r$ iff $f(x, r) = 1$.
- $g(x) = 0$ iff $x \in x$ iff $x \notin r$ iff $f(x, r) = 0$.

But as also can be checked,

- $g(r) = 1$ iff $r \notin r$ iff $f(r, r) = 0$.
- $g(r) = 0$ iff $r \in r$ iff $f(r, r) = 1$.

So from

$$g(r) = f(r, r),$$

it follows $0 = 1$. The conclusion is unacceptable, so we have an expression of the paradox (as leading to absurdity). The schema can be then applied to the liar and much more.[6]

Lawvere's theorem contraposed (called "Cantor's Theorem" in [Lawvere and Schanuel, 2009]) says that if there is a map e on Y with *no* fixed points, then for all $f : X \times X \longrightarrow Y$, there is a $g : X \longrightarrow Y$ that is *not* representable by f: for no $t \in X$ is it always the case that $g(x) = f(x, t)$. This is the standard "reductio" interpretation of the situation. Since negation has no fixed points, some mappings are not representable, and any coherent language is not semantically closed.

Lawvere's analysis of the paradoxes is a beautiful expression of the diagonal scenario, at least for those with a taste for desert landscapes. Nevertheless, the schema is so abstract, so "external" a view, that its explanatory force is lost; so it can't quite be the longed-for uniform solution. This is exemplified in Lawvere and Schanuel's own take on the schema:[7]

> Our proof of the Cantor Diagonal Theorem . . . is clearly valid in any category with products. This fact was exploited by Russell around 1900 and by Gödel and Tarski in the 1930s to derive certain results (which are sometimes described in 'popular' books as 'paradoxes'). *[Lawvere and Schanuel, 2009, p. 306]*

---

[6] Including strengthened liars, where the set of truth values is expanded to $\{0, 1, 2\}$ and $e(1) = 0$, $e(0) = e(2) = 1$ [Yanofsky, 2003, p. 371].

[7] Or Yanofsky's: "The best part of this unified scheme is that it shows that there really are no paradoxes. There are limitations. Paradoxes are ways of showing that if you permit one to violate a limitation, then you will get an inconsistent system." This is a standard classical/incompleteist interpretation, but then he says in the very next sentence, "The Liar paradox shows that if you permit natural language to talk about its own truthfulness (*as it – of course – does*) then we will have inconsistencies . . ." [Yanofsky, 2003, p. 364, emphasis added].

The most salient issue with using Lawvere's schema as an explanation of the paradoxes, then, is that it is, and seems intended to be, the proverbial "view from 10,000 feet." It gives us no clues about the underlying forces that drive the paradoxes. Looking at the Russell contradiction, for example, the Russell set is simply *given* and already has its famous properties. For g to be representable by f is for there already to be some term t given whereby $g(x) = f(x, t)$. Once that is in place, Lawvere's schema shows us what happens. But that is arriving too late, so to speak. The paradox has already happened and it is left to the logician simply to describe the aftermath.

Smith[8] warns against grouping together the paradoxes at too high a level of abstraction, especially if that grouping will be used with the principle of uniform solution to address paradoxes at lower levels of abstraction [Smith, 2000]. The principle of uniform solution may be thought of as providing a bound on the level of generality that an explanation may take.[9] A way to see that this might be a problem for Lawvere's schema is to observe how many different phenomena fit (in [Yanofsky, 2003]). Along with the set theoretic and semantic paradoxes, it also models time travel paradoxes, von Neumann's self-reproducing programs, and many other "recursive" phenomena, some of which are paradoxical and some of which, it seems, are not.

I think there is still a lot to be learned from further study of Lawvere's theorem.[10] Ultimately, though, the schema as given expresses the attitude that closure is impossible, that any isolated inconsistency reduces everything to absurdity, and, most tellingly, that there is really nothing so surprising or paradoxical about the paradoxes. This is missing some crucial elements, so we would like to enrich our view further.

2.2.2 Inclosures

Priest offers the *inclosure schema* as common pattern of the paradoxes. This section reviews the proposal and some of its virtues. There is already a book-length elaboration and defense of this in [Priest, 2002a], so I do not plan to review all that, but rather just put the main idea clearly out in the open for inspection. There are two (separate) claims to consider: first, that all the paradoxes in question fit a common pattern; and second, that they are best met with admitting true contradictions.

2.2.2.1 The Idea

Are there thoughts we cannot think? Wittgenstein observes in the preface to the *Tractatus* that

... in order to draw a limit to thinking we should have to be able to think both sides of this limit (we should therefore have to be able to think what cannot be thought).

---

[8] Critiquing Priest, not Lawvere, but the point is general.
[9] Thanks to Patrick Girard here.
[10] For further explorations of this topic, see [Pavlović, 1992], and [Pitts and Taylor, 1989].

Wittgenstein is noticing that the very existence of any limit already suggests another side. This is plausible to our spatial intuition based on common experience: for any wall or fence, there is always a far side to which the wall is blocking access. Surely even if you came to the edge of the universe, you could stick out your arm![11] And what goes for walls, goes for thoughts. Wittgenstein intends his claim as a solemn modus tollens, but a glut theorist may take the modus ponens direction. There *are* thoughts we cannot think – and we are thinking one of them *right now*.[12] The central thesis of Priest's *Beyond the Limits of Thought* is that limits to thought *can* be transcended – indeed, by virtue of recognizing one as a limit, we have already in some sense transcended it.[13] In terms of uniform solution, the intuition is that something like this dynamic – crossing an uncrossable boundary – may be what is behind all the paradoxes.

2.2.2.2 The Schema ...

The inclosure schema is proposed as the structure of the paradoxes of self-reference, first in [Priest, 1991a] following Russell.[14] The basic picture is that of a space with an operation that can escape any subregion of the space, but always stays in the overall space. The intuition is that the space is some very large totality, like the universe or set of all truths, the sort of totalities we arrive at in the broadest and most speculative of metaphysical speculations; an inclosure is formed in some attempt at collecting together in thought an "everything" that is so inclusive as to be uncollectable (except didn't we just collect it?).[15]

More formally, an *inclosure* is a pair

$$\mathcal{I} = \langle W, \partial \rangle$$

with W a set, and ∂ a function that takes subsets of W to members of W

$$\partial : \mathscr{P}(W) \longrightarrow W$$

called the *diagonalizer*, with three conditions:

Existence: $W = \{x : \varphi(x)\}$ exists, and $\psi(W)$;

and for all subsets X of W, $X \subseteq W$, such that $\psi(X)$,

Transcendence: $\partial(X) \notin X$
Closure: $\partial(X) \in W$

When all the conditions are satisfied, a contradiction is immediate, since any W is a subset of itself: thereby $\partial(W) \in W$ (by closure) and $\partial(W) \notin W$ (by transcendence). Here is a picture (following Restall's highly influential original [Priest, 2002a, p. 156]):

---

[11] As Simplicius reports Archytas arguing in the *Physics* (6th c. BCE); cf. Lucretius, *De Rerum Natura* 1.951–987.
[12] For the thinkability (or not) of contradiction, see Routley and Routley [1985, p. 210]; cf. [Priest, 2016].
[13] "The thesis of the book is that such limits are dialetheic ... the limits of thought are boundaries which cannot be crossed, but yet which are crossed" [Priest, 2002a, p. 3].
[14] "There are some properties such that, given any class of terms all having such a property, we can always define a new term also having the property in question" [Russell, 1905b, p. 142]. Cf. [Landini, 2009].
[15] Inclosure and dialetheism have been taken up by contemporary *speculative realists* in a more "continental" tradition. See Cogburn [2017, pp. 60–71].

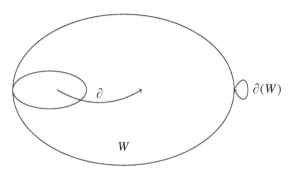

In the canonical case, Russell's paradox, the totality is the set of all sets V.[16] The diagonalizer is $r(X) = \{x \in X : x \notin x\}$; the contradiction is that $r(V) \in V$ and $r(V) \notin V$; details are in Section 2.2.3.2. It is no surprise that Russell's paradox is an inclosure. Russell abstracted the core of the inclosure schema from his paradox. Priest generalizes it in a natural way, whereby, contra Ramsey but like Lawvere, a single schema applies both to the set theoretic paradoxes *and* to semantic paradoxes such as the liar.

In the case of the liar, the carrier set is $T = \{p : p \text{ is true}\}$. The diagonal ℓ takes subsets X of truths to the sentence "this sentence is not in X," $\ell(X)$. Either $\ell(X) \in X$ or not. If $\ell(X) \in X$, what $\ell(X)$ says would be false – but every sentence of X is true, since X is a subset of T. So $\ell(X) \notin X$ by reductio (if p implies $\neg p$, then $\neg p$). Thus $\ell(X)$ is true. Therefore, it is a member of T. Contradiction at the limit: $\ell(T)$ is the liar sentence. It is both in T and not.

Inclosure is a diagnostic tool. It shows how a paradox arises, not what to do about it. One does not need to be a dialetheist, or even move to a nonclassical logic, on the basis of finding the inclosure schema appealing. Russell himself advocated for the schema as explanatory without suggesting that any contradictions are true.[17]

2.2.2.3 ... *and Its Virtues*

The inclosure analysis of the paradoxes, if it succeeds, is both wide and deep. If it succeeds, inclosure is explanatory across two dimensions. It unifies a wide range of otherwise disparate entities, and provides a clear story about what they have in common. The inclosure explanation provides a basic *unifying* pattern for the underlying mechanics of contradiction. If inclosure is right, the paradoxes arise as a result of the way diagonalizers behave in a closed space; the inconsistency is the collision at the boundary.[18] By comparison,

---

[16] For the moment, I will use roman 'V' as the generic name for the set theoretic universe, and return later to the fancy $\mathcal{V}$ mentioned in previous (and later) chapters to denote the universe; the distinction will become clear.

[17] See [Weber, 2010a].

[18] "An immovable force meets an irresistible object; and contradiction, in the shape of an inclosure, is the result" [Priest, 2002a, p. 233]. The situation is reminiscent of another aporia, described by Camus in part one of *The Myth of Sisyphus* [1942], of an "absurd" collision between the human need for an absolute versus the impossibility of ever grasping one. More positively, inclosure expresses Nietzsche's revelation:

And life itself confided this secret to me: 'Behold,' it said, 'I am that which must always overcome itself'.
[On Self-Overcoming, *Thus Spoke Zarathustra* (1883) [Nietzsche, 1976, p. 227]]

other descriptions of the paradoxes tend to be in terms of symptoms, or effects (e.g., self-reference), without getting at the root cause. Priest compares it to understanding a volcano simply as the eruption at the top, without also appreciating the geothermal activity occurring deep below [Priest, 2002a, p. 279]:

Once one understands *how* it is that a diagonaliser manages to operate on a totality of objects of a certain kind to produce a novel object of the same kind, it becomes clear *why* a contradiction occurs at the limit. *[Priest, 2002a, p. 136, emphasis in the original text]*

Inclosure suggests there is more to the paradoxes than a cardinality problem with quantitatively "overlarge" sets (Section 1.1.3.2); it points toward a more qualitative and subtle understanding; cf. [Priest, 2013a]. Paradoxes are about "shape" rather than "size."

Perhaps most pleasingly, inclosure works at a higher order of explanation. It shows what is *compulsive* about true contradictions, what makes them so hard to *solve*. That is, inclosure models revenge. For example (see Priest [2002a, §9.6]), Russell's solution to the paradoxes is captured in the slogan: "Whatever involves *all* of a collection must not be one of that collection." This solution, like most that deny closure, is self-undermining in a simple way that is itself an inclosure – it is by its own light inexpressible (transcendence) and yet it is expressed (closure).[19] Inclosures are models not just of pathological sentences or the like, but of the pathologies in our *solutions* to the paradoxes.

The story is attractive.[20] All the semantic and set theoretic paradoxes of self-reference are inclosure contradictions. More, inclosure suggests (though does not mandate) a response – dialetheism – whereby we see why the paradoxes lead naturally to contradictions and cannot be evaded (e.g., by denying closure) without triggering another inclosure-shaped contradiction. Coupled with the principle of uniform solution, this makes dialetheic paraconsistency seem highly explanatory and leaves most other solutions looking incomplete. Inclosure appears to be a candidate for the longed-for uniform solution, the explanation of the paradoxes.

2.2.2.4 Sorites as an Inclosure

Intuitively, we are in an inclosure when there is a process that both must continue and at some point cannot. The idea that the sorites paradox fits this description pushes the inclosure analysis beyond the paradoxes of self-reference.[21] The sorites is not a paradox of self-reference, so it fitting the schema expands the meaning of being an inclosure paradox. But it is a natural addition. If φ is a vague predicate, like being a heap of sand, then two quantities of sand that differ by only a grain do not differ with respect to being a heap. Vagueness means that "if one satisfies φ, so does the other – the principle of tolerance" [Priest, 2010, p. 70]. That is how closure is satisfied. TOLERANCE (Section 1.2.1) does look

---

[19] "In trying to solve the problem, Russell just succeeds in reproducing it. His theory is therefore less of a solution to the contradiction at the limits of thought than an illustration of it" [Priest, 2002a, p. 140].

[20] Or so says I, especially when I first encountered it. Of course, inclosure is not without its critics, e.g., [Abad, 2008], [Badici, 2008], [Dümont and Mau, 1998], [Zhong, 2012], among others. Cf. Priest [2002a, ch. 17] for some replies.

[21] The idea that sorites fits the inclosure was to my knowledge suggested first by Colyvan at the Australasian Association of Philosophy meeting in Armidale in 2007 [Colyvan, 2008b], inspired by Chase's "continuous sorites" (Section 1.2.1).

very much like a closure principle. It has the right "feel" for an inclosure: novel objects outside (but nearby to) a totality of φ things get pulled back in, exactly because the predicate tolerates some stretching.

Recall the sorites on Olympus Mons (Section 1.2.1). Mars and its mountain certainly exist, and W is its high-up points. Walking down the mountain acts as diagonalizing. At each high-up place, we can see that there is some lower point left out; for any subcollection X of high points, diagonal ∂ goes to the highest point not in X, the first point a_k below all of those in X. Meanwhile, since no a_k is different from its nearby points in all height respects, $\partial(X)$ is still high up, by TOLERANCE. The contradiction: the last high point is also the first nonhigh point.

As further evidence that the sorites is an inclosure problem (and, by extension, of the same sort as the set theoretic paradoxes), consider the following examples. The first example is attributed to Wang. Claim: All numbers are small. Proof: 0 is small; and it cannot be that n is small but $n + 1$ is not. So n cannot be large, for arbitrary n. The second example is attributed to Russell. Claim: There is a biggest (transfinite) cardinal. Proof: The set of all the natural numbers has a cardinal; and it cannot be that a bunch of sets have cardinals, but the collection of them does not. The set of all the transfinite numbers, then, has a cardinal. This must be the greatest cardinal. (Cf. Section 1.1.2.2.) How do these compare? The Wang argument is a straightforward sorites. The cardinality paradox is Cantor's paradox (Section 1.1.2.2), a paradox of self-reference. Juxtaposed with Wang's argument, though, Cantor's paradox (and similarly, Burali-Forti) begins to look like a vagueness problem.[22] Inclosure arguments stretch a property – being small, having a cardinal – beyond its breaking point. Such arguments point to penumbral sites where it might be expected we find paradox: boundaries.

With sorites prima facie encompassed by the inclosure schema, then, is a satisfying uniform solution to an ever wider class of paradoxes within grasp?

2.2.3 Problems with Inclosure

Two aspects of inclosure arguments jump out.[23] First, they are *informal* arguments using familiar logical steps, such as the law of excluded middle, reductio, and perhaps some other (controversial) moves such as contraposition. The arguments look valid, at least informally, but they are not laid out as formal proofs. Second, it is not clear that the contradictions established really are the right targets. In the case of the Russell paradox, the inclosure is that the Russell set is both a set and not – meaning that the contradiction is really about V, the set of all sets, and the fact that $\in$ is inconsistent over V. But this is a rather sophisticated

---

[22] Cf. Dummett: "The beginner can be persuaded to that it makes sense, after all, to talk of the number of natural numbers. Once his initial prejudice is overcome, the next stage is to convince the beginner that there are distinct transfinite cardinal numbers .... When he has become accustomed to this idea, he is extremely likely to ask: 'How many transfinite cardinals are there?' How should he be answered? He is very likely to be answered by being told, 'You must not ask that question.' But why should he not?" [Dummett, 1991, p. 315].

[23] Both are mentioned at [Priest, 2002a, §17.2].

way of seeing things. One might have thought that the basic paradox is that the Russell set is a member of *itself* and not, that the contradiction is essentially only about the Russell set, and the fact that its membership is *internally* inconsistent. The inclosure analysis makes the paradoxes *external* to the diagonal objects, ultimately a property of the totalities, as a core aspect of the explanation.

Let us take these issues – the validity of inclosure arguments, and what they argue – in turn.[24]

2.2.3.1 Informal Arguments

As Priest notes, with a little finesse almost anything can be made to fit the inclosure schema [Priest, 2002a, §9.5]. How do we determine which paradoxes really do fit?

To get a better grip on the problem, take two examples that are the right shape for inclosures, but are not genuine paradoxes. Consider a town with a barber who shaves all and only the men who do not shave themselves. This would seem to drive to the same conclusion as Russell's paradox – except that existence is not satisfied. There is not (as far as demographers have found) any such town. Or for a mathematical example, take the natural numbers $\mathbb{N}$; for any subset X of numbers, take as diagonalizer the least natural number n greater than any member of X; then the least number greater than any number would be a number and not – except that there is no reason to think that $\mathbb{N}$ is closed under any such proposed diagonalizer.[25] In both cases, we would need independent, sound arguments to make these real inclosure paradoxes.

According to Priest, satisfaction of the inclosure conditions needs to be prima facie (or perhaps a priori) plausible [Priest, 2002a, §17.2]. Prima facie validity will vary from reasoner to reasoner, though, depending on their logical training. While the informal paradoxical arguments have seemed valid to the vast majority of those who have considered the matter (that is why they are paradoxes, at least apparent ones), this only helps us classify the paradoxes in terms of what *seems* to be their structure. This raises the more important point, then: "seeming" valid is a good start, but an insufficient finish. Once one gets past initial diagnostic devices, once the paradoxes are apparently classified, then more precise and reliable instruments are needed that go beyond mere prima facie plausibility. The way to distinguish a genuine contradiction from a contradiction-shaped joke is to show that the conditions of inclosure are *genuinely* satisfied.

What does genuine satisfaction mean? At this point I will invoke some hypotheticals, to look down one potential path in answer to that question. The purpose is to ascertain whether this path, however tempting, leads to a philosophical dead-end. To that purpose, I'll allude to some results that we will see properly later in the book, but which we need to

---

[24] Similar issues were raised as serious doubts by Beall in [Beall, 2014a,b], especially about whether sorites really fits the inclosure schema (or whether it would even matter if it did). A response (with several coauthors) defending sorites as an inclosure is [Weber et al., 2014], though as will become clear later, I have shifted my position since then.

[25] As opposed to the ordinals more generally, where the existence of such a diagonalizer is essential for the existence of the ordinals – so the result, Burali-Forti's paradox, is a genuine dialetheia [Priest, 2002a, ch. 8]. But see [Priest, 1994a] for contradictions in $\mathbb{N}$.

preview now, because they make the inclosure analysis highly problematic. Let's grant the following for the sake of argument:

If the inclosure analysis of the paradoxes is correct, then a reasonable (though not the only) response is dialetheic paraconsistency. As the Introduction and Chapter 1 urged, some contradictions really are true, and logical consequence really does not validate explosion or its correlates. Moreover, if logic is *content neutral*, the same logic should always be in force (as I will argue at some length in Chapter 3). This means we should be using a fully paraconsistent toolkit, all the way up, with only paraconsistently valid consequences. If we are trying to explain paradoxes paraconsistently, *all reasoning must be paraconsistently valid*. That is, we suppose that "genuine satisfaction" of the inclosure schema reduces to the inclosure conditions being justified and true in a paraconsistent setting. If this can be done, the inclosure is genuine – a true contradiction! If every attempted argument to satisfy the inclosure conditions uses some invalid step, though, then ex hypothesi the inclosure analysis is lost, or at least severely diminished.

Whether we *ought* to hold the arguments to this standard – paraconsistent deductive validity – is the subject of the next chapter, where I will argue at length that a serious paraconsistentist should only use paraconsistent logic. Let's take this as a *question* for now: *supposing* the inclosure analysis is so wildly successful as to convince us to use nothing but paraconsistent logic, what happens to the inclosure arguments when paraconsistently formalized? Can the inclosure paradoxes be *proved* up to a dialetheic standard? To answer this rigorously, I will need to get a little ahead of ourselves and call on some of the machinery that we will get to in later chapters.[26] But the argument here is still relatively general, because while the details of different paraconsistent logics differ, a large family of them share certain core features: no matter which of these logics we adopt, the target contradiction would need to be established without using rules such as disjunctive syllogism or invalid contrapositions;[27] and it must be established without at the same time proving *triviality*. So, given these constraints, what happens to inclosure? We return to Russell's paradox, an inclosure if there ever was one, and consider how the argument works.

2.2.3.2 Russell's Paradox as Inclosure over "the Universe" Is Invalid

Let us check the details of how the argument that Russell's paradox fits inclosure is meant to go [Priest, 2002a, §9.1]. The existence condition is satisfied by the set of all sets:

$$V = \{x : x \text{ is a set}\}.$$

The property ψ is vacuously filled by $V = V$. Consider the function

$$r(X) = \{x : x \in X \text{ and } x \notin x\}$$

---

[26] The logic I have in mind is as in Section 0.2.2.3, plus a little naive set theory, too, in the vicinity of [Routley, 1977] or [Weber et al., 2016]. This will be officially presented in Chapters 4 and 5.

[27] Cf. [Priest, 1989] in [Priest et al., 1989].

that picks out all the non-self-membered members of any subset X of V. Either $r(X)$ is a member of itself or not. If $r(X)$ is a member of itself, then it is not; so by reductio, $r(X) \notin r(X)$. Then for any subset X of V,

$$r(X) \notin X$$

because otherwise $r(X)$ *would* be self-membered, which we've just showed isn't so. This establishes transcendence. But also

$$r(X) \in V$$

just because everything is in V. And that's a contradiction: when the diagonal hits $r(V) = \{x : x \notin x\}$, which is *the* Russell set, we have $r(V) \in V$ and $r(V) \notin V$. The Russell set is not a set, or so the argument is meant to show.

But when we review the reasoning, checking for (paraconsistent) satisfaction rather than prima facie plausibility, the argument breaks down. If $r(X) \in \{x \in X : x \notin x\}$, then $r(X) \notin r(X)$; so $r(X) \notin r(X)$ by reductio. This is fine – reductio will be shown to be valid, using the law of excluded middle (in Section 4.3.2.1) – but it is not enough to show that $r(X) \notin X$. For suppose otherwise that $r(X) \in X$. Then $r(X) \in X$ and $r(X) \notin r(X)$ both hold, so $r(X) \in r(X)$ by definition. That would mean $r(X)$ is inconsistent, both a member of itself and not ... but that is just a contradiction about $r(X)$, not proof that $r(X) \notin X$. We have established a conditional,

If $r(X) \in X$ and $r(X) \notin r(X)$, then $r(X) \in r(X)$ and $r(X) \notin r(X)$.

Even assuming that this conditional contraposes (falsity-preservation backward, which is doubtful),[28] which would mean that *one* of the premises is false – still, we can't isolate *which* one without using disjunctive syllogism.[29] The argument to transcendence over X fails paraconsistently.

Anticipating this, Priest says that "in case the inferences used to establish Transcendence and Closure are not always dialetheically valid, we may define the conditions more cautiously" [Priest, 2002a, p. 130, fn 7]. The idea is that disjunctive syllogism *is* paraconsistently valid, in the form

$$p, \neg p \vee q \therefore (p \,\&\, \neg p) \vee q.$$

And from there, transcendence can be (in words)

either the diagonal of X is not in X, or some (other) contradiction is true

and closure can be

either the diagonal of X is in the totality W, or some (other) contradiction is true.

And the target contradiction will then be that the diagonal of W is both in W and not, or else some other contradiction is true. (In the case of the Russell paradox, the "other"

---

[28] And ignoring any concerns about contraction, which we will be concerned about in Chapter 4.
[29] Cf. Priest [2013a, p. 1274], where working with contrapositions is treated semantically.

contradiction is that $r(X) \in r(X)$ and not, for arbitrary subset X.) This is not far off from what I will suggest in the next sections – that the true contradiction may be located somewhere other than the "edge" of the totality W. But this is not merely a more cautious formulation of the inclosure schema. It amounts to saying: *either* the paradoxes lead to contradictions at the limits of an inclosure ... *or* maybe they lead to some other contradictions somewhere else, for some other reason. Again, transcendence can fail because there may be inconsistent boundaries *somewhere else* in the totality.

2.2.3.3 *Russell's Paradox Is an Inclosure* over Itself

A simple version of the Russell paradox *can* be made to fit the inclosure schema, in a way that is paraconsistently valid [Priest, 2002a, p. 130, fn 8]. Let r itself be the totality W and the diagonalizer just be the *identity*, $\partial(X) = X$. With this rendering, a contradiction follows by paraconsistently valid reasoning: for any $X \subseteq r$,

- if $X \in X$ then $X \in r$ and so $X \notin X$; so $X \notin X$ by reductio, so $\partial(X) \notin X$ (transcendence);
- but then $X \in r$ by definition, so $\partial(X) \in r$ (closure);
- so $r \in r$ and $r \notin r$.

This is, effectively, the version of the Russell contradiction that I think is true, and the "shape" of true contradictions I endorse in what follows. Whether this is a meaningful instance of inclosure, though, is dubious. It is simply a restatement of the Russell paradox itself, not a further elucidation. The diagonalizer is doing nothing; no inevitable contradiction is "surging" toward a totality. All the features that make an inclosure distinctive are absent. The very reasons Priest gives for the importance of the underlying mechanics of inclosure [Priest, 2002a, p. 279] would suggest that this instance is degenerate, or at least not explanatory in the right sort of way. If an inclosure is supposed to be like a volcano, with the diagonalizing magma erupting at the dialetheic crater at the top, then this instance of inclosure is like the extensionless "volcano" that is the geometric point at the center of the earth.

2.2.3.4 *Russell's Paradox as Inclosure over the Universe Is Absurd*

Consider the *universe*[30]

$$\mathcal{V} = \{x : \top\},$$

where $\top$ is the True: everything entails $\top$. Set theoretically, $\top$ can be played by the formula $\exists y(x \in y)$, collecting everything that has at least one property (Section 5.2.1). And consider again the diagonal $r(X)$. Closure is automatic, $r(X) \in \mathcal{V}$, because $\mathcal{V}$ is the universe. Transcendence, though, would be

$$r(\mathcal{V}) \notin \mathcal{V},$$

---

[30] The "universe" V and the universe $\mathcal{V}$ have the same extension, but not the same antiextension. Some things are not in V, as Section 2.2.3.2 showed, whereas it would be absurd if anything were not in $\mathcal{V}$. So they are different sets. See Sections 4.2.3.4 and 5.2, and [Weber, 2010b].

which entails $\neg\top$, or

$$\bot.$$

But from $\bot$, everything follows. It can be played by the formula $\forall y(x \in y)$, stating that x has every property, and thus the property φ for any sentence φ whatsoever. $\bot$ is absurdity: "that than which no sillier can be thought" [Slaney, 1989, p. 476].

There cannot be an inclosure on the (true) universe.[31]

2.2.3.5 The Sorites Paradox as an Inclosure Is Invalid

Looking at the sorites makes it easier to understand what is going on with the Russell paradox. I just present the problem here, following [Priest, 2010, p. 70], and comment on it in Section 2.3.2.

Take a sequence A of objects $a_0, \ldots, a_n$, where $\varphi(a_0)$ and $\neg\varphi(a_n)$, for a vague predicate φ. Then the carrier set $W = \{x \in A : \varphi(x)\}$ of the inclosure is just a subset of this sequence, and a proper subset since $a_n \notin W$. That satisfies existence. Transcendence is because we are dealing with a finite sequence of objects: for any $X \subseteq W$, there is a first A not in X, call it $\partial(X)$ (bracketing the fact that "the" first $\partial(X)$ is given by a definite description, which on a paraconsistent rendering is not unique (Section 1.2.3)). But how to get closure? For a nonempty subset X, its diagonal is the first thing after some $a_i \in W$. Then because φ is vague, $\partial(X) \in W$, too, by the principle of tolerance. But how to formulate tolerance? If it is with a *detachable conditional*

$$\text{if } \varphi(a_m) \text{ then } \varphi(a_{m+1})$$

that obeys modus ponens, then satisfaction of the sorites paradox is a disaster: by mathematical induction $\forall m\varphi(a_m)$. *Everything* is φ. Since no one is prepared to accept that everything is a heap of sand, tolerance is formulated materially, as in Chapter 1:

$$\neg\varphi(a_m) \vee \varphi(a_{m+1}).$$

Disjunctive syllogism is invalid and the virulent version of sorites is stopped. (Compare this with the Russell argument in Section 2.2.3.4, where fitting the inclosure points to absurdity.) Now, though, the reasoning for closure in the inclosure schema breaks down. Take a_X to be the last member of some $X \subseteq W$. Then $\neg\varphi(a_X) \vee \varphi(\partial(X))$, by material tolerance. And $\varphi(a_X)$. But this just gives us

$$(\varphi(a_X) \,\&\, \neg\varphi(a_X)) \vee \varphi(\partial(X))$$

and, in particular, no way to conclude (without disjunctive syllogism) that $\varphi(\partial(W))$. The limit contradiction does not obtain. The sorites argument to closure is not valid. The sorites does not fit the inclosure.[32]

---

[31] This is closely related to Curry's paradox, which is the topic of the next chapter. Priest's position is that Curry paradoxes "have nothing to do with contradictions at the limit of thought" [Priest, 2002a, p. 169]; for recent debate, see [Weber et al., 2014; Priest, 2017b].

[32] Again, this is pointed out by Beall [Beall, 2014b].

Importantly, the sorites *does* give rise to a contradiction, just not necessarily the inclosure-intended one. Putting together tolerance with nontriviality and something like a least number principle (Section 3.3.3.2), we have that for *some* φ-thing, it is both φ and not – but that thing need not be at the limit. So this puts before us some options. Either sorites does not fit inclosure after all; or if it does, then either this undermines the inclosure schema as a diagnostic tool, because some inclosure paradoxes are about limits while others are not, or else it shows that other inclosure paradoxes aren't really about "limits" either.

2.2.4 Whither inclosure?

These cases show that, in a paraconsistent framework, the possibility of local contradictions – before reaching the inclosure limit – blocks inclosure arguments. And the utter impossibility of absurdity means transcendence needs to fail against some totalities. I'll spend the rest of the chapter unpacking these claims, first considering the meaning of absurdity, and then the transition to locality.

2.3 Stepping Back from the Limits of Thought

2.3.1 Absurdity

According to *non*paraconsistent logics, there is no difference between inconsistency and absurdity. All contradictory limits are equally unapproachable. A dialetheic paraconsistent view is therefore uniquely positioned to discern the following: some inconsistent limits can be surpassed, but some cannot. There can be inclosures on sets that contain everything, sets with the same extension as the universe, e.g., $\{x : x = x\}$. The one true universe, though, $\{x : \top\}$, cannot be escaped, on pain of absurdity. This claim is only *interesting* from a dialetheic paraconsistent point of view. There are totalities for which even the unstoppable force of diagonalization must stop. Full stop.

The result in Section 2.2.3.4 means that one interpretation of Priest's inclosure analysis is wrong. This is the too-good-to-be-true idea that *every* limit of thought can be overcome – that objects no matter how immovable may still be moved – and crucially that thoughts on the "other side" of the limit (like "now I am thinking the other side of the limit") are *true*. Priest has in his sights only limits of a certain kind, the Kantian/Hegelian limits broadly conceived,[33] not just any limit whatsoever; but even among this narrower class of limits, some cannot be crossed. The limits that may be broken are given by consistency. One can think the other side of the limit, whereby one is in contradiction, but still okay. For *coherence*, though, if the set $\{p : p$ is not absurd$\}$ has inconsistent membership, then absurdity follows. By the law of excluded middle, everything is either absurd or not.[34]

---

[33] "... it has seemed to people that though there be no greater than the infinite; yet there be a greater. This is, in fact, the leitmotif of the book" Priest [2002a, §2.0]).

[34] Which is *not* to say everything is either absurd or *true*; that's invalid (Section 4.3.1.2).

But if p is not absurd, that does not make it *true*. The sentence "this sentence is absurd" is simply false.

Some limits do not, as it turns out, have a far side. A two-dimensional Euclidean shape has only one side. It has no back side, as Borges used to great effect in his short story "The Disc" (1975): "... the Euclidean circle, which has but one face." You may *think* about the other side of a two-dimensional disc, of course, just as you may think about the superluminal spaceship in your garage, or the koan "everything is true." But to think a thought does not mean that those thoughts are true. Thank goodness.

In reply, someone convinced of the "ability to *break through every barrier*"[35] can offer a simple answer to the problems outlined previously. It is to deny that there is a $\perp$ particle or equivalent in our language.[36] In the truth-functional logic of LP, there is no sentence that is absurd (false-only on every valuation). Perhaps logics without $\perp$ are friendlier to dialetheism? (This question will recur in Chapter 10.) If we drop absurdity from the language entirely, there may be some interesting directions to pursue, but I suspect the problem does not go away. Notice that $\perp$ need not be an out-and-out absurdity; it just needs to be something that is in no way (actually) true.[37] There must be something that is false and in no way true, even just contingently, or there was no reason to adopt a paraconsistent logic: some things are not true, even if some contradictions are. And whatever that false-only sentence is demarcates a line that we will not cross. The negation of that sentence defines our universe.

Priest points out that rationality does not rely on consistency [Priest, 2006a, ch. 6, 7]; inconsistency is not the only reason why we might deny something. The moon is not a frog, and is in no way also a frog – not because of logic but because of the way the moon is, and the way frogs are. The main evidence that no moons are also frogs is *not* that it would be a logical contradiction for an astronomical body to be a short-bodied amphibian. Someone who needed deductive logic to make this observation is making a kind of mistake. By the same token, the reason that absurdity is beyond the ultimate horizon of thought is not that it would be *inconsistent* to cross that boundary. You can go up to the edge of absurdity and stick out your head, but you won't find enlightenment; you'll get your head cut off. Some limits to thought cannot be transcended, or even approached too closely. Absurdity is not a limit with a far side.

2.3.2 *Locality*

The ways in which paraconsistent logic recasts the inclosure reasoning point to something deeper about the paradoxes. In Section 2.2.3, I remarked that the Russell paradox should prima facie be about the Russell set, not about the universe of all sets. The paradox is

---

[35] Priest [Priest, 2002a, p. 117, emphasis in the original text] interpreting Cantor (about ordinals) via Michael Hallet.

[36] See [Casati and Fujikawa, 2019]. Cf. Priest discusses **everything** and its complement, **nothing** [Priest, 2014, §4.6, §6.13, boldface original], but these are not the totality given by $\top$ or the absolute nothingness of $\perp$.

[37] Like this not true sentence: "It was the eldritch scurrying of those fiend-born rats, always questing for new horrors, and determined to lead me on even unto those grinning caverns of earth's centre where Nyarlathotep, the mad faceless god, howls blindly in the darkness to the piping of two amorphous idiot flute-players" [H. P. Lovecraft, "The Rats in the Walls" (1924)].

not that the Russell set transcends V, but rather that it is self-inconsistent. Now we have formal evidence that this is correct. The argument in Section 2.2.3.2 establishes that Russell contradictions can occur locally. In particular, the Russell set simpliciter, $r = \{x : x \notin x\}$, generates its own simple contradiction, $r \in r$ and $r \notin r$. And *then*, as a corollary, when $V = \{x : x = x\}$ we get $r \in V$ and $r \notin V$, since by the axiom of extensionality r is a set that differs from itself with respect to membership [Restall, 1992, p. 427]; cf. Theorem 6 in Chapter 5. So the inclosure schema catches this downstream contradiction, but not at the source.

Similarly with the sorites, granting that the conditional form of tolerance must be false, then we are left with a nice version of the "line drawing" form of the paradox (Section 1.2.1). There must be a pair of objects that are very nearby each other with respect to φ, but also not, because one of them is φ and the other not. Then either the first member of the pair is both φ and not, or the second one is. A contradiction obtains, but not at any particular extremity of the sorites sequence. (This is spelled out again at Section 3.3.3.2.) The contradiction is intrinsic to the local pair.

What do these examples show? One presentation of dialetheias has them as rare and far away: spandrels, singularities that occur only at the edge of the universe, beyond the limits of thought. In everyday situations, we can assume consistency, reason with classical logic, and carry on as if the paradoxes never happened.[38] Dialetheism and paraconsistency kick in only during extreme and unusual circumstances, like at the top of the ordinals, standing before Cantor's Absolute. But the math suggests that contradictions may occur well before any limit, and cannot occur at extreme limits.

I submit, then, that once dialetheism is taken on board, then any story about the paradoxes as "singularities" or distant "anomalies" be revised. The famous paradoxes "at the limit" are only the easiest to notice. There are contradictions much closer to home, as Priest points out [Priest, 2013a, p. 1272]. As I will return to, the picture emerges in which most paradoxes are too *innocuous* to notice as paradoxes. (Anyone who has ever lived in a place with snow can tell you that it is white, but also that snow is often not all that white.) Dialetheias as limit cases perpetuates the idea that contradictions are marginal and easy to ignore. Contradictions already have a bad reputation, and thinking of them as the serpents where HERE BE DRAGONS only reinforces the bias.

The reconsideration of inclosure in light of a fully paraconsistent language shows a different way. For any data, we simply want to give the best description of it we can; dialetheism councils that sometimes, that description will be inconsistent. If this is right, then we shouldn't encourage the sentiment that an inconsistent description will be defective or bizarre, that it will involve impossible Escher towers or a psychadellic ouroboros. Dialetheic descriptions will be as mundane as bald men, heaps of sand, crossing a threshold, or comprehending a set. Dialetheism isn't a special theory for anomalies. Contradictions may occur anywhere, as local and rather banal properties of everyday objects.

---

[38] See Priest in [Priest, 1979, §4; 1989; 2006b, ch. 8]. An even more restricted place for dialetheias is outlined in [Beall, 2009]. I'll take this up in the next chapter.

2.3.3 A Reassessment

Our overarching question asks why there are paradoxes. We have previewed what an attempt to answer this question might look like from a thoroughly paraconsistent position, using some resources to be developed soon. The exercise has determined that, while Lawvere and especially Priest's schemas may be good *starting* points for explaining the paradoxes, they cannot be the *end*point of a fully dialetheic path to explanation.

Lawvere's construction is an *external* view of the paradoxes. Priest's inclosure is a more *internal* view of the situation, though it still situates the paradoxes at the level of totality. Both of the schemas, I want to say, arrive "too late." The paradoxes have already happened, in the sense that we are only seeing the end result – either fixed points for all mappings in the presence of a contradictory object (Lawvere), or the terminal node of a long process (Priest). Both still carry the sentiment that contradictions are, essentially, catastrophic. Lawvere's schema predicts that a paradox-supporting system is a trivial point; Priest's places the contradictions "at the limits," and threatens (as in Section 2.2.3.4) that the "unstoppable" diagonal will eventually drive out of even the "immovable" inconsistent universe and into the endless void of noise that lies beyond.

Previous explanations for paradoxes, from Tarskian/ZFC hierarchies through Priest's inclosure, suggest that contradictions are accounted for at the *global* level, the result of overly inclusive totalization or overly general universal quantification – some kind of mistake about the universe. Paraconsistent mathematics directs us otherwise. A thoroughgoing dialetheic account suggests that contradictions are *local*, intrinsic properties of some independent objects. We stop questioning whether we have "selected" the right universe to study, and try rather to explain why some specific objects in our universe are paradoxical. This local issue, and not some global property, is what needs to be explained.

The inclosure analysis of paradoxes, especially as it offers a unified theory, and especially when read from a classical starting point, is a compelling argument for dialetheism. If it is a successful argument for global paraconsistency, it would appear to be (self)-undermined (recalling again, though, that the initial diagnosis, at least, is independent of dialetheism [Weber, 2010a]). Dialetheism is not so undermined, though. Dialetheism follows, recall, not from an abstract paradox schema, but from extremely simple proofs that start from the naive set comprehension axiom and end in the Russell paradox or liar paradox. Nothing I've said puts pressure on those simple proofs; the paradoxes are theorems (or at least, we will show them to be in later chapters), with or without the inclosure story. Rather, what I've said in effect is that the inclosure analysis is too *classical* a way of thinking about the paradoxes – suggesting that the only way a contradiction is going to happen is as some eschatological apocalypse at the end of time. Inclosure points to contradictions, but it points *away*. Once we have climbed up over the inclosure, the direction is reversed. The paradoxes point to contradictions *inside*.

* * *

Why think that the question "why do the paradoxes exist?" can be answered? We might wish there to be an answer – "Nothing can permanently please," writes Coleridge, "which does not contain within itself the reason why it is so, and not otherwise" – but wishes are only for birthdays. Simply posing a question is not in itself much of a guarantor that it even has an answer, let alone one we might find, the principle of sufficient reason notwithstanding.[39] But the paradoxes are a special case. (They are *the* special case.) The canonical example is the liar sentence, and it has a striking property. All on its own, the liar appears to *prove its own truth*. It also appears to prove its own falsity. That's the paradox. The liar's truth status is *intrinsic*; we can see that it is both true and false without background or side assumptions. The liar can do this because it is *both* true and false – each mutually supports the other. And so it is at least worth considering whether paradoxes might be *self-explanatory*, a logician's version of Descartes's cogito, an Archimedean fixed point.[40] Perhaps the very form of these infinitely difficult problems contain within themselves the beginning of their own answer.

This chapter has brought some difficult methodological issues into view, about finding the shape of any "final" dialetheic worldview. The waters are now muddied to some extent, making our task more difficult; we cannot simply continue the previous efforts to explain the paradoxes via gluts without some reappraisal first. In the next two chapters, we consider in more detail how the emerging program is to be carried out.

---

[39] And with apologies to [Wittgenstein, 1922, §6.5], who asserts that a question can only exist when there is an answer.

[40] "The proposition that contradicts itself would stand like a monument (with a Janus head) over the propositions of logic" [Wittgenstein, 1956, p. 131]. This may have helped inspire the term "dialetheism" [Priest et al., 1989, p. xx].

3

Metatheory and Naive Theory

> Il n'y a pas de métalangage.
>
> —Lacan

The story so far .... Not long into the twentieth century, it had already become clear that there are only two options for any axiomatic theory: leave out some truths, and so be *incomplete*, or else allow some falsities in, and so be *inconsistent*. Now, in the twenty-first century, where do these options really lead?

* * *

How should a nonclassical logician reason *about* their preferred nonclassical logic? There are questions of *methodology* to be faced about *metatheory* (or sometimes, *metamathematics*) – the logic and language we use for the study of logic. This chapter extends the argument of the previous one, and pays off a debt: urging that if we need a paraconsistent logic *somewhere*, then we need it *everywhere*. I examine how this problem is handled by several nonclassical logicians, who in various ways suggest that we can continue to use classical logic. I go on to worry that holding fast to classical reasoning, even in a measured or guarded way, limits our ability to make a fair assessment of nonclassical options; such an assessment will ultimately remain beholden to classical standards. To move the nonclassical project forward, I call for totally nonclassical reasoning. That much revisionism then requires I say more about the scope of the dialetheic project as I am outlining it; I go on to discuss the role of logic in mathematical practice, the question of "classical recapture." and the more important subquestion of paradox recapture.[1]

3.1 The Myth of Metatheory

3.1.1 After Tarski: Theory, Metatheory, and the Charge of Hypocrisy

To make an assessment of the role of metatheory, let us clarify what is at stake.

There are nowadays many nonclassical logics.[2] Some have been studied extensively for both their philosophical and mathematical value. Some have even been argued for as the

---

[1] Thanks to Guillermo Badia, Patrick Girard, Toby Meadows, and Hitoshi Omori for collaboration on topics in this chapter in [Weber et al., 2016; Meadows and Weber, 2016; Omori and Weber, 2019].
[2] A plurality [Beall and Restall, 2006], or varieties [Shapiro, 2014].

correct logic. However, despite some vociferous arguments in favor of nonclassical logic, it is almost always presupposed that our informal (or not so informal) discourse *about* logic must be understood against a background of classical logic. The syntax and semantics of paraconsistent and paracomplete logics – their grammar and truth tables – are almost always taken to be "classically behaved."[3] And indeed, most prominent examples in the literature encourage this assumption, from [Kripke, 1975] to [Field, 2008] to [Beall, 2009], as we will see. Since Tarski, many have simply internalized his decision, that we "have to use two different languages in discussing . . . any problem in the field of semantics" [Tarski, 1944] (cf. Routley [1979, p. 323]), an object language and a metalanguage, and that the metalanguage is classical.

Reliance on classical theory seems peculiar, at least initially, if one takes the arguments that motivate these nonclassical logics seriously. Someone who is philosophically committed to the need for a paraconsistent logic, say because of contradictions in naive set theory, seems vulnerable to at least a nasty ad hominem if they appear to *use* an inconsistency-intolerant logic. The barbed question of *hypocrisy* is asked by Burgess about relevant logic; I've made the substitution from [Burgess, 2005, p. 740] in [Shapiro, 2005]:

> How far can a logician who *professes* to hold that [paraconsistency] is the correct criterion of a valid argument, but who freely accepts and offers standard mathematical proofs, in particular for theorems about [paraconsistent] logic itself, be regarded as *sincere* or *serious* in objecting to classical logic?

The target of this criticism can reply that they are endorsing a paraconsistent logic, e.g., for naive set theory or naive semantics, but are basing their argument *about* the logic in a classical *meta*language. The initial reason this reply seems unsatisfactory is that much of the argument for paraconsistent naive set/truth theories is that (i) other approaches appeal to expressive resources they themselves deem illicit, which is bad (Section 0.2), and that (ii) this cannot be fixed by invoking a "stronger" metalanguage, because there is no such thing as a metalanguage:

> An important advantage – perhaps the major advantage – of the dialetheic program is the possibility of a single, uniform semantics. There is no need for a separate meta-language, since the envisioned language is semantically closed. *[Shapiro, 2002, p. 818]*

This makes a resort to classical metalanguage, at least for a Priest/Routley–style dialetheist, rather awkward.

But awkwardness is not enough. Ad hominem arguments aside, why is it so important not to invoke a distinct metalanguage, with its own logic distinct from the object language? After all, do we not have different rules for different domains? E.g., addition is commutative over the natural numbers, but not over the transfinite numbers (e.g., $\omega + 1$ is a different ordinal than $1 + \omega$). So it might make perfectly good sense to have one set of logical rules L_1 for naive sets, and a different set of rules L_2 for proving theorems *about* L_1.

---

[3] E.g., Bacon [2013, p. 347, fn 12]. See [Woods, 2019] for good discussion; Woods suggests Tennant's work in nontransitive is a possible exception. For another, fully intuitionistic metatheory has been investigated since the 1960s [McCarty, 2008]. For more on reasoning *about* logic, see Meyer [1985a, p. 583].

As there are varieties of mathematics, so there are varieties of logics; why not let a thousand flowers bloom? In the next sequence of sections, let us take a closer look at the internecine justifications for this.

3.1.2 Classical Fallback

Priest is a serious dialetheist and paraconsistent logician. But he makes the suggestion that, "provided we stay within the domain of the consistent, which classical reasoning of course does (by and large), classical logic is perfectly acceptable" [Priest, 2006b, p. 222]. This fits with the prevalent methodology in nonclassical logic. Following Beall [Beall, 2011], call it default classicality or *classical fallback*. All the strategies that follow are some variation on this thought: if we are confident that the domain in which we are working is "well behaved," then, for example, disjunctive syllogism can be deployed reliably. Metametatheoretic reasoning conditions allow for all this, or so goes the idea.

3.1.2.1 The Feferman Objection

Now, it is not hard to get a *nonphilosophical* explanation for classical fallback. At the risk of making empirical claims from the armchair, I suspect that at least one reason for continued classical privilege is simply pragmatic. Classical logic is (basically) well understood and familiar, with a huge century-old cache of methods and tested results. It is generally accepted, traditional, safe. There are problems, but they are *familiar* problems. By contrast, much about the philosophical and technical aspects of nonclassical logics remains to be developed. Life is short, everyone wants to get some work done, and to date no one has known what it would mean to adopt a paraconsistent logic as a logic that we really do use in our thinking, writing, teaching, etc.

More, classical logic remains a dominant force. Gone are the days when nonclassical logic can be responsibly dismissed out of hand as "deviant" or "merely changing the subject" à la Quine, but it is fair to say that nonclassicality is still eyed with suspicion. Expressions of concern are especially pressed on dialetheic proposals, especially any suggestion that *arithmetic* is inconsistent – it "defies belief" [Field, 2008, p. 377].[4] My sense is that for many people the idea that a logico-mathematical apparatus might allow some *inconsistency* just seems to engender worries about collapsing bridges and airplanes falling out of the sky. So far, this has been mitigated by an agreement that, while we can talk about different nonclassical logics at the "object" level, we will continue to share a classical background logic at the metalevel. That way, only object-level airplanes risk falling out of the sky, which is less frightening.

Less cheekily, it has been hard to imagine *really* using paraconsistent logic because the logics involved are weak, much weaker than classical, and attempts, e.g., to reconstruct proofs with them have "not been terribly encouraging" [Priest, 2006b, p. 222]. The pragmatics begin to weigh rather heavily, and pose to the paraconsistentist a question: can we

---

[4] See also [Shapiro, 2002].

still do even basic mathematics?[5] Perhaps the most basic reason for resisting nonclassical metatheory is something like Feferman's injunction [Feferman, 1984, p. 95, emphasis added], that

nothing like sustained ordinary reasoning can be carried on

in most nonclassical logics. There are clear pragmatic reasons to want to fall back on classical logic.

I believe that the *Feferman objection*, as I will refer to it, is by far the most important problem confronting the dialetheic paraconsistent project.[6] Ultimately much of the purpose of the second half of this book is to respond to the Feferman objection: just by sustaining some ordinary reasoning in paraconsistent proofs. My purpose at the moment is not pragmatics, though. I am wondering about the philosophical justification for classical default. On to details.

3.1.2.2 Assuming Consistency

In [Field, 2008], a paracomplete theory of truth is proposed. In Beall [Beall, 2009], a paraconsistent theory is proposed. In [Priest, 2006b], a much more wide-ranging paraconsistent approach is explored. All argue that, due to paradox, classical logic is not correct. All use classical model theory in classical set theory (ZFC) in order to build models of their theories, e.g., in the Kripkean fixed point style.[7] It would appear, then, that the justification offered *for* the theory is not justified according *to* the theory itself. Without some explanation, this can seem bad, as Burgess says. All anticipate the objection, as follows.

[Field, 2008, §5.6] parses the objection in various ways, but in general he sees two prongs to it: (1) the nonclassicist should not accept their own theory (or at least their own meta-results), and (2) "this is a serious embarrassment" [Field, 2008, p. 108]. Field's response is that the nonclassicist can accept their own theory, because

the non-classical logician needn't doubt that classical logic is 'effectively valid' in the part of set theory . . . [that] suffices for giving a model-theoretic account of validity for the logic that is at least extensionally correct [Field, 2008, p. 109].

For Field, "effectively valid" means capturing the classical arguments by assuming instances of the law of excluded middle, as *nonlogical* axioms.

In a similar spirit, since his 2009 book, Beall has developed in more detail a strategy that (following Harman) distinguishes between *logical* entailment, and entailment more generally, which may use *nonlogical* resources [Beall, 2013a]. Paraconsistently, disjunctive syllogism is not valid, but

---

[5] This cuts across nonclassical traditions, e.g., Terui writes of using linear logic for set theory and arithmetic (LAST) that it "is hardly considered as a *working* system of mathematics, because the reasoning allowed by LAST is too poor to formalize proofs of mathematically interesting theorems" [Terui, 2004, p. 38].

[6] It is routinely repeated, coupled with Burgess's point that most nonclassical logicians do not actually use their preferred logic; see, e.g., Burgess and Woods, 2015.

[7] About the "proof of such fixed points, it is important that we accept classical set theory, which I do accept. This is not to say that a paraconsistent set theory couldn't provide the required 'fixed point' result; however, as far as I can see, it might not. In any event, I accept classical set theory, e.g. ZFC, which is assumed in our formal modeling" [Beall, 2009, p. 24].

$$p, \neg p \vee q \therefore q \vee (p \,\&\, \neg p)$$

is valid, because the possibility of the inconsistency of *p* is retained; then the idea is that we can "select" which disjunct of the conclusion to accept, and in particular can usually accept *q* by default classicality. This can be done systematically, via "shrieking" – producing theories by adding non-logical rules expressing consistency assumptions [Beall, 2013b]. Conclusions drawn by nonlogical inferences (instead of the theory obtained by closing under logical consequence) may be reliable, or true for other reasons, but they are not deductively certain; or, they are as deductively certain as the consistency of the subject matter at hand is; cf. [Beall, 2011, 2014a].

Both Field's and Beall's suggestions are akin to what Priest calls "quasi-valid" [Priest, 1989, 2006b, ch. 8], a nondeductive notion, which is reliable but defeasible. As with Beall's proposal, instances of invalid forms such as disjunctive syllogism may be used to obtain conclusions that are likely, but not necessarily logically, true; cf. [Priest, 1991b]. The idea is that classical consequence is reliable assuming there are no paradoxes in the offing. Statistically speaking, says Priest, we are usually not reasoning about paradox-triggering topics, so classical reasoning is, if not completely deductively valid, then at least provisionally ("perfectly"?) acceptable. (Priest does hold that *deductively valid* reasoning is only paraconsistent; he has a further proposal for the legitimacy of using classical reasoning discussed in, Section 3.1.2.6.)

If these approaches are broadly acceptable, then the nonclassical logician can get some use out of the results of classical model theory. They also (apparently) don't need to answer the Feferman objection. So what about assuming classicality for some domians? The explicit way to do this would be to conditionalize, adding as some additional premise to our reasoning that the domain in question is classically behaved. Would that work? Listen to Belnap and Dunn [Anderson et al., 1992, p. 503, emphasis in the original text], with "paraconsistency" substituted:

One might think as follows. The point of [paraconsistency] is taking seriously the threat of contradiction. But there is in this vicinity (that of fairly low level mathematics) no real such threat. So here it's O.K. to use [classical logic]. That *sounds* O.K., but is it? After all, we suppose that 'here there is no threat of contradiction' is to be construed as an added premiss. But a little thought should show that *no* such added premiss should permit the [paraconsistentist] to use [classical logic] for a very simple reason: ... one thing that is clear is that *adding* premises cannot possibly *reduce* threat. If in fact the body of information from which one is inferring *is* contradictory, then it surely doesn't help to add as an extra premiss that it is *not*. That way lies madness.

I don't see any reply to this. In terms of pure deductive logic (as opposed to defeasible evidential reasoning), a dialetheist cannot assume consistency. The idea of assuming classicality for some domains – when you have *already conceded* that classicality cannot always be taken for granted–is a chimera. There are nondeductive senses in which classical reasoning is heuristically or informally useful, just like inductive reasoning is; but falling back to a nondeductive reasoning apparatus is to concede Belnap and Dunn's point: that isn't *logic* any more. Mathematical "proofs" constructed using reliable but logically invalid steps may be convincing enough *arguments*, but they are not *proofs*; see Section 3.1.3.1.[8]

For the sake of argument, though, let's presume activities such as sudoku puzzles or finite combinatorics are consistent, and that in consistent situations classical logic is applicable. Where is the nonclassicist expecting nonclassicality? Well, *pace* Field's suggestion, the part of set theory that characterizes validity (which comes back to *truth*) is *exactly* where there are problems (Section 4.1.2). The theory of ordinals that tracks a transfinite fixed point or revision sequence construction is *exactly* where there are problems (Section 1.1.2.5), not alleviated by reassurance that we are "low enough down" in the ordinals to be safe. A nonclassicist *might* assume that some finite abelian group is consistent, and reason accordingly, without embarrassment; but when it comes to theories that focus on the notions of truth and proof, I don't see how the reliability of classical metatheory can be taken for granted.

3.1.2.3 Truth in a Model Is Not a Model of Truth

Perhaps more importantly, Field ponders whether, if the objection *is* right, then should the nonclassicist be embarrassed? He thinks not: in a similar spirit to his earlier fictionalism about mathematics, model theory is just a tool, *good without being true*,[9] and "there is no need for the notions employed in a model theory to closely relate to truth" [Field, 2008, p. 110]. Field takes an instrumentalist attitude about model theory. This appears to relieve the philosophical burden. Beall, seeing the objection here as a threat of revenge, says much the same:

[T]he formal account is instrumental; it serves as a simple but almost entirely heuristic guide. *[Beall, 2009, p. 39]*

Truth in a model can be heuristically helpful without being a model of truth.[10] And Beall's attitude toward classical ZFC is that "mathematics is free simply to axiomatize away" however mathematicians see fit [Beall, 2009, p. 112]. Invoking instrumentalism suggests that there is little constraining our theory of collections beyond usefulness, and so no reason why it cannot be classical if so desired.

These proposals have the virtues and vices of instrumentalism about science and mathematics. Taken together, Field and Beall tell us not to take the models, or even mathematics, overly seriously. This is fine as far as it goes; there are constraints on what we can hope to learn from our models. On this approach, logico-mathematical frameworks do not play the foundational role of grounding truth as originally intended. Maybe this is not so bad; maybe we should accept a "foundation without foundationalism," in Shapiro's idiom (and as seems to still be the message in [Shapiro, 2014, chs. 6, 7]). The real problem is that, once

---

[8] Beall states explicitly that he does not intend his position to cover mathematical discovery [Beall, 2009, p. 3].
[9] Lewis's gloss [Lewis, 1991, p. 58].
[10] Echoing a slogan of Hodes 1984, as quoted by [Priest, 2007, p. 230].

we have done so, *the instrumental approach does not itself justify using classical logic*; it is just as amenable to a nonclassical metatheory, provided the candidate framework can do the lifting required. Tools in themselves are just tools. If it is in part a historical accident that classical logic and ZFC set theory played dominant roles in the twentieth century, then on purely instrumental grounds, there is no reason *not* to use some other framework, if it is illuminating. If the paradoxes push us to think that classical logic is *wrong* in some important respects, then that is a positive reason for *non*classical default.

An instrumentalist may reply that I am overlooking a salient aspect of classical metatheory, namely its massive *success* as a battery of results and techniques. I don't deny any of that (setting aside the paradoxes). This line of thought brings us back to the need to answer the Feferman objection in Section 3.1.2.1 – the need for a nonclassical alternative to present as a serious rival – to be discussed at length in the next sections.

3.1.2.4 *Ladder Kicking and Gentile Preaching*

Field does make an all-else-fails suggestion: the nonclassicist can view the ZFC models as a temporary communication device. The use of classical logic could be charitably taken as the nonclassicist "preaching to the gentiles in their own tongue," to use a phrase of Meyer.[11] Classicality is no more than a ladder for climbing up over classical accounts that will eventually be kicked away. The fact that much of nonclassical mathematics can be translated or represented in classical mathematics (e.g., S4 models of intuitionistic logic, or ZFC models of paraconsistent set theory) is not in itself any point in favor of the classical apparatus. A paraconsistent logician may be trying to *communicate* to the classicist without *endorsing* classicism.

Routley's inconsistent mathematics manifesto *Ultralogic as Universal?* [Routley, 1977] is a visionary, polemical call to recast logic, mathematics, linguistics, and physics (for a start) in paraconsistent logic – what he calls "ultramodal logic" or "ultralogic." Routley uses a classical metatheory but he does emphasize the need to overcome this eventually (apparently partly in response to some substantial written criticisms from Newton da Costa). In a contemporaneous paper, he writes

It is not satisfactory to reject classical logic systematically, e.g. as involving mistakes or illegitimate assumptions, and to use it metasystematically without further ado or qualification; for to do so would be to proceed by what are confessedly mistaken paths. Such choices of system and metasystem are of

---

[11] See (if you can) his [Meyer, 1985b], redressing the use of classical metatheory. Since you probably can't, I quote from page 1: "Those of us philosophically committed to relevant logics did not require *further* evidence [in the form of classical semantics] for the justified true belief that our systems made sense.....Routley and I are entitled to feel a bit miffed about having been taken to task for preaching in Classicalese to the Classicalists. For it is they, and not we, who hold the view that *only* Classicalese makes sense; and that, for anything else to make sense, it must have a Classicalese translation. But that gut feeling won't satisfy everybody. (In fact, having tried it out on a number of people, I can candidly report that, to date, it hasn't satisfied anybody.)" Meyer was quick to discourage the notion that classical expressions of nonclassical truth are adequate, just as "those who are unwilling to make the effort to comprehend Shakespeare in his native linguistic habitat will have to make do with what can be preserved of him in a foreign language" [Meyer and Routley, 1977, p. 355]. The concern is echoed by Brady, in the midst of a nontriviality proof for naive set theory. "It may seem ironic that hierarchies are being used to prove a result which is designed to eliminate certain hierarchies. However, once our result is established, hopefully one can choose a weaker and non-hierarchical logic to work in" [Brady, 1989, p. 443].

course valuable for *limited* specific purposes ... but such choices are not generally satisfactory: they fail to cohere. *[Routley, 1980a, p. 94]*

So a classical metalanguage may be a limited-use ladder, to be kicked away.

Supposing that ladder kicking and gentile preaching are sufficient justification for classical default, though, this can only be a limited-time account. One would still want a decent theory available once the ladder is gone – a framework to work from *after* all the gentiles are converted![12]

3.1.2.5 Pluralism All the Way Up

An apparently different tack on the metatheory question is taken jointly by Beall and Restall in defense of their logical pluralism [Beall and Restall, 2006]. They argue that there is more than one correct logical consequence relation, more than one notion of validity that counts as genuine logical validity, or in their vocabulary, more than one *admissible* logic. Beall and Restall say that the question of whether an argument is valid simpliciter has no answer [Beall and Restall, 2006, p. 29]. Then what is the underlying logic of their book? The reply [p. 99]:

As to which relation we wish our own reasoning to be evaluated by we are happy to say: any and all (admissible) ones!

Assuming pluralism itself can make sense, this is an enticing thought. How does it work? E.g., here is a stretch of reasoning by disjunctive syllogism, which is admissible in classical and paracomplete logics but not paraconsistent ones. The reasoning is valid in some admissible logics, not valid in another logic. Since these are both acceptable logics, should we say that the argument is valid and not valid? The reply here should be the same again: the question of whether an argument is a valid-and-not-valid simpliciter has no answer. It will have to be pluralism "all the way up."

For a pluralist, this result is just the right thing to expect. But there are subtleties. If there is a collection of admissible logics, this is itself a set (or class) of some kind. With pluralism all the way up, this collection could be theorized in equally legitimate, but incompatible, ways. For example, if we want to say "let a thousand flowers bloom," which arithmetic, in which logic, are we using to count the flowers? If there are infinitely many admissible logics, which set theory are we using to reckon that number? There would seem to be more than one way to conceive of the admissible logics; or at least, comparing the extensions is nontrivial [Meadows and Weber, 2016]. The pluralist might keep trying to stay one step ahead – any and all admissible domains! – but this regress faces the same problem as all regresses. At some point, you have to answer the question, if a genuine explanation is, as Wittgenstein put it, one that must eventually "come to an end somewhere."

---

[12] In *Relevant logics and Their Rivals*, Routley projects sections of a second volume that never appeared as planned; cf. [Brady, 2003]. The original table of contents for that sequel as given in [Routley et al., 1982, p. ix] includes Chapter 15, "Throwing Away the Classical Ladder," with the subsection "dispensing with a classical-type metalogic." Brady tells me that in all of Routley/Sylvan's remaining unpublished materials, no drafts of these chapters are found.

3.1.2.6 Bootstrapping

Rounding out this section, Priest agrees that classical default is problematic. Priest's original argument for dialetheism in 1979 was that natural-language mathematical proof is inconsistent, due to the incompleteness theorems. So one would not expect him to rely on a classical set theory; cf. Priest [1990, p. 208].

He proposes a möbius-like workaround [Priest, 2006b, p. 257]. Here is the program. Start from the assumption that classical Zermelo–Fraenkel (ZF) plus at least one inaccessible cardinal (to be stronger than ZF) has a model. Use a (classical) model-theoretic result called the *collapsing lemma* [Dunn, 1979] to construct models of LP naive set theory. Show that one of these LP models includes inside of it a model of the cumulative hierarchy of ZF. For all we know, this model is the way things really are; at least, it is possible. And *if* this is the universe we happen to find ourselves in, then the LP set theorist can then have it all: "We may therefore establish things in ZF in the standard classical way, knowing they are acceptable from a paraconsistent perspective" (Dunn, 1979, p. 258) – including recovering the model-theoretic collapsing construction that makes this result possible![13]

For the sake of argument, presume that this approach is coherent: we use a classical framework to build up and legitimate a nonclassical framework, which then can appear to "recover" the classical framework, and conclude that the whole thing could be done nonclassically. (Presumably, we assume this procedure too for using the "model selection" strategy in minimally inconsistent LP.) Now, for this to seem convincing, the classical ZF construction must be correct. But Priest, despite having some stern words for ZF, has never been a mathematical revisionary. "The picture has always been that classical mathematics, and the reasoning that this embodies, is perfectly acceptable as long as it does not stray in to the transconsistent" [Priest, 2006b, p. 248]. So never mind that the whole idea is classical "bootstrapping" (a euphemism for "unapologetically circular"). If it worked, what would it show?

After running Priest's suggested program, a dialetheist may hope that their framework is now independently viable – free of any reliance on classical methods. The dialetheist may then be asked to justify various aspects of their formalism, to reproduce the arguments (from ZFC model theory) that got us here. Then there are two possibilities. Either the proof of, e.g., the collapsing lemma was inherently classical, involving reasoning that cannot be recast proof theoretically in paraconsistent logic – as seems overwhelmingly likely with a logic such as LP. In this case, the nonclassicist cannot justify their own position, and they accept some properly classical logic that cannot be appropriated. The underlying classical results, if only obtainable classically, will always remain classical-only, and can never be got by some circular model theoretic argument. Or *else*: the proofs could have been done in a nonclassical setting all along, in which case there never was need of classical metatheory except as a useful shortcut.

---

[13] See [Meadows, 2015] for a critique of this argument, and in reply, the end of [Priest, 2017b].

One should only kick a ladder once one is safely off of it. And if one is *safely* off, when it comes to mathematics, this amounts to having never needed the ladder in the first place.

3.1.3 The Need for Inconsistent Metatheory

There are two paths from this junction. One is to use classical logic in "consistent" domains and paraconsistent logic in the transconsistent, either by dialing down logic to defeasible quasivalidity, replacing *proof* with "acceptable and mostly reliable argument," or else by embracing a permanently open-ended, incompleteist pluralistic view. The other path is to redouble a commitment to *deductive* validity, with explanations that somewhere do come to an end; down this path, "classical default" theory switching is impossible. This latter is a commitment to the search for a *universal logic*.

3.1.3.1 Logic as Universal?

The discussion of the limitive theorems in Section 0.2.2.2 intimated how my approach to dialetheic paraconsistency is, in part, motivated by the allure of a closed, complete theory. This is to tilt toward the windmill of *universality*; as Routley puts it,

A universal logic in the intended sense, is one which is applicable in every situation whether realised or not, possible or not. Thus a universal logic is like a universal key, which opens, if rightly operated, all locks. It provides a canon for reasoning in every situation, including illogical, inconsistent and paradoxical ones. *[Routley, 1977, §1]*

This follows a long tradition, according to which logic should provide the most general schemata for laws. Frege said of logic that "these basic propositions must extend to everything that can be thought" [Frege, 1885, p. 95]. On an old conception, logic is *topic neutral*;[14] it is not about any one thing, and so it is about *everything*.[15]

To a would-be universalist, classical logic looks insufficiently *general*. It is overly specific, overly narrow, in what it can speak about – specifically, not being able to make true claims about truth, set membership, and vagueness. The paradoxes are *paradoxes* because you should fairly expect logic to be able to handle these notions. But when it comes to studies of nonclassical logic, one finds that Burgess's accusation is based on a true observation – that the nonclassical logic literature is suffused with classical logic at the top. And prima facie it looks like shifting back to a classical logic at the metalevel is a tacit admission that (1) the object-level nonclassical logic is not always the preferred logic and (2) logic is not topic neutral or universal in the old way. Rather, we end up saying that

---

[14] The term seems due to Ryle. Topic neutrality is, of course, disputed; see [Williamson, 2014]. A different conception of logic has become known as *antiexceptionalism*, and is sometimes (though not always) coupled with logical pluralism. See [Woods, 2019].

[15] NB a universalist (like Routley, or me) does *not* need to say that they themself *know* which logic is the correct or universal logic, or even that there is only one such logic; universality says only that there is *at least one* universal logic out there somewhere. I think some paraconsistent logic is a reasonably promising start, but I could certainly be wrong about that. Even if there is *some* True Logic, we may never know it – and "it remains true even if all humans should hold it to be false" [Frege, 1903a, p. xvi].

there are different logics for different tasks, e.g., classical logic is good for sudoku puzzles, fuzzy logic is good for vagueness, paraconsistent logic good for liars, and so forth.[16]

This does not work if deductive logic is about which arguments are *always* valid. There are many other related and respected disciplines about arguments that are rational and good, but do not ask so much as to be deductively valid. Informal, heuristic reasoning is crucial for understanding and getting around in the world. But when it comes to pure deductive logic, there are no concessions. Validity is a global property. If an argument form is "valid" in some domains but not others, then it is not valid. The whole point is that logic should work come what may, and we don't know what will come. We don't say that affirming the consequent, $p \rightarrow q, q$ $\therefore$ p is valid as long as we know we are in p-situations, or that "some-implies-all" is valid as long as we are in one-element domains. These forms are invalid because they have counterexamples, and at the global level they instantiate as *bad reasoning*. Dialetheic paraconsistency as I understand it is a program at the global level of deductive logic; there are any number of nondeductive ways to respond to the paradoxes.

In light of all this, if some contradictions are true, then nonparaconsistent logic is invalid, tout court. The same logic must be used everywhere – perhaps most importantly, in developing foundational mathematics. For mathematical proofs are paradigm cases where arguments are constrained by deductive logic; they are what Peano, Frege, Russell, et al. developed logic *for*. If logic is universal, and we think the correct logic is paraconsistent, then mathematics must be conducted paraconsistently.[17] Minimally, we cannot presume consistency when reasoning about notoriously paradoxical objects in set theory or semantics. I might take back what I said in the previous paragraph, and agree with Priest's *methodological maxim* [Priest, 2006b, p. 116] about "quasi-valid" principles such as disjunctive syllogism, that

Unless we have specific grounds for believing that the crucial contradictions in a piece of quasi-valid reasoning are dialetheias, we may accept the reasoning.

But there *are* grounds for believing that there are true contradictions in a great many areas. In mathematics, sets are the lingua franca. Outdoors, vagueness in particular is ubiquitous. The boundaries of ordinary objects are even more so. So the methodological maxim has the modus tollens, not the intended modus ponens, effect.

3.1.3.2 Too Much Paraconsistency?

The best argument for using a paraconsistent metatheory would be, not just the threat or possibility of inconsistency in a great many areas, but the *actual* obtaining of a great many contradictions in a great many areas. And I've now just appealed to this, turning

---

[16] Routley: "This *local* logics option soon runs into difficulties (as the geographical image suggests) at boundaries, as to how the local logics impinge upon one another and how they combine. For example what happens in a boundary area between two localities? In new (unclassified) situations? If one can't guarantee the location (e.g. because consistency isn't provable)?" [Routley, 1977, §1, emphasis in the original text]. Routley sees the "local logic" option as "a guise for classical exploitation" [p. 897]; cf. [Routley, 1975]. But enough about politics.

[17] A project like Beall's is much more modest (self-confessedly so Beall [2009, p. vii]) than my more "fundamentalist" one, which is to develop mathematical theories using (only) logical entailment. Priest now takes a pluralist view of mathematics [Priest, 2013b], though see Priest [2008, p. 585].

the discussion from our principled view of what logic is (or ought to be), to what logic *must* be, if the sort of glutty views about vagueness and the boundaries of ordinary objects I've suggested is correct. Which, in turn, brings us back to the more acute worries about airplanes falling out of the sky.

For doesn't this dialetheic account from Chapter 1 mean that almost everything is inconsistent in some sense? After all, almost every predicate is vague. Every object has a boundary. That's a lot of inconsistency! We are almost literally *swimming* in contradictions. In particular, that would undermine Priest's methodological maxim, or the related claim that most ordinary reasoning is untouched by paraconsistent considerations, if inconsistencies are statistically unusual. Priest rejects the presupposition [Priest, 2017b]:

Take a vague predicate, such as 'red'. The vast majority of objects are not red, and consistently so. It follows that the collection of objects which are red and not red are a very small proportion of the total.

True. But take almost any object; it is a borderline case of *some* vague predicate. Any red thing is a borderline case of some subcollection within the red things – e.g. the set of "red but also a bit yellowish orange" things. So the collection of objects that are φ and not φ for *some* closely related predicate is the *entire* total. If motion and change are also sites for dialethia [Priest, 2006b, chs. 11, 12], then similar comments apply.

To make the point in one fell swoop, let us look at some naive set theory, and meet for the first time a form of one of the most important objects in this study, the *Routley set* (Section 5.2.2.1). For any set X, by naive comprehension there is its "Russellization":

$$X^r = \{x : x \in X \text{ and } r \in r\},$$

which is all the members of X to the degree that the Russell set is a member of itself. Then, if we accept naive comprehension, we must accept that there has always been the following set, "out there in Aussersein":[18]

$$V^r = \{x : x = x \text{ and } r \in r\},$$

which is the universe to the degree that the Russell set is a member of itself. Now ...

- You are a member of this set, because you are self-identical, and $r \in r$.
- You are not a member of this set, because $r \notin r$, so the conjunction of $r \in r$ with anything is false.
- Therefore, *everything has at least one inconsistent property already.*

A commitment to naive comprehension does not just entail Russell contradictions; it is a commitment to inconsistency on a massive scale. This does not make everything inconsistent (Section 4.2.3). But it means there are no "safe" domains for invoking consistency-dependent logics.[19]

---

[18] A phrase from [Routley, 1977], where this set also comes from. Cf. Ripley [2015d, p. 559].

[19] Thus I go in a different direction than paraconsistency programs following da Costa [da Costa et al., 2007] (in [Jacquette, 2007]), seeking *consistency operators*, the *logics of formal inconsistency* [Carnielli et al., 2007], especially since these logics don't support naive set theory. See Carnielli and Coniglio [2016, ch. 8].

So, yes – that's a lot of inconsistency. Dialetheism begins with banalities, and ends with a radical insight into the nature of the world. It confuses the position then to say that one of the most basic assumptions in most of Western thought is false, but then to say that usually classical logic is just fine, because dialetheism is mostly irrelevant to the ordinary world.[20] "Quasi-valid" is *invalid*. Priest is correct that "counter-examples to inferences such as the disjunctive syllogism occur only in the transconsistent" [Priest, 2006b, p. 222]. But the transconsistent is when you meet your friend at "noonish," when you buy something that is just a little too expensive, when you remark that nothing your uncle says is true. The transconsistent is *everywhere*.

3.1.3.3 Nonclassical Default

That completes our review of the standard practice for nonclassical logic. For me, I persist with my (perhaps unrealistically) high hopes for (without claiming to have yet found) a universal logic. There are different rules for different games, but logic is not a game. Logic and the language we use to talk about it cannot be swapped in and out so that we can, *consistently*, say in one breath what we assert we cannot say in another. Deductive proofs cannot be conducted in a "metalanguage":

... the whole *point* of the dialetheic solution to the semantic paradoxes is to get rid of the distinction between object language and meta-language. *[Priest, 1990, p. 208, emphasis original]*

For me, using classical metatheory is out. It is time to consider alternatives.

3.2 Classical Recapture

With the plan of venturing off into inconsistent mathematics, we arrive at the question of "classical recapture." For there is, as we all know, a very large and rightly venerated body of work known as *mathematics*, a "dignified and vitally important endeavor" [Shapiro, 1997, p. 3], much of which was (re)constructed during the twentieth century using classical logic. But a rather drastic reconsideration of what logical principles are valid runs the risk of being cut off from some obviously legitimate human achievement. It invites the Feferman objection (Section 3.1.2.1). Or to put it maximally ideologically[21] – what is the place of *non*dialetheic mathematics in the (eventual) presence of an "ultralogical" rival?

In order to sustain the ultramodal challenge to classical logic it will have to be shown that even though leading features of classical logic and theories have been rejected, one can still get by. In particular,

---

[20] In a different context, Priest sounds like he would agree, responding to [Shapiro, 2002]: "Shapiro's objections stem from being half-hearted about dialetheism. If one endorses an inconsistent arithmetic, but tries to hang on to either a consistent computational theory or a consistent metamathematics of proof, one is in for trouble. The solution to Shapiro's problems is therefore not to be half-hearted, and to accept that these other things are inconsistent too" [Priest, 2006b, p. 243].

[21] From the other end of the spectrum, da Costa makes a nonideological suggestion about the place of paraconsistent mathematics (not necessarily dialetheic): "It would be as interesting to study the inconsistent systems as, for instance, the non-euclidean geometries: we would obtain a better idea of the nature of paradoxes, could have a better insight on the connections amongst the various logical principles necessary to obtain determinate results, etc. ... It is not our aim to eliminate the inconsistencies, but to analyze and study them" [da Costa, 1974, p. 498].

it will have to be shown that by going ultramodal one does not lose great chunks of the modern mathematical megalopolis. *[Routley, 1977, p. 927]*

This is a very imprecise, if compelling, image. How much is a "great chunk"?

3.2.1 Whither Paraconsistent Arithmetic?

A longstanding project is to recast *arithmetic* in a paraconsistent logic. For while set theory is a source of many of the motivating paradoxes, and itself serves as a foundation for arithmetic, the theory of natural numbers is arguably more basic, and has from Frege and Peano been a prime testing ground for logic. To show that a logic can support arithmetic is a good first step toward demonstrating *independent viability*; failure to support arithmetic calls the usefulness of any logic into question.

So for some decades, there has been an ongoing effort to fill in the details of arithmetic in a paraconsistent logic, and in particular, a *relevant logic*. Relevant logics are paraconsistent logics, those demanding a "meaningful" connection between antecedent and consequent in valid implications. Ex falso quodlibet is invalid on the grounds of being irrelevant. Dialetheic paraconsistency and relevant logics in Australasia grew up together in the 1970s.[22] Meyer [Meyer, 1976] proposed replacing classical logic with the relevant logic R, and rephrased the axioms of Peano arithmetic (PA) with relevant conditionals, to obtain $R^{\#}$ (said "Arr sharp"). He touts it for its paraconsistency:

The hope was that $R^{\#}$ offered the best of two worlds. On the one hand, its concern for relevance makes $R^{\#}$ arguably more reliable than PA. For even if, perish the thought, the Gödel formula is a theorem [thus proving a contradiction], there is no way to construct therefrom a proof of $0 = 1 \ldots$ On the other hand, early investigations suggested that $R^{\#}$ was as reliable as PA. *[Friedman and Meyer, 1992, p. 827]*

Circa 1976, Meyer appeared to show that $R^{\#}$ can prove its own consistency (nontriviality), essentially because it has finite models.[23] Priest has shown that inconsistent arithmetic based in the even weaker paraconsistent logic LP is decidable.[24] However one goes, the envisioned system is unhindered by Gödelian limitations, and, if using relevant logic, involves no reasoning concealed by implicational paradoxes.[25]

Indeed, it is not just arithmetic that the paraconsistency program has targeted. It has seemed desirable that all the theorems of classical mathematics tout court be somehow "recaptured" within a paraconsistent framework. If paraconsistency could somehow restore naive set and truth schemas without costing any results, this would vindicate a brash claim, that [Routley, 1977, p. 929]

---

[22] For history, see [Brady and Mortensen, 2014].

[23] See also work with inconsistent mathematics maestro Chris Mortensen [Meyer and Mortensen, 1984; Mortensen, 1988]. A weaker relevant arithmetic (in DKQ is laid out by Routley [Routley, 1977, §9].

[24] [Priest, 1994a, 1997a, 2000, 2006b, ch. 17]; cf. [Asenjo, 1989].

[25] [Routley, 1977, p. 904]. Including, perhaps, avoiding a mate of Curry's paradox, Löb's theorem [Smith, 2007, p. 230]. For some related work, see [Meyer and Mortensen, 1987; Meyer and Restall, 1999; Tedder, 2015].

we can do everything you can do, only better, and we can do more.

This, in very imprecise terms, is what has become known as *classical recapture*.

It is increasingly clear, though, that nothing so simple is going to happen. Much energy was expended on the so-called *gamma problem* for $R^\#$, which was the hope that disjunctive syllogism would be admissible (turn out to be valid "for free") and hence that classical recapture of Peano arithmetic would be in reach. Friedman and Meyer [Friedman and Meyer, 1992] solved the gamma problem in the negative: arithmetic in any relevant logic weaker than R will not contain all of classical arithmetic.[26] The project is thus said to have "ended in failure" [Priest, 2006b, p. 222].[27] Meyer's open question:

Given that the straightforward approach of simply grafting the first order peano postulates in to *R* has failed, is there a relevant way of thinking about the natural numbers which will produce a more satisfactory result? Hence our question: Whither relevant arithmetic? *[Friedman and Meyer, 1992, p. 825]*

3.2.2 Recapture and Rehabilitation

In retrospect, I think we should say that the project called "classical recapture" *rightly* ends in failure. A theory closed under classical logic such as ZFC or PA is a very specifically formed artifact, shaped by the accidents of classicality. It would be extremely surprising to reobtain exactly this artifact using a different, weaker logic. It would also be of dubious value if part of the purpose of "going nonclassical" is to go nonclassical. Let us distinguish, then, the question of whether general arithmetic/mathematical *practice* is still possible nonclassically from the question of whether all of classical PA or ZFC is *derivable* in a nonclassical framework. The negative answer to the latter question informs how we positively answer the former.

What is wanted is, not recapture, but classical *recovery*, in Routley's sense, a kind of rehabilitation:

Classical mathematics should be recoverable *insofar as it is correct*. *[Routley, 1977, p. 894 (emphasis added)]*

On this conception, paraconsistent mathematics never should have aimed at preserving every speck of irrelevant dust kicked up by classical PA. A motivation for paraconsistency, after all, is to *avoid* some of the more untoward classical results – to capture some naively attractive properties that are impossible in (or over) PA, such as the provability of the soundness of proof (cf. Priest [2006b, ch. 3]).

Perhaps dialetheic paraconsistency was not intended to be revisionary. But the way I am telling it, it is.[28] Some founders in this area – Meyer in particular – seemed to believe

---

[26] In [Friedman and Meyer, 1992] it is shown that the complex ring $\mathbb{C}$ is a model of relevant arithmetic in R (and hence that some classical PA theorems have counterexamples); cf. [Meyer, 1998]. It is worth revisiting the Friedman–Meyer proof, which uses classical metatheory essentially, and asking to what extent it holds nonclassically.

[27] More generally, Thomas has some convincing evidence that the classical recapture will not happen; see [Thomas, 2014].

[28] Contra Priest: "The programme of paraconsistent logic has never been revisionist in the same sense [as intuitionism]. By and large, it has accepted that the reasoning of classical mathematics is correct" [Priest, 2006b, p. 221]. All I've said, note, is that *some* part of classical mathematics is wrong. This could, for all I've said, be a very *small* part – so the scope or scale of the revisionism here is undetermined, and could turn out to be much less radical than it sounds. Or not.

at least at one point that all that would be needed to "go paraconsistent" would be to remove ex falso quodlibet from logic (which would be painless since it is "bad reasoning" and never used anyway) and everything else would carry on more or less as before, but better: "... what is to be hoped for most of all are not new routes to old truths but an expansion of the pragmatic imagination" [Meyer, 1975, p. 5]. That guess was wrong. For his part, Routley in various places is much more ready to call for revisions – he talks about "classical" mathematics like it is no more than a passing fad:

Insofar as mathematics relies on valid argument, its proper formalisation is not in terms of classical logic. *[Routley et al., 1982, p. 52]*

It strikes Routley as an almost trivial point that "alternative formalisations ... using relevant systems can undoubtedly be devised," that

the bulk of intuitive mathematics ... is not classical, except insofar as recent classical logical reconstructions have pushed it in that direction .... *[Routley, 1977, p. 903]*

Now, Routley is being far too sanguine in his expectations that reconstruction will be easy. But he is correct in his critical claim that in light of the universality of logic, true contradictions, and the invalidity of explosion, at least some of the classical analysis of mathematics is wrong.

One cliché, then, and one tautology. We don't want to throw the baby out with the bathwater – but if we want to change the bathwater, then we need to throw out some bathwater.

Farewell to the theorem-for-theorem classical recapture. Why aspire to being a weak imitation – exactly like classical mathematics, just harder and weirder? The important concern behind "recapture" is to show that real, everyday(ish) arguments are viable – answering the Feferman objection and "recovering the bulk of intuitive mathematics." What, exactly, this amounts to is hard to make precise; it would be imprudent to be too precise from the armchair. The last part of this chapter, and the second half of this book, is my small attempt at providing an answer.

3.2.3 *The Logic of "Real" Mathematics*

Surely, someone (maybe Frege) says, *the best logic for old-fashioned (platonistic) mathematics is classical logic?* Someone who asks a question like this is also likely to think that, if we must decide between nonclassical theories, surely a good criterion is to be "as close to classical logic as possible." Both of these are natural enough thoughts, but I think in light of the paradoxes they are up for a reevaluation. To be concrete – in the next chapters I present and use substructural paraconsistent logic for mathematics; before doing so, I should clarify what I think I am up to.

3.2.3.1 *Descriptivism*

Is the view being advanced here *descriptivist*? Once upon a time, in the late nineteenth and early twentieth century, logicians took themselves to be directly describing the logico-mathematical apparatus in use in mathematical practice, e.g., Peano's analysis of arithmetic (or even before that, Boole's laws of thought). Maybe Peano succeeded, or maybe not (especially in light of Gödel's theorems [Priest, 1979; Meyer, 1996]). In any case, that is not what I am doing here. I'm under no illusions; mathematical proofs "in the wild" use all sorts of reasoning that is not in general paraconsistently valid: disjunctive syllogism, most saliently, but also a host of other illicit forms, to be discussed in the next chapter. Many reasoners seem to believe in and use the naive set comprehension and truth schemas, too, which means in some sense that the overall "system" they are using is incoherent. Mathematical practice is a complicated sociological phenomenon; what results count as accepted and which do not is deeply complex;[29] and to whatever extent Peano et al. were qualified to psychoanalyze all that, I am not. This is not anthropology of mathematics or description of the logic that mathematicians "really" use. I doubt there is any one such logic, as recent work in mathematical pluralism suggests [Shapiro, 2014].

By the same token, the undeniable success of (classical?) mathematics in *applications* – in empirical science – might be held up as a reason not to call for any revision, but this would be to make a simple positivist mistake. A theory can make correct predictions without being true. So the success of mathematics is not to be written off, but honoring it does not require an eternal commitment to classicism. And if a theory makes predictions that turn out to be *false* (there is no universe?), then that really is evidence that the theory might be good without being true.[30]

3.2.3.2 Revisionism and Normativity

I have already accepted that this approach is, at least to some degree, *revisionist*. This strain of paraconsistent mathematics begins again, at the beginning, with previous foundational efforts and higher theories as a very useful *guide*, but with the possibilities for this new attempt still wide open. I am exploring a possible world where everyone from Euclid to Euler knows and uses paraconsistent logic, as a way to understand how the world could rationally contain paradox. Logic codifies the *norms* governing this activity. What are the implications?

In terms of practice, I am again under no illusions – I do not expect actual revision of actual mathematical practice. For a start, there are longstanding antifoundationalist or antilogician attitudes in the literature, to the effect that formal logic is irrelevant to real working mathematicians.[31] Indeed, many people in the naturalistic tradition share David Lewis's deferential attitude toward mathematics, when he rhetorically asks philosophers,

How would *you* like the job of telling mathematicians that they must change their ways ... ? Not me! *[Lewis, 1991, p. 59].*

---

29 See the footnotes throughout [Lakatos, 1976]. On revisionism in mathematics, see [Gilles, 1992], and for the "maverick" idea of mathematics as a social phenomenon, see, e.g., [Mancosu, 2008].

30 For more on the idea of applied inconsistent mathematics, see [Colyvan, 2009], and for consequences, [Colyvan, 2008a].

31 "If ever the day comes when the logicians find some inconsistency in arithmetic, our reaction will surely be 'Oh that's just a trick of the logicians; let them worry about it'" [Jones, 1998, p. 204]. Cf. the parable retold in [Kanamori, 1994, p. 481] comparing logicians to spiders building epic cobwebs in the vaults beneath a cathedral.

Well, not me, either! But, while I appreciate Lewis's humility, his thought experiment of marching over to the mathematics department and making demands is a misdirection, a false choice. I have no such plans, and if you do, please don't write to me. I *do* plan to follow arguments where they lead, even if they sometimes lead away from conventions of the day. One could take this as normative ("follow me!"), or simply as an alternative way of doing things ("...if you want to!").[32] And since disciplinary boundaries are largely historically and socially contingent – there is no sharp demarcation between mathematics and philosophy – if there is some (any) adoption of a inconsistent mathematics among philosophical logicians (for which I am still not holding my breath), then this would count as some change in some actual mathematical practice.

In any case, this is not a popularity contest.[33] Of more gravity is what the alternative being presented actually looks like. This is in part "establishing the ultramodal adequacy of any such formalisation" [Routley, 1977, p. 903]. What is "ultramodal adequacy"? If not simply "doing everything that [classical mathematics] can do, only better," then what? It is

a further stage in the dialectical process ... namely dialectical ascent to, or incorporation of, the metalanguage. *[Routley, 1977, p. 910]*

The end goal is the development of a theory that is independent and does not hold itself to the standard of classical or conventional approaches, a practice that is internally coherent and self-standing (cf. [Belnap and Dunn, 1981]). That *is* to do what classical mathematics has done after all. And what of "we can do more"? The theory should be *closed*, in the sense of including all the notions needed to express that theory – including full set and truth schemas, as well as predicates for validity and proof. It should then have a logic weak enough that these principles do not lead to absurdity. It should nevertheless have a logic strong enough that "the bulk" of standard mathematical results are still provable – albeit sometimes with novel arguments or techniques that are not familiar from classical practice.

3.2.3.3 Proofs

Our approach, then, will be axiomatic, focused on *proofs*, after Euclid. We assume some propositions (or propositional schemas) to be *true*, some rules to be *valid*, and see what we can see. This is the "honest toil" approach (in Russell's idiom). It is more cumbersome than the *model theoretic* approach taken by most other paraconsistent mathematicians (e.g., [Mortensen, 1995]). With models, one can say what is *not* provable, on the basis of counterexamples) – but this is all so far conducted in *classical* model theory. The tools of paraconsistent model theory are not available yet; they first need to be built, by proving things with only low-tech, elementary means. This historical progression, from axiomatic to model theoretic, is standard in mathematics; axiomatics come first, from Euclid to Hilbert to Anderson–Belnap. We start with what we *can* positively prove, to build up intuitions for any later, more sophisticated, developments.

---

[32] Akin to, e.g. *smooth infinitesimal analysis* [Bell, 2008]; cf. ch. 8.
[33] Unless inconsistent mathematics becomes more popular, of course.

On this last front, since I just admitted that "real" mathematicians are not using a paraconsistent logic, I should clarify that the intent is that, wherever a theorem is usually proved using disjunctive syllogism or other classically only valid inference, there should be an alternative proof – perhaps still needing to be discovered – that leads to the same or similar result using only paraconsistently valid inferences. In the event that there is no such alternative proof, then the theorem is *essentially* classical and, depending on the case, may not be correct. We cannot know until we try. As Slaney states:

> There is a vast amount of formal hard work to be done before we gain a clear vision of the formal sciences [in nonclassical logic]. What appears below scarcely begins that work and indicates how intricate much of the honest toil is likely to be. *[Slaney, 1982, p. 3]*

Some of the honest toil is to leave our expectations behind, and to be ready to see something new.

Von Neumann famously said, "You don't understand things in mathematics. You just get used to them." Some of us have gotten very used to a classical version of mathematics. Some of us are getting used to another.

3.3 Naive Theory

The previous section discussed what we are *not* going to attempt, namely, to imitate classical logic with a paraconsistent logic. In this section, I will set out the targets for what we will be attempting: what we will try to prove, and how to go about proving it.

The overriding concern is how to present and understand a *uniformly* paraconsistent theory, with no reliance on a classical superstructure. So, first thing first, how shall we think about the underlying logic itself?

3.3.1 There Is No Middle Value

In Section 0.2.2.3, we saw a paraconsistent logic, using two valued *relational* semantics. Standard presentations of paraconsistent logic, going back to the seminal works [Priest, 1979], are via a three-valued *functional* semantics, with values {t, f, b}, the third value read as "both." For example, the truth table for negation in LP looks like this:

So if p is a true contradiction, rather than say p is t and f as in the two-valued case, we say p is b.

Put this way, a paraconsistent logic looks like a species of multivalued logic. This may have some pedagogical advantages for presenting the idea to unfamiliar audiences, but it does not do justice to the story told in the previous chapters. For a start, it does not really capture the idea of a glut as a *contradiction*, since in either case some sui generis *new* item has been introduced, some other value b, that is not defined in terms of t or f. It is hard to see what this value has to do with truth or falsity. Maybe b is a banana, or Julius Caesar.

But maybe this problem of interpreting a middle value is just unfamiliarity, something to get used to. The more important point is that adding new truth values embraces revenge.[34] Multi-valued pictures may add texture, but they don't change the basic underlying scenario: that there is truth, and negation of truth, being "other than" truth. Multivalued logics replace truth with *designated* values, for instance, so in LP the designated values are t, b and an argument is valid iff there is no assignment that makes all the premises designated but the conclusion undesignated [Priest, 2008]. This expands the acceptable range of truth values, but in doing so basically just expands the definition of truth. Sentences with designated values are still taken as describing what is the case, even if some of them also do not describe what is the case. There is still, as Priest has observed, a binary distinction between "the good ones" and "the rest." And once you see this, you see why revenge is inevitable if you try to stratify with additional values.

A three-valued semantics makes it appear that there is after all an exclusive and exhaustive partitioning of the universe of truths, now into *three* categories: into the all-and-only truths, all-and-only untruths, and the all-and-only "boths." If the original Tarski problem was insoluble, this new, three-tiered approach will be no better. It is susceptible to revenge, in the form of a new liar: "the (only) value of this very sentence is f." With v an interpretation from sentences to exactly one of, e.g., three values $\{t, f, b\}$, then there will be a sentence q that says of itself that it is only false, q iff $v(q) = f$, and then we are no better off than we were with classical logic.[35] If the division between the "good" values and the others is both exclusive and exhaustive, then the same old scenario plays out. We should have learned by now: it remains the case that we must choose between untruth avoidance at all costs, or truth seeking, come what may.

3.3.2 Just true?

The three-valued approach encourages a common criticism of dialetheism – that some important expressive power has been lost, namely, the ability to demarcate the truths (t valued) from the true contradictions (b valued).[36] The glut theorist cannot pin down all and only the just-trues, on pain of absurdity. "But surely," says the critic, "this distinction is available – there it is in your semantics! – and yet the object language cannot express it." This is called the "just-true problem" or "true-only problem," and many people think it is an important objection to dialetheism, some kind of revenge. For example, Rossberg writes

---

[34] For instance, see Cook's "Embracing Revenge" in [Beall, 2007]. He writes that "The Revenge Problem, it turns out, is not a problem at all, but instead affords the crucial insight that . . . we can never speak a 'universal language', quantify over all sentences, or speak of all truth values at once" (p. 35). This is exactly the standard solution from the Introduction and has exactly the usual fatal flaw. The "crucial insight" does exactly what it says cannot be done.

[35] See Priest [2006b, p. 288]; but also essentially this problem as Curry's paradox, as discussed in the next chapter.

[36] Discussions include [Berto, 2008; Restall, 2010b; Rossberg, 2013; Murzi and Rossi, 2020], and essays in [Priest et al., 2004].

...[T]he [just-true] problem appears pressing because of our manifest ability to communicate that we hold a sentence to be just true (and not also false), or satisfying *our sheer desire to be able to do so.* *[Rossberg, 2013, emphasis added]*

But if the dialetheist could express "just true" or "just false" the way the critics want, it looks like there would be the same problem with a strengthened liar. Dialetheists face revenge along with everyone else. Or so goes one reconstruction of the "just-true" objection.

The "just-true" problem is not a good objection.[37] Priest points out that there is no real issue at the level of truth or falsity: a dialetheist can (truly) express that a sentence "is true and not false" with those very words; "what she cannot do is ensure that the words she utters behave consistently" [Priest, 2006b, p. 291]. The crucially important point, though, is that *neither can anybody else.* The premise of dialetheism is, in effect, that some consistency demands are impossible to satisfy; the "just-true" objection simply demands otherwise.[38]

I do think that the "just-true" problem is a *social* problem. Dialetheists are people who admit that they sometimes assert falsities, knowing that they are false – because they are true, too. This creates interpersonal problems, with trust and credibility. But the syntax of logic was never meant to carry a person's integrity. As a logic or philosophy problem, the "just-true" problem is just not about dialetheism. Indeed, it would be a devastating criticism ... if not, again, for the *original* problem, which is that *no one* can in fact make the demarcation, by Tarski, between all-and-only the truths and everything else. But since that was starting point for all parties involved, the dialetheist is at no special expressive loss. We are *all* at the same loss. The "just-true" problem is an objection to the repercussions of the limitive theorems. Either there is no universe, or it cannot be divided in two, or the division will have some overlap, whether or not you have the "sheer desire" otherwise; that's it. The "just-true" problem is an objection to the paradoxes themselves.

With this in mind, the glut theorist should use a formalism that does not invite or suggest misinterpretation. To this end, the presentation of truth values should be *relational.* (In fact, as far as possible, almost everything functional in mathematics is going to be glossed as relational [Section 5.2.2.3] – but I'm getting ahead of myself.) The truth value of a proposition will be a subset of $\{t, f\}$, and then the conditions on connectives are as given by the semantics in the Introduction. Here they are again, in a familiar form:

$\neg$		&	t	f		$\vee$	t	f
t	f		t	t	f		t	t	t
f	t		f	f	f		f	t	f

Such two-dimensional displays are often implicitly assumed to be functional look-up tables. But no such assumption is explicitly displayed or otherwise enforced by what is

---

[37] Field, no paraconsistentist, might agree: "I am embarrassed to admit that I once tentatively voiced this worry in print" [Field, 2008, p. 362, footnote].

[38] See Priest [2006a, ch. 5; 2006b, §20.3]; Beall [2009, ch. 3].

on the page. It is simply presupposing classicality to see these as functions. You get out what you put in: classical metatheory, classical object theory; paraconsistent metatheory, paraconsistent object theory.

The truth table for negation, especially, might look to be classical negation. But look again. The point is appreciated by Belnap and Dunn: gazing at the apparently standard relational clauses for de Morgan (paraconsistent) negation and Boolean (classical) negation (notation adapted),

de Morgan: $\neg A$ is t iff A is f, $\neg A$ is f iff A is t.
Boolean: $-A$ is t iff A is not t, $-A$ is f iff A is not f.

Then they observe

> These clauses certainly seem to support the distinction ... where de Morgan negation is 'internal', Boolean is 'external'. They do anyway until one stops to ask what kind of metalinguistic 'not' it is that occurs in the Boolean clauses above.
>
> This is a profound question .... If the 'not' is a de Morgan negation (as it surely would be for the relevantist [and dialetheist]), then ... the 'internal/external' distinction collapses .... *[Belnap and Dunn, 1981, p. 345]*

Without assuming consistency, two-valued truth tables are displays of semantic relations and allow overdetermination of truth value. "The truth conditions .... actually use the notion of negation ['not'] and hence are ambiguous, depending on whether this is itself [paraconsistent] or [classical] negation" [Priest, 1990, p. 207]; cf. [Sylvan, 1992]. To be very careful about it, the tables can be read as, "if t is *among* the values of φ, then f is *among* the values of $\neg\varphi$." But the more concise reading will do, too: "if φ is true, then $\neg\varphi$ is false."

We will return to discussing full-blown dialetheic semantics, especially conditions for negation, in Chapter 10.

3.3.3 Paradox Recapture!

Much more important than classical recapture, or replying to (question-begging) philosophical objections, is answering the Feferman objection, showing that some real mathematical reasoning can be supported paraconsistently. I wish to isolate a tractable part of it by focusing on *paradox recapture*. It is one thing to say that some classically valid arguments are not valid, or that some classical theorems are not true (!); but *what of the particular arguments and theorems that led us to dialetheic paraconsistency in the first place?* This is the problem we are left with from Chapter 2, and it is a serious problem.

3.3.3.1 Why Recapture Paradoxes?

Take a solution to the sorites paradox using paraconsistent logic, as sketched in Chapter 1. Ask: what becomes of the sorites when we turn the tables, and *pose* the problem using paraconsistent logic? What if the candidate logic is too weak to be able to formulate the sorites paradox to begin with? If, for example, one counts heaps of sand using a paraconsistent *arithmetic*, does any paradox still obtain? So too with the continuous and topological sorites (Section 1.2). There would seem to be something awry if a logic were not strong enough to express the very problems that logic was invoked to address. We saw the inclosure schema analysis fall apart for just this reason.

If arguments for a paraconsistent approach are fundamentally *classical*, then committed paraconsistentists cannot justify their own position, and the very raison d'être for contradiction-tolerant approaches seems to disintegrate. The nonclassical solvers are in danger of having kicked the ladder away – right out from under themselves.

A natural thought is that it is bad if a solution leaves solvers unable to justify their own solution. This has some of the symptoms of *revenge*: a solution that fails on its own terms. Solvers, having adopted a thoroughgoing paraconsistent logic, now have a weaker notion of logical consequence at their disposal, and can no longer reconstruct the memory of what led to being in this position. Solvers seem caught in a vicious circle. Disjunctive syllogism is invalid in inconsistent contexts, and vague φ is an inconsistent context – *as we have proved using disjunctive syllogism*. Without the argument using disjunctive syllogism, there is no proof that φ has inconsistent instances, and so no reason not to use the disjunctive syllogism![39] As Santayana taught us, "those who cannot remember the past are condemned to repeat it."

Perhaps, though, this objection is too quick. The paraconsistent solver can no longer *prove* that the sorites is a problem – but isn't that the point of a solution: to make the problem go away? Other purist nonclassical approaches would seem to be in the same boat as the paraconsistentist: upon describing the sorites scenario, it turns out that there is nothing paradoxical. Thoroughgoingly constructive (paracomplete) mathematics will lack the requisite "cutoff points" in sets of reals to conjure the sorites to begin with, for example. And in a way, maybe this is how it should be, if these are intended as *solutions*. A fuzzy solution couched in fuzzy mathematics is designed to be fuzzy to its core, so the problem of precise borderlines does not arise. Inexpressibility is the point, not a problem, right? So if faced with Santayana's objection, solvers could try to justify their position by way of a hypothetical. "*If* you took a standard, classical approach, *then* you would have a sorites paradox." Solvers pretend, as it were, to be classicists long enough to show that the position is untenable – preaching to the gentiles in their own tongue.

As before, I do not think this is the way to go – we have already examined and rejected solutions that cannot express or justify their own tenets. Even if the program just sketched is feasible,[40] unfortunately, the classicist can turn around and play the same trick. "You tried to convince me of p; but if your argument was right, then your argument was *not* right. The argument you've given is self-undermining on *your* terms, not mine. In pretending to be classical, all you've done is demonstrate that I can't be otherwise." The dialectic is completely symmetrical, and devolves into burden shifting. Let's not.

---

[39] This sort of logic revision cycle is discussed in [Woods, 2019].

[40] The dialog itself – pretending to take on an opponent's view in order to refute it – is steeped in *counterlogicals*. This raises the issue of transworld logical debates; see [Meadows and Weber, 2016].

I think naive set comprehension is true, and that the Russell set is both a member of itself and not; I think that there are sharp cutoff points for vague objects, inconsistent cutoff points; I think discrete material objects can overlap at their nonempty boundaries. Therefore, I use a paraconsistent logic (always and everywhere). But when I go to use paraconsistent logic, if I find that the mathematics does *not* lead to paradoxes after all – say if the Burali-Forti paradox of the greatest ordinal could not be derived – then on my preferred approach, these conclusions are *not* inconsistent. I've discovered that I made a mistake (and not in a paraconsistently "okay" way where I also did not make a mistake [ha ha]). It's not just a bad faith/ad hominiem/tu quoque sort of problem that worries. It is the problem of simply being *wrong*, on one's own terms. Again: one ought to be able to express the justification for one's position.

Something more is required. And in providing it, the paraconsistent approach then could be at an *advantage* over other nonclassicists, insofar as it overcomes the dialectical problems just outlined and retains the ability to describe the problem to which it is a solution. Returning to the question of classical rehabilitation, and what needs to be provable to count as "the bulk of intuitive mathematics," then a minimal condition is: at least enough mathematics to explain why one is doing mathematics this way. That's the goal. Let's list what we will be looking for.

3.3.3.2 The Sorites Reasoning

To express the sorites problem, we presented arguments that appealed to various mathematical principles; these (or something in the vicinity) will need to be recaptured in paraconsistent settings. The move to a paraconsistent logic should still make available – rehabilitate – the basic mathematical infrastructure of set theory, arithmetic, real analysis, and topology: especially (though perhaps perversely) those parts that seem essential for generating famous paradoxes.

Let us set the targets, looking at three versions of sorites. This is what we need:

The **discrete sorites** is the canonical version – the sorites paradox of the heap, from Section 1.2.1: it requires the following:

(1) A linear well-order $\{0, 1, 2, \ldots n\} \subseteq \mathbb{N}$ exists.
(2) TOLERANCE: $\forall n(\neg\varphi(n) \vee \varphi(n+1))$.
(3) NONTRIVIALITY: $\varphi(0)$ but $\neg\varphi(k)$ for some $k > 0$.

To get a contradiction, then, we call on the (finite) *Least Number Principle.* If $\neg\varphi(k)$ for some $k > 0$, then there is a *least* such: $\neg\varphi(k)$ and $\varphi(n)$ for all $n < k$. There is some least k where

$$\varphi(k - 1) \,\&\, \neg\varphi(k)$$

but by tolerance,

$$\neg\varphi(k - 1) \vee \varphi(k).$$

Conjoining these and distributing the conjunction over the disjunction, this is a contradiction:

$$(\varphi(k-1) \mathbin{\&} \neg\varphi(k-1)) \vee (\varphi(k) \mathbin{\&} \neg\varphi(k)),$$

that is, either $k-1$ or k are inconsistent with respect to φ (as in Section 2.2.3.5).

In a **continuous sorites**, the following is given:

(1) A real interval $[0,1] \subseteq \mathbb{R}$ exists.
(2) TOLERANCE: for vague φ, it is not the case that all members of a subset of an interval are φ but their least upper bound is not φ.
(3) NONTRIVIALITY: $\varphi(0)$ and $\neg\varphi(1)$.

So the action is at the *least upper bounds*: any nonempty set bounded from above has a least upper bound. The contradiction is that the least upper bound of the φs is both φ and not φ.

In a **topological sorites, the following is given:**

(1) There is a connected topological space X.
(2) TOLERANCE: for a vague φ, the subset $A = \{x \in X : \varphi(x)\}$ is topologically closed, and the complement of A in X is closed.
(3) NONTRIVIALITY: but both A and its complement are not empty.

To get the contradiction here, we need a small raft of results about connected spaces and topological boundaries, with the upshot that A shares a boundary with its complement, and that boundary is nonempty because X is connected. Then since A is both open and closed, points on the boundary are both in A and not in A. The boundary of A is inconsistent.

These arguments, or something like them (perhaps with some suitably modified definitions), need to be established in paraconsistent mathematics. Similarly, the key paradoxes of set theory – Russell, Cantor, Burali-Forti – listed in Chapter 1 need to be theorems. So this generates our "to-do" list: build up enough mathematics to make Chapter 1 true.

Beyond that horizon, our ultimate goal is a result that, in some sense, unifies all these (as gestured at in the Introduction), a version of Brouwer's fixed point theorem. Ascending the mathematical mountain high enough to attain this theorem (at least in two-dimensional point-set topology, in a parameterized form) will count for today as a significant "chunk" of mathematics, pace the Feferman objection; and, it is hoped, give us a new vantage point for explaining the paradoxes.

3.3.3.3 Structure in the Cracks

In the pages ahead, the idea is to work basically within the ordinary mathematical universe, but use our more sensitive logic to discern more, Meyer's "expansion of the pragmatic imagination." When a new definition or theorem looks strange, or even false, ask: is this in fact classically equivalent to something that is classically *true*? The answer will often be yes. For example, since classically there are no non-self-identical objects, anything we say conditionally about a non-self-identical object will be classically true, by ex falso quodlibet ("there are no φs, so all φs are ψs"). Or for another example, we will be working with relations where in conventional mathematics we would be using functions, but the intent is that most of the claims about, say, "continuous relations" are true in the special case where the relation is a function. It will be (mostly) a new look at old objects ... plus, of course, some entirely new and magnificently transconsistent objects, only visible on this sort of approach.

We will need to begin from the beginning, to recast set theory, arithmetic, real analysis, and topology. (Other important areas, such as computation, will have to wait until your book.) In the next chapter, we pack our supplies in preparation to begin.

4

Prolegomena to Any Future Inconsistent Mathematics

> We have always held that, in these permissive days, no rule is sacrosanct
> – except modus ponens.
> —*Meyer, Routley, and Dunn [Meyer et al., 1978, p. 128]*

There is one more thicket of paradox to chop through before we can get cracking on the mathematical ultraverse. This is Curry's paradox, and related problems to do with extensionality. In this chapter, we have a look at them, with the main goal of using these paradoxes as a guide to pinning down the logic to be used ahead for mathematical proofs. The symbols $\rightarrow$ and $\vdash$ stand for an implication connective and a logical consequence relation, respectively. Their meanings until Section 4.3 of this chapter are underdetermined; the purpose of the discussion is to determine them.

The paradoxes in this chapter are hard. They turn our attention away from issues of simple negation consistency, which paraconsistency has a good control over, to issues of absolute triviality in just the positive (negation-free) fragment of the language. The problems that these paradoxes uncover are deep in the logical machinery itself, and they can start to seem very daunting even for a dauntless would-be inconsistent set theorist. The guiding principle is that, insofar as logic is the laws of truth (Section 0.2.3), we must be prepared to (re)formulate our theory of logic according to the needs of truth (though not without any limit whatsoever, as epigram at the beginning of this chapter indicates). We don't plan to *save* truth from paradox, but we need to find out how truth can *bear* the paradoxes.[1]

4.1 Curry's Paradox

So far, we have been focusing on truth, falsity, and some logical connectives. Curry's paradox applies at that level – but does not stop there. Facing up to the Curry paradox will lead us from basic nonclassical logic to *substructural* logic.[2]

---

[1] "What is difficult [*schwer*]? asks the spirit that would bear much, and kneels down like a camel wanting to be well-loaded. What is most difficult,... that I may take it upon myself and exult in my strength?" [Nietzsche, 1976, *Zarathustra*, p. 138]

[2] Thanks to Colin Caret for collaboration on this topic in [Caret and Weber, 2015].

4.1.1 Operators

Reviewing: in the face of the paradoxes, at first you might think that they can be accommodated just by dropping either of the following principles:

Excluded Middle: $\vdash \varphi \vee \neg\varphi$
Explosion: $\varphi, \neg\varphi \vdash$

giving a paracomplete or paraconsistent theory. (A sentence on the right of a consequence relation $\vdash$ with nothing on the left is a theorem; premises on the left with nothing on the right are absurd.) Having tinkered with the logical *operators*, the material conditional $\varphi \supset \psi := \neg\varphi \vee \psi$ will no longer be adequate, because for a paracompleteist without excluded middle,

$$\vdash \varphi \supset \varphi$$

fails, and for a paraconsistentist to keep explosion from being valid, material modus ponens (disjunctive syllogism)

$$\varphi, \varphi \supset \psi \vdash \psi$$

fails. As flagged in Section 0.2.2.3, though, this just shows that the *material* conditional is not a *conditional*: modus ponens is "analytically part of what implication is" [Priest, 2006b, p. 83]. So either way, parconsistent or paracomplete, you will need a new account of the conditional – but you probably should have wanted that anyway, given the "paradoxes of material implication" [Routley et al., 1982, 1§1.2]. Decades of work have been expended on refounding a theory of the conditional.[3] But even if this project were completely successful, it would not be enough.

4.1.1.1 Curry

From naive comprehension, there is a *Curry set* (after Haskell Curry [Curry, 1942]):

$$c = \{x : x \in x \rightarrow \bot\}.$$

This is the set of all things such that, *if* they were self-membered, then absurdity, $\bot$, would follow. It is the set of all non-self-membered-on-pain-of-incoherence sets. If the conditional $\rightarrow$ is the material conditional, then the Curry set just reduces to the Russell set. (If $x \in c$ then $x \in x \rightarrow \bot$, so $x \notin x \vee \bot$, and then, since $\bot$ implies anything, it follows by cases, $x \notin x$.) If the conditional is *stronger* than the material conditional, though ($p \rightarrow q$ implies $\neg p \vee q$, but not vice versa), then the Curry set is a *proper subset* of the Russell set. The Curry set is inside the Russell set, as its hard inner core.[4] The expression "$p \rightarrow \bot$" implies

---

3 See Priest [2008, part 1]. Beall calls it, somewhat provocatively, a "quest." A different option is to forgo any implication connective, and stick with the material "conditional." This is sometimes called the "Goodship project," after [Goodship, 1996]; see Beall [2011, 2013a]; Omori [2015]; Priest [2014, 2017b]; cf. Section 3.1.2.2.

4 See Priest [2002a, p. 168]. Curry can be stated with an arbitrary formula φ in the place of $\bot$. Priest focuses on this general form, to keep it from fitting the inclosure schema [Weber et al., 2014; Priest, 2017b]. Here that debate does not matter much, since I set aside inclosure in Chapter 2, and the $\bot$ version is what needs most urgent attention, no matter what you call it.

that p is not (at all) true; it is a kind of strong negation or rejection of p, which I will sometimes call *annihilation*.

From the definition of c, we have

$$c \in c \leftrightarrow (c \in c \rightarrow \bot).$$

Reasoning informally, suppose $c \in c$. Then $c \in c \rightarrow \bot$. Putting this together with the supposition $c \in c$, by modus ponens, $\bot$. So, assuming $c \in c$ has led to $\bot$, that would seem to prove that $c \in c \rightarrow \bot$. But that is exactly what we cannot conclude, since then it proves $c \in c$ and $\bot$ really does follow, not just on supposition but "in real life."

Absurdity?

Look again, using the truth predicate. There is a Curry sentence

$$C \leftrightarrow (T(\ulcorner C \urcorner) \rightarrow \bot), \tag{4.1}$$

which, again if $\rightarrow$ is material, is just the liar sentence, and if $\rightarrow$ is stronger than material, is stronger than the liar. It says, "if this sentence is true, then absurdity follows." Reasoning informally, then, the Curry sentence must not be true – on pain of absurdity. But that is exactly what it says of itself.

Absurdity?

This is where we need to start paying very close attention to each step in the reasoning.

4.1.1.2 Implication

Suppose the principle of *contraction*

$$\varphi \rightarrow (\varphi \rightarrow \psi) \vdash (\varphi \rightarrow \psi)$$

is valid for the operator $\rightarrow$. This says, roughly, that using φ twice to get ψ is the same as using it once. As an accounting rule, it says that arbitrarily many φs need only be recorded as a single φ.[5]

By the T-schema, from (4.1) we have

$$C \leftrightarrow (C \rightarrow \bot), \tag{4.2}$$

so by contraction on (4.2) we have

$$C \rightarrow \bot. \tag{4.3}$$

This along with (4.2) right to left implies

$$C, \tag{4.4}$$

and then modus ponens on the previous two lines yields

$$\bot.$$

---

[5] In the literature, it is sometimes called *absorption*, e.g. Routley et al. [1982, p. 39].

The message I take from this, following others, is that naive truth/set comprehension is not compatible with $\rightarrow$ contraction, as has been known in one way or another for some time.[6] This shows that there are problematic principles at the level of genuine (nonmaterial) implication, not only material implication and negation.[7] Naive set/truth theory requires more than dropping ex falso quodlibet (and with it, disjunctive syllogism) to be coherent. Contraction is at least mechanically speaking, and according to one paraconsistency tradition – wrong for the implication connective.[8]

4.1.1.3 "Pseudo Modus Ponens"

It is even more disconcerting to find that Curry problems go all the way down to *conjunction*. Suppose conjunction is "extensional," or "additive";[9] that is, consider an *idempotent* conjunction $\wedge$, where

$$\varphi \leftrightarrow \varphi \wedge \varphi$$

is valid, and in particular in the direction $\varphi \rightarrow \varphi \wedge \varphi$, i.e., any proposition implies two instances of itself. Then the (apparently) unimpeachable axiom of modus ponens with this conjunction

$$\varphi \wedge (\varphi \rightarrow \psi) \rightarrow \psi$$

leads to trouble. Take the instance of axiom modus ponens

$$C \wedge (C \rightarrow \bot) \rightarrow \bot. \tag{4.5}$$

Then from (4.2), since C is equivalent to $C \rightarrow \bot$, by substitution on (4.5),[10] we have

$$C \wedge C \rightarrow \bot \tag{4.6}$$

and then by idempotence

$$C \rightarrow C \wedge C \tag{4.7}$$

and transitivity on (4.6) and (4.7),

$$C \rightarrow \bot \tag{4.8}$$

---

[6] See [Geach, 1955; Myhill, 1984; Restall, 1994]; cf. [Priest, 2006b, ch. 6; Beall, 2009, 2§2]. Clearly contraction is not the only principle used in these and other derivations that follow. There are liberal dollops of transitivity and reflexivity, among others. As ever, I am following one possible series of choices. See, e.g., work by [French, 2016] and [Ripley, 2013] for alternatives at the substructural level (Section 4.1.2.3); see [Ripley, 2015a] for comparisons.

[7] Conditionals that solve the paradoxes of material implication, such as *strict implication*, still contract [Priest, 2008, ch. 4].

[8] Slaney says, "It is in some ways more faithful to the motivating thoughts of relevant logic to count repetitions strictly, requiring . . . that premises be used *exactly* as many times as they are assumed" [Slaney, 1982, p. 4, emphasis in the original text]. A common interpretation of noncontractive logics is "resource sensitivity," that once a premise is assumed, it is "used up" [Girard, 1998]. Other suggestions are based around information [Mares and Paoli, 2014]. See also Rosenblatt on structural contraction and why it fails, which includes failures of cut (transitivity) as well [Rosenblatt, 2021].

[9] "Intensional" or "multiplicative" conjunction is sometimes called *fusion*. See the 130-page fifth chapter "And" in [Humberstone, 2011] for encyclopedic discussion. For relation to Curry's paradox, see [Priest, 2015].

[10] Or without substitution, but using $p \wedge q \rightarrow r, p \rightarrow q \vdash p \wedge p \rightarrow r$.

which from (4.2) is

$$C \tag{4.9}$$

and therefore from (4.8) and (4.9)

$$\bot$$

again.

This leads Meyer, Dunn, and Routley to conclude that naive theories are incompatible with even modus ponens, and that the project is over:

> One may, of course, still cleave to the rule of modus ponens without the modus ponens axiom . . . But the conclusion is none the less clear; unless we are prepared to give up a great deal of logic – not only of classical logic but of intuitionist and even relevant logic as well – a naive set theory is untenable *[Meyer et al., 1978, p. 128]*.

Nevertheless, naive theorists are thick skinned sorts, and have persisted. Priest [Priest, 1980] takes Meyer et al.'s first suggestion, keeping modus ponens as a valid *argument* form, but not as a valid *sentence* (theorem); then line (4.5) is not instantiated. Axiom modus ponens, as phrased with an indempotent conjunction connective $\wedge$, came to be called *pseudo modus ponens*. Routley and Brady's set theories [Routley, 1977; Brady, 2006], and then Beall's theory of truth [Beall, 2009], are based on a conditional that respects a large number of restrictions, including dropping pseudo modus ponens.[11]

Even if we get to keep modus ponens as a rule, giving it up as an axiom is heavy. Curry puts extreme pressure on the would-be ultralogician.

4.1.1.4 Stratification?

Most urgently, let us see what happens if we accept that *axiom* ("pseudo") modus ponens is invalid,

$$\nvdash (\varphi \wedge (\varphi \to \psi) \to \psi),$$

but the *rule* of modus ponens

$$\varphi, \varphi \to \psi \vdash \psi$$

is valid. Then the rule of $\to$-introduction,

$$\varphi \vdash \psi \qquad \text{only if} \qquad \vdash \varphi \to \psi,$$

a.k.a. the *deduction theorem*, must fail. Accepting this package would mean that modus ponens is valid *over* the language, but not *in* the language.

---

[11] They also give up *assertion*, $\varphi \to ((\varphi \to \psi) \to \psi)$, *permutation*, $\varphi \to (\psi \to \chi) \vdash \psi \to (\varphi \to \chi)$, and *residuation* $(\varphi \wedge \psi \to \chi) \dashv\vdash (\varphi \to (\psi \to \chi))$. For further constraints on implication for naive set/truth theory, see [Slaney, 1989; Øgaard, 2016].

While there are many nonclassical logics without a deduction theorem, from the standpoint of naive closure this seems extremely bad. It means that there is an object-level/metalevel distinction after all, that cannot be avoided. There are things we can informally say that cannot be formally expressed: *we* can say the sentence "*q* follows from *p* and *p* → *q*," while our *theory* cannot. At the very least, it brings back a typing structure and all the concomitant problems with hierarchies.[12] It is no small thing, as it turns out, to retain naive set theory. And it becomes well nigh impossible if you also insist on having a closed theory, with no recourse to a metalanguage.[13]

4.1.2 Structural Rules

The upshot of the previous section is unsatisfactory from the standpoint of closure. It left us with an object language/metalanguage distinction after all. A closed logical system mandates that true logical statements about the system show up in the system. This means not only that an implication connective respects some conditional-introduction rule, but more generally, there must be some sort of predicate expression of validity, as we will now see.

4.1.2.1 A Validity Predicate

Validity is necessary truth preservation from premises to conclusion, in virtue of form.[14] Validity, like truth, corresponds to a predicate, either two-place ("*y* follows validly from *x*") or one place ("*z* is a valid argument"). Logical *operators* take sentences, while a *predicate* takes names (possibly of sentences). Predicates, like a truth predicate, let you say things like "all the theorems of arithmetic are true," or "some sentences of Peano Arithmetic, I know not which, may be false," allowing quantification into sentence position. A predicate allows for (infinite) generalizations over an operator [Shapiro, 2011, p. 326].[15] Validity is a predicate; claims like "not all valid arguments are sound" are essential for learning and teaching – expressing – truths about logic.

In parallel with notion of semantic closure for a truth predicate, a naive formal system is expected to be capable of expressing its own consequence relation, toward a closed or universal theory. A *naive validity schema* for a two-place predicate $\mathsf{Val}(x, y)$ looks something like

$$\mathsf{Val}(\ulcorner\varphi\urcorner, \ulcorner\psi\urcorner) \text{ iff } \varphi \vdash \psi$$

---

[12] Myhill observes that, in dropping contraction, we immediately have "levels of implication" [Myhill, 1975]. See also [Whittle, 2004; Weber, 2014].

[13] See Field [2008, ch. 26], where this is leveraged into a substantial criticism of the paraconsistent approach.

[14] At least, on a mainstream account that marries semantic and syntactic considerations; cf. [Cook, 2014]. We will have enough trouble with this (apparently) moderate definition without querying the foundations of logical consequence; see the seminal [Etchemendy, 1990] and, more recently, essays in [Caret and Hjortland, 2015].

[15] A truth operator $T(\varphi)$ could just be defined as, e.g., φ, which gives us a completely paradox-free (but also useless) version of the T-schema. See Beall [2009, pp. 1–3]. Along these lines, *deflationism* about truth lends itself easily to a similar sort of deflationism about validity: it is "no more problematic than extending deflationism from truth to *falsity*" [Shapiro, 2011, p. 332].

for single-premise arguments. For arguments with finitely many premises, a reasonable generalization[16] would be $\mathsf{Val}(\ulcorner \varphi_0 \& \ldots \& \varphi_n \urcorner, \ulcorner \psi \urcorner)$ iff $\varphi_0, \ldots, \varphi_n \vdash \psi$. And a validity predicate must obey some kind of "modus ponens," in the shape

$$\varphi, \mathsf{Val}(\ulcorner \varphi \urcorner, \ulcorner \psi \urcorner) \vdash \psi,$$

suggesting natural introduction and elimination rules for Val.

4.1.2.2 A Validity Curry

Now, paradox. As with a truth predicate, direct self-reference or even self-referential loops, like

Student: All of Teacher's arguments are valid.
Teacher: Everything Student says is true. Therefore, we are doomed!

(cf. [Kripke, 1975]) will generate paradoxes. More simply, an expressive device that can vindicate validity will allow a sentence like

Absurdity follows validly from this very sentence,

which is a Curry sentence. More precisely, the instance $\mathsf{Val}(x, \ulcorner \bot \urcorner)$ has a fixed point (cf. Section 1.1.2.1), some C that is fully intersubstitutable with $\mathsf{Val}(\ulcorner C \urcorner, \ulcorner \bot \urcorner)$:

$$C \leftrightarrow \mathsf{Val}(\ulcorner C \urcorner, \ulcorner \bot \urcorner).$$

Then – disturbingly – the mere *assumption* of C appears to derive $\bot$, absurdity, from no assumptions [Beall and Murzi, 2013]:

(1) Suppose C.
(2) Then $\mathsf{Val}(\ulcorner C \urcorner, \ulcorner \bot \urcorner)$, from (4.1) and the definition of C.
(3) So $\bot$, from (4.1), (4.2), and Val-modus ponens.
(4) Therefore, $\mathsf{Val}(\ulcorner C \urcorner, \ulcorner \bot \urcorner)$ by (4.1)–(4.3).
(5) Therefore, C, from (4.4) and the definition of C.
(6) Then $\bot$, from (4.4), (4.5), and Val-modus ponens.

The argument is terribly simple: supposing C, then absurdity follows; but that is what C says, so C is true; so absurdity follows. "I must be rejected, or else!" must be rejected, or else. But doing so is to *accept* the very thing being rejected.

Absurdity?

When it came to a conditional operator, insofar as we were inclined to take the advice from tradition and accept that contraction (on $\to$) is invalid, the same strategy can be carried forward here. The previous derivation of absurdity involves *repeated use* of a single assumption. Looking again at a subproof of the Val–Curry, line (2) is justified by line (1), and line (3) is in part justified by line (1). So line (1), the assumption that C, is

---

[16] Assuming that collections of formulas may be conjoined into a single sentence – the rule of *adjunction*, $\varphi, \psi \vdash \varphi \& \psi$. Some traditions in paraconsistency deny adjunction [Jaśkowski, 1969].

used (inevitably) twice. This proves that *two* uses of C lead to $\bot$, which means (4) in the derivation should say

$$\mathsf{Val}(\ulcorner C \ \& \ C \urcorner, \ulcorner \bot \urcorner).$$

If we then go on to "contract," treating two uses as one, treating C as logically identical to $C \ \& \ C$, then disaster ensues. But if not, then not.[17]

4.1.2.3 Structural Contraction

Whatever the merits of rethinking the logical operators, the moves in Section 4.1.1 do not make any changes to the *structural* rules of a logic. Structural rules concern logical consequence itself. Structural rules express proof theoretic "metavalidities," of the form, "if this argument is valid, then that argument is valid." Standard structural rules, presented Gentzen-style, are

$$\frac{\Gamma \vdash \varphi}{\Gamma, \psi \vdash \varphi} \ weakening \qquad\qquad \frac{\Gamma \vdash \psi \quad \Delta, \psi \vdash \chi}{\Gamma, \Delta \vdash \chi} \ cut$$

$$\frac{\Gamma, \varphi, \varphi \vdash \psi}{\Gamma, \varphi \vdash \psi} \ contraction \qquad\qquad \frac{\Gamma, \varphi, \psi \vdash \chi}{\Gamma, \psi, \varphi \vdash \chi} \ exchange$$

where $\Gamma, \Delta \ldots$ are collections of premises and the horizontal line represents a valid step. The proposal then is that contraction is not a valid structural rule. It does not preserve validity, as Val–Curry teaches us (again). A logic for a genuinely closed theory needs to be *substructural*.

Presented this way, there are two types of conjunction, at least as exemplified by introduction rules:

$$\frac{\Gamma \vdash \varphi \quad \Gamma \vdash \psi}{\Gamma \vdash \varphi \wedge \psi} \ extensional \ \text{``and''} \qquad\qquad \frac{\Gamma \vdash \varphi \quad \Delta \vdash \psi}{\Gamma, \Delta \vdash \varphi \ \& \ \psi} \ intensional \ \text{``and''}$$

Extensional $\wedge$ contracts; two uses of Γ are recorded as one. Since $\varphi \vdash \varphi$, this rule will have as a consequence that $\varphi \vdash \varphi \wedge \varphi$. Meantime, for intensional &, even in the case where $\Gamma = \Delta$, we have

$$\frac{\Gamma \vdash \varphi \quad \Gamma \vdash \psi}{\Gamma, \Gamma \vdash \varphi \ \& \ \psi},$$

but one does not go on to conclude $\Gamma \vdash \varphi \ \& \ \psi$, at least not without assuming contraction. Dropping structural contraction means: don't use assumptions more than once without telling someone. I will follow this strategy, of dropping (or, more staunchly, never having accepted in the first place) structural contraction.[18]

---

[17] A proof that dropping *structural* contraction renders naive truth theory coherent is spelled out by Zardini [Zardini, 2011], with suggestions about the metaphysics of noncontraction; cf. [White, 1993]. For doubts about the validity Curry, see [Cook, 2014].

[18] Alternatively, dropping cut instead – denying the transitivity of consequence – has similarities with dropping modus ponens (since $\vdash p$ and $p \vdash q$ can hold without $\vdash q$) and so is like working in LP [Dicher and Paoli, 2019].

4.1.2.4 Modus Ponens Regained

This provides a means to resolve the modus ponens problems from the last section. For, going down the substructural path, we arrive at a clear choice:

- If conjunction is "extensional," then the axiom form of (pseudo) modus ponens is invalid.
- If axiom modus ponens is valid, then conjunction is "intensional."

If modus ponens – as an "object-level" sentence – is sacrosanct, then the appropriate notion of conjunction is *intensional*; the true sentence-level form of modus ponens is formulated with &. This, I would suggest, is the natural move, consonant with our previous choices. For if we accept that operator contraction on → is invalid, then we already accept that using a premise twice is different than using it once; then we should not accept the principle "if φ, then φ and φ." But then, the previous plan of accepting axiom modus ponens as merely "pseudo" looks like an overreaction. The step in Section 4.1.1.3 from $C \& C \rightarrow \bot$ to $C \rightarrow \bot$ is invalid, because $C \rightarrow C \& C$ isn't true. And even if it were, line (8) is used twice, without being properly substructurally accounted for. Without structural contraction, these derivations are halted before the question of questioning axiom modus ponens arises.

One might ask if a choice between formalizations of conjunction is really necessary. Would it be possible to have both an extensional ∧ and intensional & in our language? The answer looks to be "no" in the presence (always in the presence) of naive set comprehension. Assuming that extensional conjunction validates the two principles (see the appendix of this chapter),

$$(p \rightarrow q) \wedge (q \rightarrow r) \rightarrow (p \rightarrow r)$$

and

$$(p \rightarrow q) \wedge (p \rightarrow r) \rightarrow (p \rightarrow q \wedge r),$$

then the following "mixed" Curry is found in*[Routley et al., 1982, p. 367] (and is there reckoned to be an argument against &, called "fusion"):[19]

1	$C \leftrightarrow (C \& C \rightarrow \bot)$	(Curry sentence)
2	$C \rightarrow (C \rightarrow C \& C)$	(axiom for &)
3	$C \rightarrow ((C \rightarrow C \& C) \wedge (C \& C \rightarrow \bot))$	(1, 2, axiom for ∧)
4	$C \rightarrow (C \rightarrow \bot)$	(3, axiom for ∧, transitivity)
5	$C \& C \rightarrow \bot$	(4, axiom for &)
6	C	(1, 5, modus ponens)
7	$C \& C$	(6, repetition)
8	$\bot$	(5, 7, modus ponens)

This proof uses structural contraction (lines 1, 5, and 6 are all used more than once), so perhaps there is still some room for both conjunctions. Nevertheless, for the remainder we will take the mirror-image response as Routley et al. following this proof, that extensional

---

[19] Cf. Øgaard [2016, p. 253].

conjunction, "though not nearly as classical as classical negation it is too classical … for relevant dialectical purposes" [Routley et al., 1982, p. 367]

Dropping structural contraction – and concomitantly endorsing intensional conjunction & as conjunction – avoids the near-fatal position of having to deny axiomatic modus ponens, or that in general some conditionals and validities are inexpressible. It turns away from the bad faith idea that there is some "meta" truth that cannot be expressed at the object level. It avoids, as Beall and Field are (independently) driven to do, denying that soundness is expressible (what is provable is true) or that validity is definable.[20] It keeps us on the path to closure.

4.1.3 Into the Substructural Ultraverse

We have a straightforward *counterexample* to contraction, at the operator and structural levels. For the validity Curry sentence C, it is true that $C, C \vdash \bot$. It is not true that $C \vdash \bot$. In fact, if $C \vdash \bot$, then $\vdash \bot$, so by reductio, $C \vdash \bot$ is simply false (which, as we'll see, isn't to say that C is true). Without contraction,

if the Curry sentence were true, then absurdity would follow

is just *false*. It is true, rather, that

if the Curry sentence were true, and the Curry sentence were true, then absurdity would follow.

But that is *not* what the Curry sentence says. Once contraction is dropped, C and C & C are not the same.

We already saw (Section 1.2.3) how Russellian definite descriptions allow "unique" objects to be multiply realized in paraconsistent contexts – there can be more than one "last" moment of the day. Similarly, there can be more than one "the Curry sentence," more than one unique "the premises" of an argument. Each thing is different; some things are even *different from themselves*. As we are about to see in the next section and throughout, *identity is not*, as Lewis said, "utterly simple and unproblematic" [Lewis, 1986, p. 192]. The project of a universal theory, combining a charactaristica universalis and calculus ratiocinator, a complete and computable theory, arrives at a logic that, without contraction, must take each thing on its own, one at a time – sometimes, even more than one time.[21] To be universal, we must be particular.

There is much more to say, one day, about Curry's paradox, contraction, gluts, and paraconsistency. But there are more pressing pragmatic issues. Working in logics without contraction turns out to be overwhelmingly more profound a difference than "just" trying to make do without disjunctive syllogism. And there are further paradoxes to face.

---

[20] See Beall [2009, pp. 35–41]; Field [2008, pp. 286–290]; Shapiro [2011, p. 335].

[21] "[L]ack of contraction implicitly amounts to a shift of consideration, viz. from formulas as *types* to formulas as *tokens*" [Petersen, 2000, p. 369]. So maybe Lewis is right after all: "there is never any problem about what makes two things identical; two things never can be identical" [Lewis, 1986, p. 192]. Or maybe we should have a poem: "When Spring returns … / Not even the flowers or green leaves return. / There are new flowers, new green leaves. / There are new balmy days. / Nothing returns, nothing repeats, because everything is real" (Pessoa 7-XI-1915).

4.2 Grišin's Paradox and Identity

The paradoxes of set theory are almost always attributed to – blamed on – the naive set comprehension principle. There are two axioms of naive set theory, though, and the discussion in Section 1.1.3.1 of Chapter 1 suggests that we could equally say that comprehension is not the culprit, but rather, it is the axiom of *extensionality* – the assumption that membership determines identity (Section 1.1.1). The strongest evidence for this possibility comes from Grišin's paradox [Grišin, 1982; Cantini, 2003].

The arrow $\rightarrow$ and consequence relation $\vdash$ in this section continue to be *generic* placeholders.

4.2.1 Substitution and Extensionality

Comprehension gives existence conditions for sets. What about uniqueness? Two principles are considered. *Leibniz equality*

$$x = y \leftrightarrow \forall z(x \in z \leftrightarrow y \in z) \tag{4.10}$$

encodes the identity of indiscernibles and the indiscernibility of identicals. It says that identical things are in all the same sets – and by comprehension, then, have all the same properties; so Leibniz equality in the presence of comprehension amounts to *substitution of identicals*:

$$x = y \rightarrow (\varphi(x) \rightarrow \varphi(y))$$

for any φ. Meanwhile, *extensional equality*

$$x =_e y \leftrightarrow \forall z(z \in x \leftrightarrow z \in y)$$

says that identical sets have the same members. For either notion, assuming that the $\leftrightarrow$ connective (presumably, the conjunction of two conditionals) is reflexive, symmetric, and transitive, then the corresponding identity will inherit the properties of an equivalence relation (cf. Section 5.1.2).[22]

4.2.2 Paradox

Set theory is the theory of extensions. Naive set theory makes the uniqueness condition on sets the axiom of extensionality – a principle asserting that extensionally identical sets are really identical:

$$x =_e y \rightarrow x = y.$$

We thought earlier that dropping structural contraction, and taking conjunction to be intensional rather than extensional, would free up our implication connective from the

---

[22] A dialetheic theory of identity in which some or all of the classical properties of identity fail (by taking $\leftrightarrow$ to be a material biconditional) is in [Priest, 2014].

iatrogenic disorder of "pseudo" modus ponens and the like. But without being very careful about the properties of the conditional underlying identity, the extensionality axiom leads to absurdity.

4.2.2.1 Grišin's Paradox

We have seen in Section 4.1.1.3 that Curry's paradox, and the sanctity of modus ponens, together push us to say that conjunction is not idempotent: $\varphi \rightarrow \varphi \& \varphi$ is not true in every instance. But instances may still be provable. Note that

$$x = x \& x = x \tag{4.11}$$

always holds, since $x = x$ is true and $x = x$ is true. Then consider the instance of Leibniz identity (4.10):

$$x = y \rightarrow ((x = x \& x = x) \rightarrow (x = y \& x = y)).$$

By permutation (rearrangement) of antecedents,

$$x = x \& x = x \rightarrow (x = y \rightarrow (x = y \& x = y)). \tag{4.12}$$

By modus ponens on (Eqs. 4.11 and 4.12), we have contraction for $=$,

$$x = y \rightarrow x = y \& x = y. \tag{4.13}$$

This becomes a major problem, since comprehension generates a set term for every formula, so "contracting" on $=$ will trickle back to "contracting" on $\in$, as we now see.[23]

Assume that $\rightarrow$ obeys *weakening*,

$$\varphi \rightarrow (\psi \rightarrow \varphi).$$

(A relevant arrow will not weaken, but other reasonable arrows for substructural logics will.) Consider the sets

$$\mathcal{U}_0 := \{x : x = x\}$$
$$\mathcal{U}_1 := \{x : x = x \& \varphi\}.$$

The first is the set of all self-identical things; the second is the set of all the things that are self-identical and φ. If φ is true, then it would be intuitive to think that these are the same set (Section 4.2.3.4). Using weakening and permutation, reason as follows:[24]

---

[23] It may look like contraction on identity, $x = x \rightarrow x = x \& x = x$, is wrong from the standpoint of a noncontractive theory. Perhaps there is a way to block it, and avoid the paradoxes that follow? Shirahata tells us why this idea is doomed: "It is true that substitutivity allows the duplication of terms in substitution, which is responsible for the contraction over equalities, and this seems to violate the spirit of [noncontraction]. However, the same is true for the comprehension, which allows us to abstract over multiple occurrences of a variable at once. In fact, it is no use to prohibit the duplication in substitution without changing the comprehension at the same time, since the former can be recovered through the latter. We simply abstract over the multiple occurrences of a term t in a formula φ so that we have the formula $t \in \{x : A[x/t]\}$, and substitute t on the left by s. We then move the term s back inside φ. Hence, unless we want to make a change in comprehension, we have to accept [Grišin's paradox]" [Shirahata, 1994, p. 137]. His analysis seems unimpeachable. The whole aim of the enterprise is not to meddle with comprehension.

[24] This streamlined proof is based on one by Terui. See also [Cantini, 2003, p. 355].

1    $\varphi \leftrightarrow (x = x \leftrightarrow x = x \,\&\, \varphi)$.
2    $\varphi \leftrightarrow \forall x (x \in \mathcal{U}_0 \leftrightarrow x \in \mathcal{U}_1)$.
3    $\varphi \leftrightarrow \mathcal{U}_0 = \mathcal{U}_1$.
4    $\varphi \leftrightarrow \mathcal{U}_0 = \mathcal{U}_1 \,\&\, \mathcal{U}_0 = \mathcal{U}_1$.
5    $\mathcal{U}_0 = \mathcal{U}_1 \,\&\, \mathcal{U}_0 = \mathcal{U}_1 \leftrightarrow \varphi \,\&\, \varphi$.
6    $\varphi \leftrightarrow \varphi \,\&\, \varphi$.

If each step is justified (by some fiddling), that would prove contraction for *arbitrary* formulas, which is bad. This is Grišin's paradox. We would have, for instance, $c \in c \rightarrow c \in c \,\&\, c \in c$, allowing the Curry proof to complete. So even in a contractionless logic, the conditional governing extensionality needs to be tightly controlled, or else triviality still follows.

4.2.2.2 More Paradox

When it rains, it pours. Variants of the following are sometimes called the *Hinnion–Libert* paradox.[25] Let $\varnothing = \{x : \bot\}$, the empty set. The *Hinnion–Libert* set is $\{x : \{y : x \in x\} = \varnothing\}$. It is very like the Curry set: informally, the problem is that, if this set were self-membered, it would by definition be empty (on pain of $\bot$); being self-membered *and* $\bot$-empty is absurd, so it isn't self-membered, on pain of absurdity; but then by definition it is self-membered after all. What is new is that this trades on equality rather than implication (or validity), so, since $=$ contracts (4.13), the general failure of contraction does not block the reasoning.

Let's work with an even spikier version of it, that skips to the fixed point; Cantini [Cantini, 2003] calls this Gordeev's trick. Let

$$g = \{x : g = \varnothing\}.$$

Then

$$x \in g \rightarrow g = \varnothing. \tag{4.14}$$

Since identity contracts,

$$g = \varnothing \rightarrow g = \varnothing \,\&\, g = \varnothing. \tag{4.15}$$

But using the definition of g,

$$g = \varnothing \,\&\, g = \varnothing \rightarrow x \in g \,\&\, x \in g. \tag{4.16}$$

Therefore, by transitivity of (4.14)–(4.16),

$$x \in g \rightarrow x \in g \,\&\, x \in g. \tag{4.17}$$

Now, assuming substitution holds in the form

$$a = b \,\&\, \varphi(a) \rightarrow \varphi(b),$$

---

[25] After [Hinnion and Libert, 2003; Libert, 2005]; cf. Restall's arguments in [Restall, 2013], and also [Øgaard, 2016; Field et al., 2017].

then we have as a valid instance of it that

$$g = \varnothing \ \& \ x \in g \rightarrow x \in \varnothing.$$

But $x \in g \rightarrow g = \varnothing$, so

$$x \in g \ \& \ x \in g \rightarrow \perp. \tag{4.18}$$

Putting together (4.17) with (4.18), we get

$$x \in g \rightarrow \perp$$

for any x. Now, by $=_e$, this appears to prove that g is extensionally equivalent to $\varnothing$. But if $g = \varnothing$, then the defining condition of g is *true*, and $\forall x (x \in g)$. Then $\forall x (x \in \varnothing)$ and $\perp$ and we can all go home.

4.2.2.3 Prognosis: Extensionality Lost?

There are lots of moving parts here. Key in both the Grišin and Hinnion–Libert paradoxes would seem to be the assumption that $=_e$ entails $=$, that extensionality is identity. Are we driven to say that "the failure of weak extensionality can be sharpened to the point of destroying all illusions about possibilities of finding a realm of objects which is well behaved" [Petersen, 2000, p. 384], that "it is hopeless to combine extensionality and abstraction ..." [Libert, 2005, p. 31]?[26] Clearly we cannot have everything; the paradoxes are paradoxes. There are morals to draw and choices about the fine details of the logic to be made.

4.2.3 Extensionality Regained: Varieties of Implication

The base of logical connectives $\&, \vee$ and $\neg$ in Box 0 (Section 0.2.2.3) is insufficient for mathematical practice because it does not include a conditional. The purpose of the chapter up to this point has been to isolate desiderata for extending the connectives with an implication connective. Insofar as a conditional will play key roles in the set theoretic axioms – especially in underwriting the extensionality axiom and a theory of identity – we now have some fairly rigid constraints. To satisfy them, let's meet two logics, a descendant of linear logic called BCK, and a weak relevant logic called DK, which will form the core of our official logic in Section 4.3.

4.2.3.1 BCK

Given the Grišin/Hinnion–Libert paradoxes, one option is to drop the axiom of extensionality, "$x =_e y$ implies $x = y$," entirely. An extensional *equivalence* is established between sets that, if the equivalence were identity, it would turn out to be absurd; so this equivalence is insufficient for identity. On this idea, in Section 4.2.2.2, while g is in some

---

[26] See also [Brady, 1971; White, 1979, 1993].

sense "coextensive" with the empty set, it is not *identical*, in the Leibniz sense, to the empty set. So being "coextensive," in the scare quoted sense, does not imply identity. This (schematically) solves the problem: we proved $g =_e \varnothing$ but not the defining condition for g, that $g = \varnothing$. If we now tried to run the same paradox from Section 4.2.2.2 with

$$g' = \{x : g' =_e \varnothing\},$$

but without assuming that $=_e$ implies $=$, then we would find only that the proof fails at the point were substitution is used, and not even get to $g' =_e \varnothing$. Grišin's paradox is similarly avoided, because $\mathcal{U}_0 =_e \mathcal{U}_1$ but not $\mathcal{U}_0 = \mathcal{U}_1$. A venerable response, then, is that extensionality in the presence of set comprehension must fail.[27]

In this tradition, the *light affine logic* BCK stands out as a logic without contraction that can accommodate naive comprehension. Introduced by Imai and Iseki in 1966, and brought to some prominence by Ono and Komori in 1985, it is notable for lacking contraction and so is consistent with comprehension, by a cut elimination argument.[28] This logic has a *flexible* conditional, making it easy to work with. (It is named after principles from combinatory logic corresponding to properties of its conditional: transitivity [or prefixing] [B], permutation [C], and weakening [K].) The logic BCK is presented in the appendix to this chapter. We will start writing BCK's conditional with

$$\Rightarrow .$$

Part of the idea behind introducing this arrow is that, as we'll see, it closely mirrors properties of logical consequence $\vdash$ and so will have nice conditional-introduction/deduction theorem properties.

Thus, for the BCK arrow, $\forall z(z \in x \Leftrightarrow z \in y)$ can hold (as in Grišin/Hinnion–Libert) without entailing identity between x and y. (If $x = y$, then substitution *is* possible, but this will effectively only happen in the case of syntactic identity.) One consequence is the failure of formula substitution: the plausible principle that logical *extensional* equivalents are intersubstitutable in any context χ,

$$(\varphi \Leftrightarrow \psi) \Rightarrow (\chi(\varphi) \Rightarrow \chi(\psi)),$$

is incoherent. Otherwise, like identity, a biconditional will contract:

$$(p \Leftrightarrow q) \Rightarrow (p \Leftrightarrow q) \& (p \Leftrightarrow q).$$

That would be bad, in light of a Curry sentence $C \Leftrightarrow (C \Leftrightarrow \bot)$. For then we would have

$$C \Rightarrow (C \Leftrightarrow \bot) \& (C \Leftrightarrow \bot)$$

and then $C \Rightarrow C \& C$ and the original (bad) Curry reasoning would hold.

---

[27] The idea goes back in some way to [Chang, 1963] and [Gilmore, 1974]; cf. [Dunn, 1988] in [Austin, 1988]. This is adopted in versions of fuzzy class theory [Běhounek and Cintula, 2005; Běhounek and Haniková, 2015]. A tempting thought in the paraconsistent context is that extensionality could be repaired by adding a clause that says if sets have the same extension *and* the same *anti*extension, *then* they are identical. This conjecture fails; see Istre's example in Section 4.3.1.2, from [Istre, 2017].

[28] As observed in [Grišin, 1982], and taken up by by Girard in [Girard, 1998]. See Petersen [2000, §3]; Cantini [2003, p. 348]; Terui [2004, p. 35]; cf. [Komori, 1989]. For BCK in general, see Ono and Komori [1985]; Galatos et al. [2007, p. 101]; Humberstone [2011, 1§29, p. 1110]. For connections between BCK and other logics (such as FLew, intuitionistic logic without contraction), see [Ono, 2010]. For connections with relevant logic and BCK, see [Meyer and Ono, 1994].

"Negation" in BCK is the *annihilation* $p \Rightarrow \bot$, not a paraconsistent negation $\neg$. It explodes (p and $p \Rightarrow \bot$ entail $\bot$), but is not exhaustive ($p \vee (p \Rightarrow \bot)$ isn't true). It "contraposes," in the sense that if $p \Rightarrow q$, then $q \Rightarrow \bot$ implies $p \Rightarrow \bot$, by transitivity. Hence modus tollens in the form

$$\varphi \Rightarrow \psi, \psi \Rightarrow \bot \vdash \neg\varphi$$

is valid. (Anything that implies an absurdity is absurd, and therefore false.) See Section 4.3.2.1. But since $\Rightarrow$ weakens, for negation $\neg$, the BCK conditional will *not* contrapose (Section 0.2.2.3). BCK on its own won't help us with the glutty program set out in previous chapters.

So BCK has a lot going for it, but not enough. Without an extensionality axiom, BCK delivers *naive property theory*, not naive set theory. Any *set* theory will have some notion of extensional identity between sets. That's half the "naive set" package I was praising in Chapter 1. I think the above (sinister) versions of Curry's paradox means we need to be very *careful* about how the axiom of extensionality is phrased, and with which conditionals; but there must be some connective and sense in which it *is* coherent. What needs to be accepted is that extensionality will just not be quite the way we once imagined, before we learned the grown-up truth about living in Cantor's paradise.

4.2.3.2 DKQ

Before going too far down the path of foregoing extensionality, observe that the triviality proof in Section 4.2.2.1 relies on $\rightarrow$ obeying *weakening*, in the instance $\varphi \rightarrow (x = x \rightarrow \varphi)$. The moves in Section 4.2.2.2 similarly require a flexible conditional So perhaps questioning this principle, $p \rightarrow (q \rightarrow p)$, provides another option?

Relevant logics are well-understood paraconsistent logics without weakening.[29] Relevant logics require there to be a meaningful connection between the antecedent and consequent of a (relevant) conditional; weakening is invalid because it is irrelevant. Relevant logics are then naturally paraconsistent since $p \& \neg p \rightarrow q$ is invalid on relevance grounds. Basic relevant implication does not contract, and so can support both comprehension *and* extensionality, for naive *set* theory: the logic DKQ, suggested by Routley (following DL in [Routley and Meyer, 1976]), is one of many proven nontrivial for full naive set theory.[30] This is achieved by having a conditional that is completely *inflexible*, as mentioned in footnote 13 at the end of Section 4.1.1.4. The arrow introduction rule fails for it: $p \rightarrow q$ is properly stronger than $p \vdash q$. We will fix

$$\rightarrow$$

as a relevant (meaning preserving) conditional. It *does* contrapose: if $p \rightarrow q$, then $\neg q \rightarrow \neg p$. DKQ is also presented in the appendix to this chapter.

The Grišin/Hinnion–Libert derivations carry through (up to extensionality) in BCK because it is an easy arrow to work with. Proving statements of the form $\forall z(z \in x \Leftrightarrow z \in y)$

---

[29] See [Dunn and Restall, 2002; Mares, 2004a]; cf. Section 3.2.1.
[30] See [Brady and Routley, 1989; Brady, 1989, 2006]. The result began with [Brady, 1971].

for certain x, y is not too hard, but you get what you pay for; you don't get identity. In DKQ, $\forall z(z \in x \leftrightarrow z \in y)$ *does* establish full identity, because $\leftrightarrow$ is not at all easy to work with; the preceding triviality proofs break down for DKQ. Relevant implication is very strong, much stronger than a deduction. I like an early interpretation of Routley [Routley, 1977, p. 895–896], there calling relevant logic *ultramodal*:

The central deducibility relation of ultramodal logics, entailment, is intended to capture the notion of sufficiency. This means, in particular, sufficiency of the antecedent of an entailment on its own, without any additional imported truth, especially imported logical truths. ... Sufficiency is a go-anywhere notion, which is not limited by the fact that the situation in which it operates is somehow classically incoherent, e.g. inconsistent or paradoxical. If A is sufficient for B then it does not matter what else goes on; logical laws may go haywire but nothing subtracts from A's sufficiency.

The idea is that there is such a thing as an entailment connection that can survive in an almost unlimited amount of ambient chaos, and this "absolute sufficiency" notion delivers a notion of being *really* coextensive.[31]

Let's see how this "absolute-sufficiency" connective works for identity.

4.2.3.3 Relevant Identity 1: Substitution

The extensionality axiom says that when x has the same members as y, then x shares all properties with y. Naive set theories like in DKQ can include both extensionality and Leibniz identity (Section 4.2.1); Brady gives them in the form

$$x =_e y \,\&\, z =_e z \rightarrow (x \in z \rightarrow y \in z)$$

or as a rule

$$x =_e y \vdash \varphi(x) \rightarrow \varphi(y).$$

These are both effectively noncontraposable.[32] In DKQ set theory, $x =_e y$ is defined as $\forall z(z \in x \leftrightarrow z \in y)$, relevantly: if this arrow is satisfied, then being a member of x *intrinsically means* being a member of y, and vice versa. So they really are identical.

Why does Brady make these principles noncontraposable, when $\rightarrow$ can contrapose? Let us focus on formulations that are not workable for naive set theory. The most salient candidate to be considered for substitution is the relevant formulation,

$$x = y \rightarrow (\varphi(x) \rightarrow \varphi(y)).$$

This may look acceptable, but it is not. A first clue is that, letting $\varphi(x)$ be some fixed propositional constant p, then we have the irrelevant $x = y \rightarrow (p \rightarrow p)$. That's not a

---

[31] "Relevance" isn't the key point, then. Routley: "[L]ogical sufficiency ... is what is really fundamental to the logics being promoted. Relevance of consequence to antecedent, though a hallmark of an adequate implicational relation, is strictly a by-product of a good sufficiency notion; for if B has nothing to do with A then A can hardly be sufficient for B. But relevance is not of the essence" Routley [1977, p. 896]. Cf. Routley et al. [1982, p. x], that "main concern is not really relevance at all."

[32] The first does give $\neg(x \in z \rightarrow y \in z) \rightarrow x \neq_e y \vee z \neq_e z$, but the second disjunct of the consequent can't be eliminated without disjunctive syllogism. And the relation $\vdash$ of DKQ is based off that of LP; it does not contrapose because $q \vdash p \vee \neg p$ is valid, but the contrapositive of that is explosion. See Brady [1989, p. 440]; Brady [2006, p. 242].

disaster, but it is an early indicator that something is wrong.[33] Now, instead of an arbitrary fixed proposition, pick the liar, ℓ. Then our candidate "relevant Leibniz" has as an instance

$$x = x \rightarrow (\ell \rightarrow \ell).$$

Now, if we have a *counterexample* principle, $p \& \neg q \vdash \neg(p \rightarrow q)$, as in Sections 0.2.2.4 and 3.3, then since $\ell \& \neg\ell$, we would have

$$\neg(\ell \rightarrow \ell).$$

By the contrapositive of relevant Leibniz,

$$\neg(\ell \rightarrow \ell) \rightarrow x \neq x.$$

And then since x was arbitary,

$$\forall x (x \neq x).$$

Although that is not complete triviality, it says that *everything* is non-self-identical, which is a rather profligate conclusion to draw from the liar paradox (cf. (Section 3.1.3.2)). This result seems like strong evidence that an appropriate substitution principle does not contrapose, and that a principle such as $\exists z(x \in z \& y \notin z) \rightarrow x \neq y$ isn't true. And it makes sense in an intensional context – sets may differ in some of their *properties*, without differing in their *members*. Identity is an *internal* matter, intrinsically determined "downward" by what is on the left of $\in$, not an external matter extrinsically determined "upward" by what is on the right of $\in$.[34] Paired with extensionality, the identity of indiscernibles and the indiscernibility of (extensional) identicals hold, but the non-identity of discernibles fails.

When $\forall z(z \in x \leftrightarrow z \in y)$, when being an x is absolutely sufficient for being a y and vice versa, then $x = y$ and they share all their properties. This just doesn't happen very often, and when it does, *not* sharing all properties does not track back to $x \neq y$. Along these same lines, the relevant biconditional, unlike the BCK biconditional, *can* enjoy formula substitution, in noncontraposable rule form:

$$\varphi \leftrightarrow \psi \vdash \chi(\varphi) \rightarrow \chi(\psi).$$

This will yield $p \leftrightarrow q \vdash (p \leftrightarrow q) \& (p \leftrightarrow q)$, but unproblematically so, with respect to Grišin's paradox. Given a relevant Curry, $C \leftrightarrow (C \leftrightarrow \bot)$, we will get $C \vdash C \& C$, and since from $C \& C$ one can derive $\bot$, we will have that $C \vdash \bot$, but emphatically *not* the intensional $C \rightarrow \bot$. So the Curry derivation – that if C were true, then one could prove $\bot$ – does not validate what C says of itself, which is that C *intrinsically means* $\bot$. An absolutely sufficient biconditional can intersubstitute exactly because it is so hard for such a biconditional to be true. Relevance!

---

[33] See Routley [1977, p. 920]; Dunn [1987].

[34] Cf. standard Fregean puzzles about "transparent" versus "opaque" contexts, e.g., maybe Pythagoras loved the Morning Star but did not love the Evening Star, without that somehow causing Venus $\neq$ Venus. See Priest [2008, ch. 17] and Section 4.2.3.4.

4.2.3.4 Relevant Identity 2: Doppelgängers

Relevant logics are in many ways ideally suited for the purpose of an intensional and extensional naive set theory (Section 1.1.3.1). But, some would argue, relevant logic goes too far. An axiom of extensionality formulated relevantly can be preserved, but prima facie it is too stringent, because there are extensionally identical sets that are not relevantly identical.[35]

In Section 4.2.2.1, the two sets $\mathcal{U}_0 = \{x : x = x\}$ and $\mathcal{U}_1 = \{x : x = x \ \& \ \varphi\}$ will fail to be identical using relevant arrows, because $x = x \rightarrow (x = x \ \& \ \varphi)$ is irrelevant; the antecedent is insufficient – even if φ is a logical truth. This is sometimes called the "doppelgänger problem" because if φ is always true, then it seems intuitive that $\mathcal{U}_0$ and $\mathcal{U}_1$ are both "the" universal class. This phenomenon recurs at any level, e.g.:

$$\text{horse} = \{x : x \text{ is a horse}\}$$
$$\text{horse}^{p \rightarrow p} = \{x : x \text{ is a horse} \ \& \ (p \rightarrow p)\}.$$

A requirement of relevance prevents these sets from being identified together, because being a horse does not relevantly imply arbitrary tautologies. That you have the property horse does not *mean*, absolutely, that you have the property $\text{horse}^{p \rightarrow p}$.

I would argue that the lesson of the extensionality paradoxes is that the doppelgänger problem is not a problem. This is the intensional/extensional duality underlying the set concept itself resurfacing. Doppelgängers are not a *problem*, because these are not, in fact, the same sets. Relevant biconditionals as used in an extensionality axiom will tie set identity to property identity – as Frege's Basic Law V demands (Section 5.1.1): sets are identical *exactly* when they are extensions of the same property. Since properties are fine-grained (the property of being an equilateral triangle is not the same as the property of being an equiangular triangle), identity is fine-grained, too; identicals must be the *same thing*. That's what relevant conditionals deliver. And I want to suggest that, since this is the only notion of identity that is workable for a naive *set* theory (as opposed to property theory), that is good evidence that it is the *correct* notion of identity for naive set theory. There are weaker notions available, under which doppelgängers are "coextensive" in an extensional sense, e.g., $(x \text{ is a horse}) \Leftrightarrow (x \text{ is a horse} \ \& \ (p \rightarrow p))$, and that's useful; it's just not *identity*. So $\mathcal{U}_0$ and $\mathcal{U}_1$ are not identical because they are not extensions of the same property; *they are not the same set*. When only absolute sufficiency will do, as in the case of identity, only relevance will suffice.

So, DKQ has a lot going for it, but not enough. For many practical purposes, $\rightarrow$ is too stringent to represent a conditional. (Whatever the conditional is in the sentence, "If you are going to the kitchen, then be careful not to slip on the wet floor," it is certainly not a deep meaning connection between antecedent and consequent.) Ultralogical absolute sufficiency is very unforgiving, by design, but then as the few who have tried to do relevant mathematics can attest, it does not allow various moves that are (apparently) essential

---

[35] For more on the topic of this section, see [Weber, 2010b; Field et al., 2017; Field, 2020; Weber, 2020].

in many unproblematic mathematical arguments [Istre, 2017, chs. 2, 3]. And even for a purist relevantist, willing to kiss off any "irrelevant" mathematical proofs, there is still the philosophical problem from Section 4.1.1.3. The consequence relation of ⊢ of DKQ *does* permit weakening and permutation as structural rules (to say nothing of modus ponens, as in Section 4.1.1.4). So one way or the other, the relevant logician is still using these inferential moves – *over* the language. And not admitting it at the sentence level, by not having a connective in the language that represents ⊢, violates closure. For both practical and principled reasons, I see a BCK-shaped hole in DKQ's heart.

4.2.3.5 The Plan

After this somewhat breathless charge through Curry's paradox and its ilk, my solution to a logic for paraconsistent set theory and beyond is relatively simple. Naive set theory is an intensional theory of extensions, as I said in Chapter 1; that is why it is inconsistent. Its logic represents all these features. There will be an "intensional" conditional, $\rightarrow$, that delivers set theory, and an "extensional" conditional, $\Rightarrow$, that picks up the deductive slack, obeying conditional introduction and (recalling the validity predicate) generally providing a sentence-level representation of the consequence relation ⊢. Neither connective, note, is reducible to the other or to combinations of $\vee, \&, \neg$. The intensional arrow is stricter than the extensional one, so to connect them we have the one-way principle:

$$\text{If } p \rightarrow q \text{ then } p \Rightarrow q.$$

Relevant set theory DKQ and light affine property theory in BCK are both known (classically) to be coherent. So, let's put them together. That's what happens in the next section.

4.3 Logic

Recapping: there are two countervailing demands on a logic (or logics) for *inconsistent mathematics*. It must be *weak* enough to support set comprehension, without triviality, but *strong* enough to be able to support actual proofs and carry through to familiar results. Since the primary feature of most nonclassical logics is to avoid certain classical principles, which makes carrying through proofs difficult, the aim is to retain as strong a logic as possible without trivializing, e.g., naive comprehension.

An effective conditional will have the following properties:

 i. *Not* be the material conditional
 ii. *Not* obey contraction
iii. Either contrapose, or weaken, but not both
 iv. Have a reasonable introduction rule/deduction theorem
  v. Support a theory of identity compatible with the axiom of set extensionality

Our solution is that this list can be satisfied by two conditionals. The main one, $\Rightarrow$, meets all the criteria except contraposing and compatibility with set extensionality. The other, $\rightarrow$, does not weaken but allows substitution of equivalents.

An effective *negation* is needed, too: it will

vi. Be exclusive, or exhaustive, but not both
vii. Not support ex contradictione quodlibet
viii. Support reductio ad absurdum arguments
ix. Have full de Morgan properties

Our solution is that this list can be met with two 'negations'. The main one, $\neg$, is exhaustive and not explosive. Another negation, call it annihilation, based on $\bot$, is explosive and supports reductio ad absurdum arguments. For de Morgan laws, these will be in terms of $\neg$, making conjunction and disjunction dual.

The idea is that all the work done by a few classical connectives can be done by *more* non-classical connectives. The difference is that classical logic overloads, asking one connective to do more than it reasonably should. In the nonclassical case, the labor is divided. The weaker the logic, the more distinctions become available.

The logic in itself is not primarily designed as an object of intrinsic beauty. As Peano and Frege abstracted their logics from perceived mathematical practice (in arithmetic) so our logic is "reverse engineered" from experience in substructural inconsistent mathematics. It is comprised of the tools that seem required to carry through proofs and shorn of the extras that cause trouble.

4.3.1 The Logic

Substructural dialetheic logic with quantifiers subDLQ, a first-order logic with identity, is now presented.[36]

One way to think of subDL is as the logic for the connectives conjunction, disjunction, and negation as given in the logic LP, from the Introduction, and adding to it two implication connectives that together add up to everything you want a conditional to do (and nothing you don't want it to do) – which also ends up making the conjunction connective "intensional." Another way to think of subDL is as obtained by taking the relevant logic DK, dropping the structural rule of contraction, dropping any axioms that imply $\varphi \to \varphi \& \varphi$, and adding a connective $\Rightarrow$ that reflects the properties of the consequence relation. And yet another, perhaps simplest, way to think of subDLQ is as based on the light affine logic BCK, and adding to it axioms for the connectives, including a de Morgan negation, and stapling on a relevant conditional too.

4.3.1.1 subDLQ

The language has connectives $\to$, $\Rightarrow$, &, $\vee$, $\neg$, $\forall$, $\exists$, $=$, brackets, and propositional atoms; well-formed formulae $\varphi, \psi, \ldots$ are built in the usual way. To save on brackets, the assumed

---

[36] This logic (or fragments thereof) is studied in [Badia and Weber, 2019] and deployed in [Weber et al., 2016; Girard and Weber, 2019]. Thanks to Badia and Girard. For classical semantics of logics in this vicinity, see [Allwein and Dunn, 1993; MacCaull, 1996].

Box 4.1 **The logic** subDLQ

Axioms

(1) $\varphi \rightarrow \varphi$

(2) $(\varphi \rightarrow \psi) \,\&\, (\psi \rightarrow \chi) \rightarrow (\varphi \rightarrow \chi)$

(3) $(\varphi \rightarrow \psi) \rightarrow (\neg\psi \rightarrow \neg\varphi)$

(4) $\varphi \,\&\, \psi \rightarrow \varphi$

(5) $\varphi \,\&\, \psi \rightarrow \psi \,\&\, \varphi$

(6) $\varphi \,\&\, (\psi \,\&\, \chi) \rightarrow (\varphi \,\&\, \psi) \,\&\, \chi$

(7) $\varphi \rightarrow \varphi \vee \psi$

(8) $\psi \rightarrow \varphi \vee \psi$

(9) $\varphi \vee \neg\varphi$

(10) $\neg\neg\varphi \leftrightarrow \varphi$

(11) $\varphi \vee \psi \leftrightarrow \neg(\neg\varphi \,\&\, \neg\psi)$

(12) $\varphi \,\&\, \psi \leftrightarrow \neg(\neg\varphi \vee \neg\psi)$

(13) $\varphi \,\&\, (\psi \vee \chi) \leftrightarrow (\varphi \,\&\, \psi) \vee (\varphi \,\&\, \chi)$

(14) $\varphi \vee (\psi \,\&\, \chi) \leftrightarrow (\varphi \vee \psi) \,\&\, (\varphi \vee \chi)$

(15) $\forall x\varphi \leftrightarrow \neg\exists x\neg\varphi$

(16) $\exists x\varphi \leftrightarrow \neg\forall x\neg\varphi$

(17) $\forall x\varphi \rightarrow \varphi_t^x$, for any t

(18) $\forall x(\varphi \vee \psi) \rightarrow (\varphi \vee \forall x\psi)$, x not free in φ

(19) $(\varphi \rightarrow \psi) \Rightarrow (\varphi \Rightarrow \psi)$

(20) $\neg(\varphi \Rightarrow \psi) \Rightarrow \neg(\varphi \rightarrow \psi)$

(21) $\varphi \,\&\, \neg\psi \Rightarrow \neg(\varphi \Rightarrow \psi)$

(22) $(\psi \Rightarrow \chi) \Rightarrow ((\varphi \Rightarrow \psi) \Rightarrow (\varphi \Rightarrow \chi))$

(23) $(\varphi \Rightarrow (\psi \Rightarrow \chi)) \Rightarrow (\psi \Rightarrow (\varphi \Rightarrow \chi))$

(24) $\varphi \Rightarrow (\psi \Rightarrow \varphi)$

(25) $(\varphi \Rightarrow \chi) \Rightarrow ((\psi \Rightarrow \chi) \Rightarrow (\varphi \vee \psi \Rightarrow \chi))$

(26) $\varphi \Rightarrow (\psi \Rightarrow \varphi \,\&\, \psi)$

(27) $(\varphi \Rightarrow (\psi \Rightarrow \chi)) \Rightarrow (\varphi \,\&\, \psi \Rightarrow \chi)$

(28) $\forall x(\varphi \Rightarrow \psi) \Rightarrow (\exists y\varphi_y^x \Rightarrow \psi)$
    x not free in ψ

(29) $\forall x(\psi \Rightarrow \varphi) \Rightarrow (\psi \Rightarrow \forall y\varphi_y^x)$
    x not free in ψ

(30) $\forall x(\varphi(x) \,\&\, \psi(x)) \Rightarrow \forall x\varphi(x) \,\&\, \forall x\psi(x)$

Rules

Modus Ponens: If φ, and $\varphi \Rightarrow \psi$, then ψ.
Universal Generalization: If φ, then $\forall x\varphi$.
Substitution (1): If $x = y$, then $\varphi(x) \rightarrow \varphi(y)$.
Substitution (2): If $\varphi \leftrightarrow \psi$, then $\chi(\varphi) \leftrightarrow \chi(\psi)$.

binding order is $\neg$, $\&$, $\vee$, $\rightarrow$, $\Rightarrow$ (so $\neg p \vee q \Rightarrow r$ is $((\neg p) \vee q) \Rightarrow r$). Biconditionals are defined: $\varphi \leftrightarrow \psi := (\varphi \rightarrow \psi)\&(\psi \rightarrow \varphi)$, and $\varphi \Leftrightarrow \psi := (\varphi \Rightarrow \psi)\&(\psi \Rightarrow \varphi)$. The logic is presented as a first-order axiomatic Hilbert system (with redundancies, valuing transparency over economy), in Box 4.1.

As a Hilbert system, the logic has many axiom schemas (some of them, again, superfluous) and three rules. The substitution rule is for intensionally equivalent formulas (Section 4.2.3.3).[37] A similar rule is *not* assumed for extensional equivalence $\Leftrightarrow$, in light of paradoxes.

We say that φ is *derivable* from formulas Γ, and we write

$$\Gamma \vdash \varphi$$

---

[37] Following [Routley et al., 1982, p. 292], where the "substitutivity of coimplicants" is explicitly added to the axiomatization of basic relevant logic. Thanks to Tore Fjetland Øgaard here.

when there is a sequence of formulas $\varphi_0, \ldots, \varphi_n$ such that φ is φ_n, and for every φ_i, either $\varphi_i \in \Gamma$, φ_i is an instance of an axiom, or φ_i or results from application of a rule on previous lines that haven't been used in other applications of rules. When φ is derivable from Γ, then and only then we say that $\Gamma \vdash \varphi$ is *valid*. A *theorem* is a valid sentence, derivable from the empty set.

The consequence relation $\vdash$ is reflexive, transitive, monotonic, and insensitive to order of premises (cf. the Excursus in the middle of Chapter 9):

- $\Gamma, \varphi \vdash \varphi$
- If $\Gamma \vdash \varphi$ and $\Delta, \varphi \vdash \psi$, then $\Gamma, \Delta \vdash \psi$
- If $\Gamma, \varphi \vdash \psi$, then $\Gamma, \varphi, \chi \vdash \psi$
- If $\Gamma, \Delta \vdash \varphi$, then $\Delta, \Gamma \vdash \varphi$

These are the salient "structural" properties of the logic. They show what deductions *preserve validity*, and are sometimes called "meta-inferences." But contraction is not on the list. How is it avoided?

Contraction fails by taking due care with the implicit meaning of the clause "results from" in the definition of derivation – a definition that is stated in an already non-contractive metalanguage![38] In the course of a derivation from Γ to φ, we associate with each member φ_k of the sequence of formulas a collection Γ_k made up of exactly the previous members of the sequence that φ_k followed from (so Γ_k is empty if φ_k is an axiom or member of Γ). When all the members $\gamma_0, \gamma_1, \ldots$ of all these Γ_k are taken together, we say that a derivation is valid *only if* every one of the the γs is already a member of Γ. This ensures that repeated assumptions are accounted for [Badia and Weber, 2019].[39]

Example [Badia]. In the derivation

1	p	premise
2	$p \Rightarrow (p \Rightarrow p \& p)$	axiom
3	$(p \Rightarrow p \& p)$	1, 2, modus ponens
4	p	premise
5	$p \& p$	3, 4 modus ponens

the repetition of premise p at line 4 is required to arrive at 5. In general, the valid $\psi, \psi \vdash \psi \& \psi$ (shown later) does not reduce to $\psi \vdash \psi \& \psi$, because a collection of premises that contains both ψ, ψ is not subsumed by a collection that contains only ψ.

Any valid derivation will respect this "tracking" of occurrences, because contraction is not valid in the language – this one – in which we are stating the definition of the Hilbert system.

---

[38] Meyer: "I am not going to fight with the C[lassical]-partisan about what goes on in the 'metalanguage.' ... [T]hat would just give us another formal system to talk about ... but eventually the escalator ride has got to stop.... The solution, already presented, is not to ride escalators" [Meyer, 1975, p. 8] As in Section 3.3.2, the "object" logic reflects the "metalogic."

[39] In standard metatheory, this would be accomplished by treating collections of premises as *multisets*, which are sensitive to repetitions of members [Blizard, 1989]. For some problems with contraction and consequence, see [Ripley, 2015b], for replies, and more on multisets, see [Cintula and Paoli, 2016].

See the appendix for BCK as a Gentzen system. I'll almost always be using this logic as an axiomatic system, except to make a couple of points in this chapter.

4.3.1.2 Properties of the Logic

The extensional arrow $\Rightarrow$ obeys the rule of conditional introduction, or a *deduction theorem*,

$$\Gamma, \varphi \vdash \psi \text{ iff } \Gamma \vdash \varphi \Rightarrow \psi,$$

so that "metalinguistic" facts about what-follows-from-what can be expressed in the language. (The deduction theorem proper is the left-to-right direction; right-to-left is modus ponens.) Conditional introduction *preserves validity*.[40] To prove a conditional, as long as a proof of q from assumptions $p_0, \ldots, p_n$ has not used any assumption more than once, it suffices for the conditional $p_0 \& \ldots \& p_n \Rightarrow q$ on no assumptions; or more generally, however many times assumptions are used, they are recorded as repeated antecedents of the derived conditional.

In particular, from the valid rule of modus ponens

$$\varphi, \varphi \Rightarrow \psi \vdash \psi$$

using conditional introduction, we have as a theorem

$$\varphi \Rightarrow ((\varphi \Rightarrow \psi) \Rightarrow \psi)$$

and then by axiom (27) of subDL, it is a theorem that

$$\varphi \& (\varphi \Rightarrow \psi) \Rightarrow \psi$$

as is good and beautiful and true. The quantified version

$$\varphi(a) \& \forall x(\varphi(x) \Rightarrow \psi(x)) \Rightarrow \psi(a)$$

holds too: supposing $\forall x(\varphi(x) \Rightarrow \psi(x))$, then $\varphi(a) \Rightarrow \psi(a)$ by $\forall$-elimination; supposing $\varphi(a)$, then $\psi(a)$ by modus ponens; and then conditional introduction completes the proof.

The rule of *adjunction*

$$\varphi, \psi \vdash \varphi \& \psi$$

is valid. *Proof:* Using modus ponens,

1	φ	premise
2	$\varphi \Rightarrow (\psi \Rightarrow \varphi \& \psi)$	axiom
3	$\psi \Rightarrow \varphi \& \psi$	1, 2, modus ponens
4	ψ	premise
5	$\varphi \& \psi$	3, 4, modus ponens

---

[40] See Badia and Weber [2019, theorem 1] for proof, by induction on formula complexity. For problems, see [Chvalovský and Cintula, 2012].

This shows that conjoined premises can be separated: if $\varphi \,\&\, \psi \vdash \chi$, then $\varphi, \psi \vdash \chi$ by transitivity of consequence. On the other hand, finitely many premises can be combined into a single sentence: if $\varphi, \psi \vdash \chi$, then $\vdash \varphi \Rightarrow (\psi \Rightarrow \chi)$, and then $\vdash \varphi \,\&\, \psi \Rightarrow \chi$, and then $\varphi \,\&\, \psi \vdash \chi$, using conditional introduction, logic, and modus ponens. This shows that that

$$\varphi_0, \varphi_1, \ldots, \varphi_n \vdash \psi \qquad \text{iff} \qquad \varphi_0 \,\&\, \varphi_1 \,\&\, \cdots \,\&\, \varphi_n \vdash \psi$$

by repeating the preceding arguments n times.

Disjunction is "additive" or "extensional" – in order to support *argument by cases* (axiom (25)), which is vital for mathematical proofs. It proves

$$\varphi \vee \varphi \Rightarrow \varphi$$

as a special case. Conjunction, though, as we've seen is, "multiplicative" or "intensional" – and in particular we *cannot* have $\varphi \Rightarrow \varphi \,\&\, \varphi$. Given that the subDLQ does have de Morgan laws for negation (axioms (11) and (12)), asserting that & and $\vee$ are dual to each other, this looks like a tension. But let the tension be a *frisson*. In nonclassical logic, we take notions that are equivalent according to orthodoxy, break them up into independent parts, and individually add back in those bits as needed. Here, we take this $\vee$ / & / $\neg$ package in order to have as much strength as possible within the bounds of coherence. (E.g., de Morgan laws preserve duality between finite unions and finite intersections; cf. Section 4.3.2.1.) This is coherent because $\Rightarrow$ does not contrapose, and more generally $\Leftrightarrow$ is not a congruence (it does not substitute). Consider:

Example [Ripley]. The derivation

1	$\varphi \,\&\, \varphi \Rightarrow \psi$	premise
2	$\neg(\neg\varphi \vee \neg\varphi) \Rightarrow \psi$	1, de Morgan
3	$\neg\varphi \vee \neg\varphi \Leftrightarrow \neg\varphi$	theorem
4	$\neg(\neg\varphi) \Rightarrow \psi$	2, 3, substitution
5	$\varphi \Rightarrow \psi$	4, double negation

would appear to prove contraction – except the step from 3 to 4 is invalid, an illicit substitution under negation.

There is no corresponding $\leftrightarrow$ version of line 3 that would validate a substitution. Argument by cases is only over the extensional arrow; if it were over $\to$, then because $\to$ contraposes this, plus de Morgan laws would give back $(\varphi \to \psi) \,\&\, (\varphi \to \chi) \to (\varphi \to \psi \,\&\, \chi)$, which we have deleted from DKQ, since it proves $\varphi \to \varphi \,\&\, \varphi$. On the other hand, since conjunction is commutative and associative by axioms 5 and 6, then by contraposition and de Morgan laws, so is disjunction.

Since the law of excluded middle (LEM) is a theorem, $\varphi \vee \neg\varphi$, then as forecast in the Introduction, the law of noncontradiction

$$\neg(\varphi \,\&\, \neg\varphi)$$

is a theorem, by de Morgan laws. All contradictions are false. It is, however, *not* a theorem that $\varphi \vee (\varphi \Rightarrow \psi)$ (where ψ might be $\bot$), which would be an LEM for annihilation negation. Paraconsistency means that $\neg\varphi$ does not imply $\varphi \Rightarrow \bot$. Not all falsity is absurdity.

What is the relationship between the two arrows, $\to$ and $\Rightarrow$? If a $\to$ implication holds, then a $\Rightarrow$ holds; in the other direction, $\Rightarrow$ is *not* sufficient for $\to$. What is the "gap" between them? A natural conjecture would be that if both $\varphi \Rightarrow \psi$ *and* $\neg\psi \Rightarrow \neg\varphi$, then $\varphi \to \psi$. (The converse is true.[41]) That is, maybe having both positive and negative extensional information suffices for intensional? This would be wrong, however. There is more to $\to$ than this. Else we could prove that all contradictions are intensionally equivalent:

Example [Istre]. Suppose $p\&\neg p$ and $q\&\neg q$. From p, we have $q \Rightarrow p$. From $\neg q$, we have $\neg p \Rightarrow \neg q$. If that were sufficient to establish $q \to p$, then a similar argument would prove $p \to q$, and then $p \leftrightarrow q$.

This wipes out the purpose of undertaking a careful study of the transconsistent.[42] An arrow that weakens is missing more connection between antecedent and consequent than can be supplemented by negation.

What is the relationship between conditionals and counterexamples? We have an axiom asserting that an extensional *counterexample* falsifies an implication: if $p \& \neg q$, then $\neg(p \Rightarrow q)$ (and then $\neg(p \to q)$, too). But the absence of a counterexample is not sufficient for entailment; that would be to fall back into material implication. Nor does a false implication indicate the existence of a counterexample; both $\neg(p \to q)$ and $\neg(p \Rightarrow q)$ are inferential dead-ends. Conditionals are more than the dual of counterexamples; denying that conditionals, or validity for that matter, can be subjected to "extensional reduction" is a hallmark of this kind of work.[43] To say this as a slogan that we will have occasion to repeat many times ahead: *intensionality is good for sameness, and extensionality is good for difference.* Less pithily: not being the same does not mean being different, and not being different does not mean being the same.

We collect here some useful derived theorems. I'm going to prove a few of them in some detail, to give more of a feel for how this logic works.

Antecedent strengthening, $(\varphi \to \psi) \Rightarrow (\varphi \& \chi \to \psi)$, indicates that this is not a nonmonotonic logic. If an antecedent of a conditional suffices for the consequent, then adding further assumptions cannot defeat the implication. It is shown as follows.

1	$p \to q$	premise
2	$\neg q \to \neg p$	1, contraposition
3	$\neg p \to \neg p \vee \neg r$	$\vee$-intro axiom
4	$\neg q \to \neg p \vee \neg r$	2, 3, transitivity
5	$\neg(\neg p \vee \neg r) \to q$	4, contraposition
6	$p \& r \to \neg(\neg p \vee \neg r)$	de Morgan law
7	$p \& r \to q$	5, 6, transitivity

---

[41] At least, up to contraction: $(\varphi \to \psi) \Rightarrow (\varphi \Rightarrow \psi)$, and $(\varphi \to \psi) \Rightarrow (\neg\psi \to \neg\varphi) \Rightarrow (\neg\psi \Rightarrow \neg\varphi)$.

[42] Istre goes on to show that, if we tried to adopt an axiom of extensionality of the form

$$\forall z(z \in x \Leftrightarrow z \in y) \& \forall z(z \notin x \Leftrightarrow z \notin y) \Rightarrow x = y,$$

i.e., that sets are identical when their extensions and antiextensions coincide (in the sense of $\Leftrightarrow$), then triviality results as in Grišin's paradox [Istre, 2017, p. 89]. Cf. Priest [2006b, p. 254].

[43] I highly recommend, again, [Meyer and Routley, 1977].

Conditional introduction on lines 1 and 7 complete the proof. (Line 3 calls on a background axiom so we don't have antecedent strengthening as a $\rightarrow$ statement – which, given its mild appearance of irrelevance, seems about right.) If φ is *enough* – "absolutely sufficient" – for ψ, then φ along with something else χ is (more than) enough, too.

The relevant arrow is *transitive*, by axiom (2), but only in this form.[44] For other forms of transitivity, we have *prefixing* for the $\Rightarrow$ arrow, in axiom (22), and this yields *suffixing*

$$(\varphi \Rightarrow \psi) \Rightarrow ((\psi \Rightarrow \chi) \Rightarrow (\varphi \Rightarrow \chi))$$

just by permutation of antecedents (axiom (23)).

Residuation is extremely useful:

$$(\varphi \,\&\, \psi \Rightarrow \chi) \Leftrightarrow (\varphi \Rightarrow (\psi \Rightarrow \chi)).$$

The left-to-right direction is called *exporting*, or sometimes just "residuating left to right," and the right-to-left direction is *importing*, or sometimes just "residuating right to left." Right to left is an axiom of the Hilbert system. To prove left to right,

1	$\varphi \,\&\, \psi \Rightarrow \chi$	(premise)
2	$\varphi \Rightarrow (\psi \Rightarrow \varphi \,\&\, \psi)$	axiom
3	$(\psi \Rightarrow \varphi \,\&\, \psi) \Rightarrow ((\varphi \,\&\, \psi \Rightarrow \chi) \Rightarrow (\psi \Rightarrow \chi))$	axiom
4	$(\varphi \,\&\, \psi \Rightarrow \chi) \Rightarrow ((\psi \Rightarrow \varphi \,\&\, \psi) \Rightarrow (\psi \Rightarrow \chi))$	3, permutation
5	$(\psi \Rightarrow \varphi \,\&\, \psi) \Rightarrow (\psi \Rightarrow \chi)$	1, 4, modus ponens

And then

6	$(\varphi \Rightarrow (\psi \Rightarrow \varphi \,\&\, \psi)) \Rightarrow$ $((\psi \Rightarrow \varphi \,\&\, \psi) \Rightarrow (\psi \Rightarrow \chi)) \Rightarrow (\varphi \Rightarrow (\psi \Rightarrow \chi))$	axiom
7	$((\psi \Rightarrow \varphi \,\&\, \psi) \Rightarrow (\psi \Rightarrow \chi)) \Rightarrow (\varphi \Rightarrow (\psi \Rightarrow \chi))$	2, 6, modus ponens
8	$\varphi \Rightarrow (\psi \Rightarrow \chi)$	5, 7, modus ponens

This means that weakening for $\Rightarrow$ is equivalent to &-elimination. Thus it is worth emphasizing that residuation is *not* assumed (or derived) for relevant $\rightarrow$, since otherwise the true axiom $p \,\&\, q \rightarrow q$ would give us the irrelevancies $p \rightarrow (q \rightarrow p)$ and $p \rightarrow (q \rightarrow q)$.

Residuation is a chief reason $\Rightarrow$ is *flexible*: background information may be used invisibly in the course of an argument – what relevant logicians chidingly call "suppression."[45] For example, let p be a *logical law*. Then p may be used in to chain conditionals together, as in

$$q \Rightarrow r, \, r \,\&\, p \Rightarrow s \vdash q \Rightarrow s.$$

Proof: if $r \,\&\, p \Rightarrow s$, then $p \Rightarrow (r \Rightarrow s)$ by exporting. Since p is (we suppose) a law, $r \Rightarrow s$ by modus ponens. Then if $q \Rightarrow r$, then $q \Rightarrow s$ by transitivity. This is used in the next derivation.

---

[44] Adding "hypothetical syllogism," e.g. $(p \rightarrow q) \rightarrow ((r \rightarrow p) \rightarrow (r \rightarrow s))$, to DKQ amps it up to TKQ [Brady, 2006, p. 246].
[45] For example in [Routley et al., 1982, ch. 1].

Factor is what Routley calls the principle

$$(\varphi \Rightarrow \psi) \;\&\; (\chi \Rightarrow \xi) \Rightarrow (\varphi \;\&\; \chi \Rightarrow \psi \;\&\; \xi). \tag{4.19}$$

After residuation, this is one of the best features of the $\Rightarrow/\&$ connective package. To prove it, first we prove that

$$(p \;\&\; r) \;\&\; (p \Rightarrow q) \Rightarrow (r \;\&\; q), \tag{4.20}$$

which is handy in itself, showing that we can work "inside" conjunctions. Proof: $p \;\&\; (p \Rightarrow q) \Rightarrow q$ is modus ponens and $q \Rightarrow (r \Rightarrow q \;\&\; r)$ is axiom (26), so by transitivity, $p \;\&\; (p \Rightarrow q) \Rightarrow (r \Rightarrow q \;\&\; r)$. Then by exporting, $((p \;\&\; (p \Rightarrow q)) \;\&\; r) \Rightarrow (q \;\&\; r)$, and we have (4.20) by the associativity and commutativity of conjunction.

Now to prove factor: an instance of (4.20) is

$$(p \;\&\; (r \;\&\; (r \Rightarrow s))) \;\&\; (p \Rightarrow q) \Rightarrow ((r \;\&\; (r \Rightarrow s)) \;\&\; q). \tag{4.21}$$

Another instance of (4.20) is

$$((r \;\&\; (r \Rightarrow s)) \;\&\; q) \;\&\; (r \;\&\; (r \Rightarrow s) \Rightarrow s) \Rightarrow (q \;\&\; s),$$

from which it follows, since modus ponens is a law, that

$$(r \;\&\; (r \Rightarrow s)) \;\&\; q \Rightarrow (q \;\&\; s). \tag{4.22}$$

So by transitivity on (4.21) and (4.22),

$$(p \;\&\; (r \;\&\; (r \Rightarrow s))) \;\&\; (p \Rightarrow q) \Rightarrow (q \;\&\; s).$$

Now again by the commutativity of conjunction, we rearrange to get

$$(p \Rightarrow q) \;\&\; (r \Rightarrow s) \;\&\; (p \;\&\; r) \Rightarrow (q \;\&\; s).$$

And then finally by residuation again, now left to right,

$$(p \Rightarrow q) \;\&\; (r \Rightarrow s) \Rightarrow ((p \;\&\; r) \Rightarrow (q \;\&\; s))$$

as required.[46]

A corollary for $\rightarrow$ is that

$$(\varphi \rightarrow \psi) \;\&\; (\chi \rightarrow \xi) \Rightarrow (\varphi \;\&\; \chi \Rightarrow \psi \;\&\; \xi)$$

since $\rightarrow$ implies $\Rightarrow$. Factor is effectively just conjunction introduction, but it is a recurring theme of contraction-free arguments. Contractive versions of many arguments (e.g., Lemma A in Chapter 6) start with the conjoined premises and reuse them to get the conjuncts, which would require the invalid inference from $\varphi \;\&\; \chi \Rightarrow \psi$ and $\varphi \;\&\; \chi \Rightarrow \vartheta$ to $\varphi \;\&\; \chi \Rightarrow \psi \;\&\; \vartheta$. Contraction freedom can be as simple as working in a more circumspect order.

---

[46] For the relevant $\rightarrow$, it isn't clear whether one gets anything beyond antecedent strengthening – if $(\varphi \rightarrow \psi)$ and $(\chi \rightarrow \xi)$, then $(\varphi \;\&\; \chi \rightarrow \psi)$ and $(\varphi \;\&\; \chi \rightarrow \xi)$. Usually in DKQ, one would get factor with axiom V (see the appendix).

We have assumed *distribution* of & over ∨ for the relevant →, and therefore it holds for ⇒, though this form of distribution would be derivable for the extensional arrow anyway. To show this,

1	$p \& q \Rightarrow ((p \& q) \vee (p \& r))$	axiom
2	$p \& r \Rightarrow ((p \& q) \vee (p \& r))$	axiom
3	$q \Rightarrow (p \Rightarrow (p \& q) \vee (p \& r))$	1, residuate
4	$r \Rightarrow (p \Rightarrow (p \& q) \vee (p \& r))$	2, residuate
5	$q \vee r \Rightarrow (p \Rightarrow (p \& q) \vee (p \& r))$	3, 4, ∨-left
6	$(q \vee r) \& p \Rightarrow ((p \& q) \vee (p \& r))$	5, residuate

The other direction is

1	$p \Rightarrow p$ and $q \Rightarrow q \vee r$	axioms
2	$p \& q \Rightarrow p \& (q \vee r)$	1, factor
3	$p \Rightarrow p$ and $r \Rightarrow q \vee r$	axioms
4	$p \& r \Rightarrow p \& (q \vee r)$	3, factor
5	$(p \& q) \vee (p \& r) \Rightarrow p \& (q \vee r)$	2, 4, ∨-introduction

On the other hand, it looks like we *cannot* independently derive distribution of ∨ over & for ⇒ without contraction; the closest we come is $(p \& p) \vee (q \vee r) \Rightarrow (p \vee q) \& (p \vee r)$. We've just made it an axiom, though we could get it instead by having assumed distribution for the relevant →, which contraposes, so we have

1	$\neg(p \& (q \vee r)) \leftrightarrow \neg((p \& q) \vee (p \& r))$	axiom, contraposition
2	$\neg p \vee \neg(q \vee r) \leftrightarrow \neg(p \& q) \& \neg(p \& r)$	1, de Morgan
3	$\neg p \vee (\neg q \& \neg r) \leftrightarrow (\neg p \vee \neg q) \vee (\neg p \vee \neg r)$	2, more de Morgan

and then, since this holds for anything, let p be $\neg\varphi$, let q be $\neg\psi$, and let r be $\neg\chi$. Then use double negation elimination. And the ⇒ version follows.

Here are a few further derived laws:

[Exercise] Show $(\varphi \Rightarrow \psi) \Rightarrow \neg\varphi \vee \psi$.

[Hint: use the law of excluded middle.]

[Exercise] Use residuation to show that modus ponens on subformulas of an antecedent is allowed, e.g.,

$$(p \& q \Rightarrow r) \& q \Rightarrow (p \Rightarrow r)$$

in the case of two formulas in the antecedent.[47]

[Exercise] Use residuation to show that argument by cases on subformulas of antecedents is allowed, e.g.,

$$(p \& q \Rightarrow r) \& (s \Rightarrow r) \Rightarrow (p \& (q \vee s) \Rightarrow r)$$

---

[47] One solution: since $(p \& q \Rightarrow r) \Rightarrow (p \Rightarrow (q \Rightarrow r))$, use permutation to get $p \Rightarrow ((p \& q \Rightarrow r) \Rightarrow (q \Rightarrow r))$, and then residuate back for the desired schema.

in the case of two formulas in the antecedent.[48]

[Exercise] Show $(\varphi \Rightarrow (\psi \Rightarrow \xi)) \Rightarrow ((\xi \Rightarrow \chi) \Rightarrow (\varphi \Rightarrow (\psi \Rightarrow \chi)))$

And for all that, we've said virtually nothing about quantifiers – mainly because, following the Australasian paraconsistency tradition, there is already so much to focus on at the propositional level. Some basic principles are as follows:

- $\forall x (\varphi(x) \Rightarrow \psi(x)) \Rightarrow (\forall x \varphi(x) \Rightarrow \forall x \psi(x))$.
- $\forall x (\varphi(x) \Rightarrow \psi(x)) \,\&\, \forall x \varphi(x) \Rightarrow \forall x \psi(x)$.
- $\forall x (\varphi(x) \Rightarrow \psi(x)) \,\&\, \forall y (\chi(y) \Rightarrow \xi(y)) \Rightarrow \forall x \forall y (\varphi(x) \,\&\, \chi(y) \Rightarrow \psi(x) \,\&\, \xi(y))$.
- $\forall x \varphi(x) \,\&\, \forall x \psi(x) \Rightarrow \forall x (\varphi(x) \,\&\, \psi(x))$.

For the first, suppose $\forall x (\varphi \Rightarrow \psi)$. Then axiomatically, $\varphi \Rightarrow \forall x \psi(x)$. And it is an axiom that $\forall x \varphi \to \varphi$, so by transitivity, $\forall x \varphi \Rightarrow \forall x \psi$. Then the second is by residuating left. The third is by factor. For the $\to$ arrow,

- $\forall z (\varphi \to \varphi)$.
- $\forall z (\varphi \to \psi) \,\&\, \forall z (\psi \to \varphi) \Leftrightarrow \forall z (\varphi \leftrightarrow \psi)$.
- $\forall z (\varphi \to \psi) \,\&\, \forall z (\psi \to \chi) \Rightarrow \forall z (\varphi \to \chi)$.

Standard dual properties of $\exists$ are as expected, such as

$$\varphi_t^x \to \exists x \varphi$$

for any t, as the dual (contrapositive) to the elimination axiom for $\forall$.[49]

Box 4.2 is a summary of some of what we've derived in this section, collected together for reference.

Box 4.2 **Some derived laws.**

$\varphi \,\&\, (\varphi \Rightarrow \psi) \Rightarrow \psi$

$\forall x (\varphi(x) \Rightarrow \psi(x)) \,\&\, \varphi(a) \Rightarrow \psi(a)$

$\varphi \vee \varphi \Rightarrow \varphi$

$\neg(\varphi \,\&\, \neg\varphi)$

$(\varphi \to \psi) \to (\varphi \,\&\, \chi \to \psi)$

$(\varphi \,\&\, \psi \Rightarrow \chi) \Leftrightarrow (\varphi \Rightarrow (\psi \Rightarrow \chi))$

$(\varphi \Rightarrow \psi) \,\&\, (\chi \Rightarrow \xi) \Rightarrow (\varphi \,\&\, \chi \Rightarrow \psi \,\&\, \xi)$

---

[48] One solution: suppose $(p \,\&\, q \Rightarrow r)$. Then $p \Rightarrow (q \Rightarrow r)$. Now, argument by cases and residuation says $(q \Rightarrow r) \Rightarrow ((s \Rightarrow r) \Rightarrow (q \vee s \Rightarrow r))$, so by transitivity, $p \Rightarrow ((s \Rightarrow r) \Rightarrow (q \vee s \Rightarrow r))$. Then by permutation and modus ponens, if we assume $s \Rightarrow r$, then $p \Rightarrow (q \vee s \Rightarrow r)$. Residuate back for the result.

[49] A reasonable suggestion is that if a universal quantifier is like a "big" conjunction, where $\forall x (\varphi(x))$ means $\varphi(a) \,\&\, \varphi(b) \,\&\, \dots$ for every thing $a, b, \dots$ then the quantifier should inherit or display noncontractive properties. See [Zardini, 2011] and [French and Ripley, 2015] and, from a different angle, [Běhounek and Cintula, 2005]. In the current setting, $\exists$ would like $\vee$ remain extensional, and then the "de Morgan"-ish quantifier dualities would be imposed. This awaits future work.

4.3.1.3 Underivables

To round off this list, it is worth noting a couple of formulas that *cannot* be derived, on pain of reintroducing contraction and hence triviality. (The first half of this chapter already identified the most salient of these.) If an argument cannot be valid, on pain of absurdity, we'll write $\nVdash$. Obviously, "square increasingness" $\varphi \Rightarrow \varphi \mathbin{\&} \varphi$ is out, as would be the contrapositive of argument by cases,

$$\nVdash (\varphi \Rightarrow \psi) \mathbin{\&} (\varphi \Rightarrow \chi) \Rightarrow (\varphi \Rightarrow \psi \mathbin{\&} \chi),$$

since $\varphi \Rightarrow \varphi$ as both conjuncts of the antecedent and modus ponens would give back "square increasingness." More subtly,

$$(\nVdash \varphi \Rightarrow (\psi \Rightarrow \chi)) \Rightarrow ((\varphi \Rightarrow \psi) \Rightarrow (\varphi \Rightarrow \chi)).$$

This might look like a form of transitivity, but if it were a valid schema, then we could instantiate by letting φ be p, also ψ be p, and χ be q to get $(p \Rightarrow (p \Rightarrow q)) \Rightarrow ((p \Rightarrow p) \Rightarrow (p \Rightarrow q))$. Then permutation and modus ponens give back contraction.

4.3.2 The Logic in Practice

Let's turn to some issues about the practical applications of this logic (since that is all it is intended for).

4.3.2.1 Reductio

Developing mathematics without contraction or classical negation is hard going. How does an indirect proof work, for example – a reductio ad contradictione, especially one where some assumptions are reused? Taking a cue from Slaney [Slaney, 1989], using the law of excluded middle helps. Assume that $\bot$ has the property that it implies everything (to be confirmed in the next chapter). We can prove *reductio*

$$(\varphi \Rightarrow \bot) \Rightarrow \neg\varphi$$

as follows:

1	$p \Rightarrow \bot$		premise
2	$\bot \Rightarrow \neg p$		$\bot$
3	$p \Rightarrow \neg p$		1, 2, transitivity
4	$\neg p \Rightarrow \neg p$		axiom
5	$(p \vee \neg p) \Rightarrow \neg p$		3, 4, $\vee$-intro
6	$p \vee \neg p$		LEM
7	$\neg p$		5, 6, modus ponens

Similarly, we have *consequentia mirabilis*

$$(\neg\varphi \Rightarrow \varphi) \Rightarrow \varphi$$

and reductio in the form

$$(\varphi \Rightarrow \neg\varphi) \Rightarrow \neg\varphi.$$

We also have a derivable *instance* of contraction:

$$\varphi \& \varphi \Rightarrow \neg\varphi \vdash \varphi \Rightarrow \neg\varphi.$$

(Proof: Assume $\varphi \& \varphi \Rightarrow \neg\varphi$; then residuating, $\varphi \Rightarrow (\varphi \Rightarrow \neg\varphi)$. Then $(\varphi \Rightarrow \neg\varphi) \Rightarrow \neg\varphi$ as before, so by transitivity, $\varphi \Rightarrow \neg\varphi$.) Just because contraction is invalid in general does not mean we can't have safe instances of it. In fact, what this shows is that in reconstructing mathematical arguments, contraction may sometimes be safely replaced by argument by cases, through the law of excluded middle. From de Morgan laws, we have $\neg\varphi \Leftrightarrow \neg(\varphi\&\varphi)$, so the difficulty of working with intensional conjunction can sometimes be mitigated by working with extensional disjunction.

Indeed, arguments in which repeated use of premises that drive to an absurdity may be discharged. If

$$\varphi, \ldots, \varphi \vdash \perp,$$

then

$$\vdash \neg\varphi$$

follows after as many steps as there are occurrences of φ on the first line. One just needs to invoke the law of excluded middle as many times as is needed – and that's fine, since LEM is a theorem (Section 4.3.2.2). For instance, suppose we have used p three times to prove $\neg p$ (or what comes to the same thing, $\perp$, starting one line earlier and using $\perp \rightarrow \neg p$):

1	$p \& p \& p \Rightarrow \neg p$	premise
2	$\neg p \Rightarrow \neg p$	axiom
3	$p \& p \& (p \vee \neg p) \Rightarrow \neg p$	1, 2, $\vee$-elim
4	$p \vee \neg p$	LEM
5	$p \& p \Rightarrow \neg p$	3, 4, modus ponens

Then repeat the argument until you reduce to $\neg p$. So again, $(\neg\varphi \Rightarrow (\neg\varphi \Rightarrow \cdots \Rightarrow (\neg\varphi \Rightarrow \perp))) \Rightarrow \varphi$ is valid as an acceptable "contraction." Multiple uses of a premise that turns out to be absurd are reduced to zero. A general pattern to prove a theorem is: use exhaustive negation to check cases, and exclusive "annihilation negation" to eliminate them.

A similar sort of thought shows how forms of disjunctive syllogism may be reinstated:

$$p \& ((p \Rightarrow \perp) \vee q) \Rightarrow q$$

holds by using argument by cases. And similarly, a form of contrapostion, *antilogism*

$$(p \& q \Rightarrow \perp) \Rightarrow (p \Rightarrow \neg q)$$

holds in this specific instance. (It must hold not in the general form $(p \& q \Rightarrow r) \Rightarrow (p \& \neg r \Rightarrow \neg q)$, since the antecedent could just be made conjunction elimination, $(p \& \neg q \Rightarrow p)$, thus yielding explosion $(p \& \neg p \Rightarrow q)$.)

This is how the naive theorist can hope to reconstruct a lot of mathematics. To prove φ, argue by cases. First assume either φ or $\neg\varphi$. If φ, that's great. Otherwise, use $\neg\varphi$ and background axioms/theorems to derive $\bot$. Either way, the theorem holds. In this way, the noncontractive approach seems in keeping with traditional proof methods, as a way to mathematical truths.[50]

4.3.2.2 Metacontraction

The strategy just outlined for reductio – not to mention several of the proofs for derived laws of the logic – only works if we have as many instances of the law of excluded middle and other axioms on hand as are needed to finish arguments. That is, we wish to *contract on theorems*. And there would appear to be a saving grace on this front. Here is the rule for conjunction introduction, as we met it in Section 4.1.2.3:

$$\frac{\Gamma \vdash \varphi \qquad \Delta \vdash \psi}{\Gamma, \Delta \vdash \varphi \,\&\, \psi}.$$

Now, a theorem "follows from no premises," i.e., $\vdash \varphi$. Then contraction on *theorems* looks acceptable, as an instance of the &-in rule, and the reflexivity of consequence:

$$\frac{\vdash \varphi \qquad \varphi \vdash \varphi}{\varphi \vdash \varphi \& \varphi}.$$

So if φ is a theorem, then $\varphi \Rightarrow \varphi \,\&\, \varphi$. We have *contraction by cuts*,

$$\frac{\vdash \varphi \qquad \dfrac{\vdash \varphi \qquad \Gamma, \varphi, \varphi \vdash \psi}{\Gamma, \varphi \vdash \psi}}{\Gamma \vdash \psi}.$$

When φ is *proven*, two instances of φ reduce to one without further ado: in subDLQ, we can easily derive

$$\varphi \vdash (\varphi\&\varphi \Rightarrow \psi) \Rightarrow (\varphi \Rightarrow \psi).$$

Bad instances of contraction are those with *bad* (nontheorem) assumptions, e.g., Curry sentences, which devolve upon repetition into absurdity. The axioms for, e.g., $\mathbb{N}$ and $\mathbb{R}$ and their consequences, we believe, are not bad, and can be appealed to as part of the fixed inferential background. Theorems are "renewable" resources.[51]

---

[50] Regarding negation, there are issues that arise about the Ackermann constants, little t and little f [Dunn and Restall, 2002, p. 12; Slaney, 1989; Beall et al., 2011; Øgaard, 2016], e.g., to what extent $p \Rightarrow f$ can be thought of as the negation of p. (It can't, because if it could, then $(p \Rightarrow q)$ would be "contraposable" by transitivity, $(q \Rightarrow f) \Rightarrow (p \Rightarrow f)$. And contraposition plus weakening yields explosion.) Ackermann constants are omitted from in the present study.

[51] With validity predicate Val and constant $\top$ (the negation of $\bot$), we could represent that φ is a theorem as $\mathsf{Val}(\ulcorner\top\urcorner, \ulcorner\varphi\urcorner)$. Then one can have a principle that says "if something is a theorem, then two uses of it are like one," as in

$$\mathsf{Val}(\ulcorner\top\urcorner, \ulcorner\varphi\urcorner) \Rightarrow ((\varphi\&\varphi \Rightarrow \psi) \Rightarrow (\varphi \Rightarrow \psi))$$

More simply, $\top \Rightarrow \varphi$ could represent theoremhood, whence $(\top \Rightarrow \varphi) \Rightarrow (\varphi \Rightarrow \varphi \,\&\, \varphi)$ is derivable. (Proof: from $\top \Rightarrow \varphi$ and $\varphi \Rightarrow \varphi$, then by factor, $\top \,\&\, \varphi \Rightarrow \varphi \,\&\, \varphi$; then by exporting, $\top \Rightarrow (\varphi \Rightarrow \varphi \,\&\, \varphi)$; and then the result follows by modus ponens and the truth of $\top$). See Section 5.2.1.

So, that is (finally) some good news. But tread carefully. Didn't Belnap and Dunn tell us in the last chapter that "*adding* premises cannot possibly *reduce* threat," that "that way lies madness" [Anderson et al., 1992, p. 503, emphasis in the original text]? In Section 3.1.2.2, I was very judgmental about other people helping themselves to special regions of "quasi-valid safety" in otherwise invalid environs. What, then, is this locution "if φ is a theorem, then it can be used safely"?

Since a naive theory should satisfy closure, if we allow contraction on theorems, then the closed theory should express and validate "contract on theorems," *as a thesis of the theory itself*. The same would go for any of the true principles of the system. For this, we do have contraction by cuts – it is derivable in the logic, not as an extralogical assumption – but still, how does it encode? The threat of some new revenge would be lurking if we try to cheat and contract in the "metatheory." For example, it would be cheating to invoke a principle like: if p is a theorem, and $p \& p \Rightarrow q$, then q, or an *incautious cut* rule,

$$\frac{\vdash \varphi \qquad \Gamma, \varphi, \varphi \vdash \psi}{\Gamma \vdash \psi},$$

which reintroduces contraction by making two cuts on φ but accounting for it only once. Written across, we do *not* have

$$\varphi \vdash (\varphi \& \varphi \Rightarrow \psi) \Rightarrow \psi$$

without assuming some kind of contraction in the "metatheory." We already know that contraction will lead to disaster; incautious cut is contraction, and it does not work, on pain of triviality.[52]

We may contract on theorems, but we have to do it right: since we are contraction-free in the metatheory, all the way up, then the acceptable version of the principle will be expressed

if p is a theorem, and p is a theorem, then we can use p twice,

or in general, if p is a theorem and ... and p is a theorem (n times), then p may be used n times. In the derivations in Section 4.3.2.1, every time we needed the LEM, we invoked it *again*. We can do that "for free," but it still needs to be recorded. To say that theorems are renewable does not mean that they can be said once and used indefinitely; it means that they can be said indefinitely, and *then* used indefinitely. That's what it means to be contraction free *all the way up*.

4.3.2.3 Restricted Quantification

Paraconsistency allows *unrestricted quantification*: there is a universe of all objects, unlike in orthodox classical theory, and we can quantify over it. A problem then arises about *restricted* quantification, where the domain of discourse is not the entire universe but rather limited, e.g., to all the people, all the sets, all the natural numbers, etc. Classically, this

---

[52] See [Terui, 2004] for the impossibility of adding an S4 "provability" modality to BCK set theory. See [Wansing and Priest, 2015; Weber, 2016b].

is done using material conditionals, writing, e.g., $\forall n(\varphi(n))$ for $\forall x(x \in \mathbb{N} \supset \varphi(x))$, and $\exists n\varphi(n)$ instead of $\exists x(x \in \mathbb{N} \,\&\, \varphi(n))$. This then delivers (the appearance of) Aristotelian "square of opposition" dualities: not all numbers are φ iff some number is not; there is no number that is φ iff all numbers are not. Without disjunctive syllogism, though, these do not work as intended, since $\varphi(a) \,\&\, \forall x(\neg\varphi(x) \vee \psi(x))$ does not yield $\psi(a)$. But neither can we get better restricted quantifiers by simply replacing $\supset$ with either of our conditionals $\rightarrow$ or $\Rightarrow$, since, on pain of explosion,

$$\neg\exists x(\varphi(x) \,\&\, \psi(x)) \not\vdash \forall x(\varphi(x) \Rightarrow \neg\psi(x)) \text{ and}$$
$$\neg\forall x(\varphi(x) \Rightarrow \psi(x)) \not\vdash \exists x(\varphi(x) \,\&\, \neg\psi(x)).$$

While the unrestricted quantifiers are dual, $\exists = \neg\forall\neg$ and $\forall = \neg\exists\neg$, the restricted versions are not, exactly because $\rightarrow$ and $\Rightarrow$ are not material conditionals; cf. Priest [2006b, p. 76].

The problem of restricted quantification for nonclassical logics is very hard and to my knowledge is not solved for the glutty paraconsistent case.[53] Solving it is out of range for today. In the pages ahead, I will, where needed, use informal primitive restricted quantifiers, saying, e.g., that either all n are even or else some n is not, as an instance of the law of excluded middle. In doing this, I am following standard practice in this area.[54] I hope to do better one day, but, just as one can make valuable contributions in ethics without first solving the problem of free will versus determinism,[55] there are too many other interesting problems to get going on to wait.

4.3.3 Nontriviality?

Both BCK and DKQ are nontrivial for naive comprehension, according to classical proofs, and DKQ is nontrivial for extensionality, too. Further, the hybrid logic subDLQ itself is nontrivial, and in particular does not validate contraction, as shown by a (classical) four-valued algebraic model that validates all the axioms but in which $\varphi \vdash \varphi\&\varphi$ fails.[56]

As mentioned in Section 3.2.3.3, though, we don't yet have a paraconsistent model theory – obviously, since model theory is built out of set theory, and that is developed in the next chapter. Nor do we have a formal theory *of* proofs, with cut elimination theorems and the like; those require developed machinery under them, too. The proof-theoretic methods used for BCK-type logics break down when we add excluded middle. The model theoretic proofs as used by Brady, and extended by [Field et al., 2017] may be workable, but for two conditionals become very complicated. So we have a proof-theoretic apparatus, which can show what is (positively) provable; but we don't have precise formal ways of saying

---

[53] See, e.g., [Beall and Restall, 2006; Weber, 2010b; Field, 2020]. A proposal by Brady for relevant logic is in [Brady, 2003, pp. 321–331]. A proposal by Priest in [Priest, 2017b, §8] uses a consistency operator.

[54] As in [Routley, 1977; Dunn, 1980; Meyer and Mortensen, 1984; Mortensen, 1995; Restall, 2010a].

[55] Mates groups determinism and moral responsibility in with the liar, the Russell paradox, and the existence of the external world, as "absolutely insoluble" [Mates, 1981].

[56] As shown by a model devised by the program PROVER9, and checked by Badia; see [Badia and Weber, 2019] for details.

when something is *not* provable, at least not in the sense of soundness and completeness theorems (e.g., "there is a semantic counterexample, so there must not be any proof"). We tread carefully around known danger areas, outlined in the first part of this chapter, but the emphasis is on what *can* be shown. On a case-by-case basis, this logic (plus comprehension and extensionality, with those phrased relevantly) avoids all the known problems, up to the devilish [Øgaard, 2016]. But that's obviously only an inductive inference, complicated by worries we will face at Theorem 13 in Chapter 5; so time will tell.

In any case, how much should a committed ultralogician care about a metatheoretic proof of nontriviality? A nontriviality proof is *relative* to classical ZFC, or some other nonparaconsistent system. I have already argued that a committed nonclassicist should not place any weight on such systems. What, then, about an internal nontriviality proof, conducted in subDLQ itself, proving its own coherence? That turns out to be so easy as to be a joke, to close the chapter:

Lemma (Soundness of subDLQ **+ set theory).** *All the theorems of the system are true.*

Proof The axioms are all true, and whatever follows from true premises by valid rules is true. □

Theorem (Nontriviality of subDLQ **+ set theory).** *The system is nontrivial.*

Proof Either the system is trivial or not. If not, that is great. If it is, then it is a theorem of the system (in which we are now arguing) that "the system is nontrivial." Either way, by soundness, the result follows. □

* * *

In this chapter, we've seen the constraints on a logic for naive set theory under closure, and offered a (putatively) workable apparatus, subDLQ, in response. These have been serious challenges; I hope at least this gives pause to anyone who writes off paraconsistency as the "easy" way out. The motivation was to honor intuitions about sets and truth, and pay for that by adjusting our theory of logic. The logic we've arrived at is a tool, and that's okay, because, as I asserted back in the Introduction, logic in itself does not constitute a "solution" to the paradoxes. Paraconsistency and dialetheism are a method, a decision about *how* to proceed with the paradoxes; it is the first step, not the last. To this method we have now added more texture: *substructural* logic, and with it an entry to a highly intensional "ultramodal" universe. That ultraverse is the thing to try to understand, by using logic to draw a mathematical picture of it. And that is what we are going to spend the rest of the book doing.

Appendix: BCK and DKQ

As our logic subDLQ is made from DKQ and BCK, it may be useful to have these in view individually, with BCK presented as a Gentzen system.

BCK

The logic BCK is presented as a Gentzen system. It may equivalently be presented as a Hilbert system [Ono and Komori, 1985]; cf. Galatos et al. [2007, cor. 2.21]: theorems of the Hilbert system are exactly the provable sequents of the Gentzen system.

A *sequent* is of the form $\Gamma \vdash \varphi$, obtained in the following ways.

$$\frac{}{\varphi \vdash \varphi} \; id$$

$$\frac{\Gamma \vdash \varphi}{\Gamma, \psi \vdash \varphi} \; weakening \qquad \frac{\Gamma \vdash \psi \quad \Delta, \psi \vdash \chi}{\Gamma, \Delta \vdash \chi} \; cut \qquad \frac{\Gamma, \varphi, \psi \vdash \chi}{\Gamma, \psi, \varphi \vdash \chi} \; exchange$$

$$\frac{\Gamma \vdash \varphi}{\Gamma \vdash \varphi \vee \psi} \qquad \frac{\Gamma \vdash \psi}{\Gamma \vdash \varphi \vee \psi} \qquad \frac{\Gamma, \varphi \vdash \chi \quad \Gamma, \psi \vdash \chi}{\Gamma, \varphi \vee \psi \vdash \chi}$$

$$\frac{\Gamma \vdash \psi \quad \Delta \vdash \chi}{\Gamma, \Delta \vdash \psi \,\&\, \chi} \qquad \frac{\Gamma, \varphi, \psi \vdash \chi}{\Gamma, \varphi \,\&\, \psi \vdash \chi}$$

$$\frac{\Gamma, \varphi \vdash \psi}{\Gamma \vdash \varphi \Rightarrow \psi} \qquad \frac{\Gamma \vdash \varphi \quad \Delta, \psi \vdash \chi}{\Gamma, \Delta, \varphi \Rightarrow \psi \vdash \chi}$$

$$\frac{\Gamma, \varphi_t^x \vdash \psi}{\Gamma, \forall x \varphi \vdash \psi} \; t \text{ any term} \qquad \frac{\Gamma \vdash \varphi_y^x}{\Gamma \vdash \forall x \varphi} \; y \text{ not free in } \varphi$$

$$\frac{\Gamma \vdash \varphi_t^x}{\Gamma \vdash \exists x \varphi} \; t \text{ any term} \qquad \frac{\Gamma, \varphi_y^x \vdash \psi}{\Gamma, \exists x \varphi \vdash \psi} \; y \text{ not free in } \varphi$$

To think of our subDLQ as a Gentzen system, take (quantified) BCK and add to it as initial sequents all the axioms for $\rightarrow$ and principles for a de Morgan negation [Badia and Weber, 2019].

DKQ

The logic DK is based on Routley and Meyer's 1976 *dialectical logic* DL, which has the axiom $(\varphi \rightarrow \psi) \rightarrow \neg(\varphi \wedge \psi)$. For DK, this is weakened to VIII in the following listed schemes, and the "L" in the logic's name goes back one letter in the alphabet. Definitions: $A \vee B$ for $\neg(\neg A \wedge \neg B)$; $A \leftrightarrow B$ for $(A \rightarrow B) \wedge (B \rightarrow A)$; $\exists$ is $\neg\forall\neg$. Note here that $\wedge$ is additive conjunction. Crucially, to obtain subDLQ, we delete axiom V.

All instances of the following schemata are theorems:

I	$\varphi \to \varphi$	
IIa	$\varphi \wedge \psi \to \varphi$	
IIb	$\varphi \wedge \psi \to \psi$	
III	$\varphi \wedge (\psi \vee \chi) \to (\varphi \wedge \psi) \vee (\varphi \wedge \chi)$	(*distribution*)
IV	$(\varphi \to \psi) \wedge (\psi \to \chi) \to (\varphi \to \chi)$	(*conjunctive syllogism*)
V	$(\varphi \to \psi) \wedge (\varphi \to \chi) \to (\varphi \to \psi \wedge \chi)$	
VI	$(\varphi \to \psi) \leftrightarrow (\neg\psi \to \neg\varphi)$	(*contraposition*)
VII	$\varphi \leftrightarrow \neg\neg\varphi$	(*double negation elimination*)
VIII	$\varphi \vee \neg\varphi$	(*excluded middle*)
IX	$(\forall x)\varphi \to \varphi(a/x)$	
X	$(\forall x)(\varphi \to \psi) \to (\varphi \to (\forall x)\psi)$	(with x not free in φ)
XI	$(\forall x)(\varphi \vee \psi) \to \varphi \vee (\forall x)\psi$	(with x not free in φ)

The following rules are valid:

I	$\varphi, \psi \vdash \varphi \wedge \psi$	(*adjunction*)
II	$\varphi, \varphi \to \psi \vdash \psi$	(*modus ponens*)
III	$\varphi \vdash (\forall x)\varphi$	(*universal generalization*)
IV	$\varphi \to \psi, \chi \to \delta \vdash (\psi \to \chi) \to (\varphi \to \delta)$	(*hypothetical syllogism*)
V	$x = y \vdash \varphi(x) \leftrightarrow \varphi(y)$	(*substitution*)

The following "metarules" preserve validity:

$$\frac{\varphi \vdash \psi}{\varphi \vee \chi \vdash \psi \vee \chi} \qquad\qquad \frac{\varphi \vdash \psi}{\exists x \varphi \vdash \exists x \psi}$$

Note how the metarules essentially invoke working with the consequence relation as an operator; this is the motivation for introducing a conditional that obeys arrow introduction into subDLQ.

Part III

Where Are the Paradoxes?

*It is just as hard to draw a very small square circle
as it is to draw an enormous one.*
– Max Black

5

Set Theory

Wherein the basic theory of sets is developed axiomatically in a paraconsistent logic. The two main goals are (1) to establish a toolkit for elementary mathematics and (2) to prove the main antinomies of naive set theory. The two goals come together in proving the Burali-Forti paradox for the theory of ordinals, along the way getting acclimated to working in the substructural/paraconsistent atmosphere.[1]

The presentation is in mathematical English, with the background logic subDLQ of the previous chapter. In general, here and in chapters ahead the words "if ... then" or cognates will correspond to the "deductive" $\Rightarrow$ conditional. Where the "law-like" $\rightarrow$ is used, we say so explicitly.

5.1 Elements

5.1.1 Axioms

Naive set theory has two nonlogical axioms,[2] which effectively define the two nonlogical connectives $\in$ and $=$.

Axiom 1 (Abstraction). $x \in \{z : \varphi(z)\} \leftrightarrow \varphi(x)$.

Axiom 2 (Extensionality). $\forall z(z \in x \leftrightarrow z \in y) \leftrightarrow x = y$.

These are formulated with intensional, contraposable biconditionals, which assert a law-like connection – absolute sufficiency, in Routley's sense (Section 4.2.3.4) – between their right and left sides. For x to be φ just means that x is in the set of φs. And x is identical to y exactly when being an x just means being a y. These are extremely strong connections. (That is why they still hold, even in the face of inconsistency.) The axioms are equivalent to $x \notin \{z : \varphi(z)\} \leftrightarrow \neg\varphi(z)$ and $\exists z\neg(z \in x \leftrightarrow z \in y) \leftrightarrow x \neq y$, the contrapositives, as part of the naive set concept. Here and throughout, "$x \notin y$" means $\neg(x \in y)$.

---

[1] This chapter refits some material from [Weber, 2010d, 2012] into subDLQ.
[2] The language includes term-forming set abstracts, $\{\cdot : \cdot\}$. For adding such symbols conservatively to the language via skolemization, see [Øgaard, 2017]; but also see Section 5.2.3.2 of this chapter.

The abstraction axiom immediately implies, by existential generalization, the naive comprehension principle:

$$\exists y \forall x (x \in y \leftrightarrow \varphi(x)).$$

If you read footnote 3 from back in Section 1.1, you might remember that in abstraction and comprehension, there is no restriction on φ, so the set being defined may appear free in its own defining description. *Every* predicate determines a set. So there are "circular" sets; but that will be the case anyway following the fixed point result (Theorem 13), that follows from "restricted" comprehension alone.

Since these first-order axioms are reconstructed from Frege's *Grundgeztze*, it is good that his nefarious original principle can be recovered:

Theorem 3 (Frege's Basic Law V). $\{x : \varphi(x)\} = \{x : \psi(x)\} \Leftrightarrow \forall x (\varphi(x) \leftrightarrow \psi(x)).$

Proof Left to right:

1	$\{x : \varphi(x)\} = \{x : \psi(x)\}$	premise
2	$\varphi(z) \leftrightarrow z \in \{x : \varphi(x)\}$	Axiom 1
3	$z \in \{x : \varphi(x)\} \leftrightarrow z \in \{x : \psi(x)\}$	1, Axiom 2
4	$\varphi(z) \leftrightarrow z \in \{x : \psi(x)\}$	2, 3, transitivity of $\leftrightarrow$
5	$z \in \{x : \psi(x)\} \leftrightarrow \psi(z)$	Axiom 1
6	$\varphi(z) \leftrightarrow \psi(z)$	4, 5, transitivity of $\leftrightarrow$

Right to left can be proved using formula substitution, or without it by running the preceding derivation more or less in reverse:

1	$\varphi(z) \leftrightarrow \psi(z)$	premise
2	$z \in \{x : \varphi(x)\} \leftrightarrow \varphi(z)$	Axiom 1
3	$z \in \{x : \varphi(x)\} \leftrightarrow \psi(z)$	1, 2, trans.
4	$\psi(z) \leftrightarrow z \in \{x : \psi(x)\}$	Axiom 1
5	$z \in \{x : \varphi(x)\} \leftrightarrow z \in \{x : \psi(x)\}$	3, 4, trans.
6	$\{x : \varphi(x)\} = \{x : \psi(x)\}$	5, Axiom 2

as required.   □

Thus $x \in y$ means that $y = \{z : \varphi(z)\}$ & $\varphi(x)$, for some φ, as Frege originally defined it [Frege, 1903a, I, §34].

5.1.2 Identity, Subsets, and Singletons

Identity is an *equivalence relation*: reflexive, symmetric, and transitive, as in

$$x = x$$
$$x = y \Rightarrow y = z$$
$$x = y \mathrel{\&} y = z \Rightarrow x = z$$

due to Axiom 2 and properties of the underlying (bi)conditional (Section 4.3.1.2):

$$\forall x (\varphi \leftrightarrow \varphi)$$
$$\forall x (\varphi \leftrightarrow \psi) \Rightarrow \forall x (\psi \leftrightarrow \varphi)$$
$$\forall x (\varphi \leftrightarrow \psi) \;\&\; \forall x (\psi \leftrightarrow \chi) \Rightarrow \forall x (\varphi \leftrightarrow \chi).$$

For proof, recall that a biconditional is defined as the conjunction of two conditionals. For reflexivity, $\varphi \rightarrow \varphi$ is a theorem, and theorems may be repeated; so $(\varphi \rightarrow \varphi) \;\&\; (\varphi \rightarrow \varphi)$, so $\varphi \leftrightarrow \varphi$. For symmetry, $\varphi \;\&\; \psi \rightarrow \psi \;\&\; \varphi$, so $(\varphi \leftrightarrow \psi) \rightarrow (\psi \leftrightarrow \varphi)$. For transitivity, $(\varphi \leftrightarrow \psi) \;\&\; (\psi \leftrightarrow \chi) \Rightarrow (\varphi \leftrightarrow \chi)$ by factor (4.19),[3] then use $\forall x \varphi \;\&\; \forall x \psi \Rightarrow \forall x (\varphi \;\&\; \psi)$.

Intensionality is good for sameness and extensionality is good for difference. Pursuing the analogy, define *subsethood* intensionally, like identity, and *proper subsethood* as subsets with extensional difference; respectively,

$$x \subseteq y := \forall z (z \in x \rightarrow z \in y)$$
$$x \subset y := x \subseteq y \;\&\; \exists z (z \in y \;\&\; z \notin x).$$

This way of defining proper subsets implies, but is stronger than, the claim that $x \subseteq y \;\&\; x \neq y$ (see Section 5.1.3), which is often how proper subsethood is phrased. Subsets form a *partial order* – reflexive, antisymmetric, and transitive:

$$x \subseteq x$$
$$x \subseteq y \;\&\; y \subseteq x \Rightarrow x = y$$
$$(x \subseteq y) \;\&\; (y \subseteq z) \Rightarrow x \subseteq z$$

due to the underlying arrow and universal quantifier.

Terminology: when x is a (proper) subset of y, it is said that x is (properly) *included* in y; when x is a member of y, it is said that x is *contained* in y; cf. Halmos [1974, pp. 2–3]. Thus x is included in y just in case everything contained in x is contained in y, and x is properly included in y when it is a subset of y but something contained in y is not contained in x.

The axiom of extensionality may be expressed as the antisymmetry of $\subseteq$, the proposition that x is identical to y exactly when $x \subseteq y$ and $y \subseteq x$. And not only are sets with exactly the same *members* identical, so too are sets with exactly the same *parts*,

$$\forall z (z \subseteq x \Leftrightarrow z \subseteq y) \Leftrightarrow x = y, \tag{5.1}$$

showing that this is extensional set theory.

With naive comprehension, we can state a substitution principle using first-order quantifiers.

Theorem 4 (Leibniz's Law). $x = y \Leftrightarrow \forall z (x \in z \leftrightarrow y \in z)$.

---

[3] If $(p \rightarrow q) \;\&\; (q \rightarrow r) \Rightarrow (p \rightarrow r)$ and $(r \rightarrow q) \;\&\; (q \rightarrow p) \Rightarrow (r \rightarrow p)$, then $(p \rightarrow q) \;\&\; (q \rightarrow p) \;\&\; (r \rightarrow q) \;\&\; (q \rightarrow r) \Rightarrow (p \rightarrow r) \;\&\; (r \rightarrow p)$.

Proof Left to right, if $x = y$, then use $\forall x(x \in z \leftrightarrow x \in z)$ and the fact that $=$ substitutes. Right to left, instantiate z by $\{u : u = x\}$, and do modus ponens on the fact that $x = x$. $\square$

For $\Rightarrow$, there are corresponding notions of *extensional equivalence* and *extensional parthood*,

$$x \equiv y := \forall z(z \in x \Leftrightarrow z \in y)$$

$$x \sqsubseteq y := \forall z(z \in x \Rightarrow z \in y).$$

Extensional equivalence is an equivalence, and extensional parthood is a partial order,

$x \equiv x$	$x \sqsubseteq x$
$x \equiv y \Rightarrow y \equiv x$	$x \sqsubseteq y \Rightarrow (y \sqsubseteq x \Rightarrow y \equiv x)$
$(x \equiv y) \,\&\, (y \equiv z) \Rightarrow x \equiv z$	$(x \sqsubseteq y) \,\&\, (y \sqsubseteq z) \Rightarrow x \sqsubseteq z,$

using the underlying properties of $\Rightarrow$. In terms of our two arrows, the axiom of extensionality implies

$$x = y \Rightarrow \forall z(z \in x \Leftrightarrow z \in y)$$

because $\rightarrow$ implies $\Rightarrow$. Crucially, though, the step down from intensional to extensional implication is one-way; we do *not* have an identity, $x = y$, from $\forall z(z \in x \Leftrightarrow z \in y)$. The extensional $\equiv$ is handy, but should not be mistaken for the real thing, as is shown by Grišin's paradox (Section 4.2.2.1); on the basis of that, we don't assume that $\equiv$-equivalent sets can be intersubstituted. Care must be taken to distinguish between genuine *identity*, as defined by the axiom of extensionality and governed with an intensional $\rightarrow$, and which does substitute, versus the weaker *extensional coincidence* governed with a more permissive deduction relation $\Rightarrow$, and which does *not* always intersubstitute. As we saw in Section 4.2.3.4, sets with membership connected by $\Leftrightarrow$ are not necessarily the same set.[4]

Identity contracts,

$$x = y \Rightarrow x = y \,\&\, x = y, \tag{5.2}$$

because it is a theorem that $x = x$, so it is a theorem that $x = x \,\&\, x = x$; so by substitution, if x has this property, and $x = y$, then y has this property too. But we won't assume, and nothing seems to force us to accept, that extensional equivalence $\equiv$ contracts, because we haven't assumed that $\Leftrightarrow$-equivalent formulas are intersubstitutible. From $x \equiv y$, it does not appear to follow that $x \equiv y \,\&\, x \equiv y$. Formulas that are equivalent according to $\Leftrightarrow$ may be different enough that they cannot be intersubstituted *salva veritate*. Formulas that are equivalent according to $\leftrightarrow$ *may* be intersubstituted, but there are fewer occasions when such equivalences occur.

---

[4] Compare this to the material biconditional and corresponding notion of identity in [Priest, 2014], which is not substitutable, but also not transitive (and not detachable, of course). Cf. Priest [2017b, §10].

Tying identity to set membership, there are several notions of *singleton* to chose from, given the sensitivity of the language.[5] We will work with this one, giving it the familiar notation:

$$\{a\} := \{x : x = a\}.$$

Lewis [Lewis, 1991] points out that this is where the set concept comes distinctly into view: the singleton $\{a\}$ is a new thing, something over and above the object a. Or many objects – as notation, members of a set – can be written

$$\{a, b, \ldots, c\} := \{x : x = a \lor x = b \lor \ldots \lor x = c\}$$

as usual.

Singletons are used in mathematics, among other things, to represent *points*, as something like $\{\bullet\}$. We will be thinking a lot about points in the pages ahead, so it is worth noting that the properties of identity show up at the "point" level. Namely, the singleton $\{x : x = \bullet\}$ contracts: using the notion of intersection $\cap$ from (Section 5.1.4),

$$\{\bullet\} \subseteq \{\bullet\} \cap \{\bullet\}.$$

The singleton defined $\{x : x \equiv \bullet\}$, on the other hand, does not contract any more than $\equiv$ does. The intersection of singletons with the same point inside may be *structured* (Section 5.1.6.2). More to come on that front.

On a different but also baffling note, there is (from comprehension) a set that is its own singleton, $S = \{x : x = S\}$, whence by substitution

$$S = \{S\}$$
$$= \{\{S\}\}$$
$$\vdots$$
$$= \{\cdots\}.$$

There's also a set $S' = \{x : x \equiv S'\}$. It is not the same as S. To be in S', one need only *coincide* with S', but to be in S requires *being* S.

In general, circularly defined sets are not necessarily unique. Running with the example at hand, given $a = \{x : x = a\}$ and $b = \{x : x = b\}$, the membership of a and b does not determine whether or not $a = b$ [Barwise and Moss, 1996, p. 15, ex. 2.4], without some further axioms, e.g., asserting that bisimilar sets are equal, as in non-well-founded set theory [Aczel, 1988]. We have no further axioms, just lots and lots of sets. The universe is large; it contains multitudes.

---

[5] For example, $\{x : \forall z(a \in z \Rightarrow x \in z)\}$ for the singleton of a, as in [Petersen, 2000; Terui, 2004]. As far as I can tell, these singletons and the ones I am using coincide, because $=$ contracts. Not so, perhaps, for *intensional* singletons, $\{x : \forall x(a \in z \rightarrow x \in z)\}$. The pair $\{x : \forall z(a \in z \rightarrow (b \in z \rightarrow x \in z)\}$ would be different from the pair $\{x : \forall z(b \in z \rightarrow (a \in z \rightarrow x \in z)\}$, so that the intensional pair $\{a, b\}$ is not $\{b, a\}$, and all these different again from $\{x : \forall z(a \in z \& b \in z \rightarrow x \in z)\}$.

When x is a member of X, one expects that the singleton of x will be a subset of X. This can be confirmed in two ways, one for each notion of subset. First, for intensional subsethood, we could introduce the notion of a *relevant singleton*, $\{x\}_X := \{u : u \in X \& u = x\}$. Then automatically, $\{x\}_X \subseteq X$. This is somewhat artificial, since it is not contingent on whether or not $x \in X$, but it is generally the best that we can do for $\rightarrow$, because in general identity facts will not suffice (intensionally) for membership facts. It is more natural to use the extensional notions, whence from Leibniz's law and some permutation,

$$x \in X \Leftrightarrow \forall u (u = x \Rightarrow u \in X)$$

giving the property

$$x \in X \Leftrightarrow \{x\} \subseteq X \tag{5.3}$$

as desired.

5.1.3 Nonidentity

Difference of membership entails difference of sets, thanks to our "counterexample" axiom, $p \& \neg q \Rightarrow \neg(p \rightarrow q)$. Extensional sameness is not sameness, but extensional difference is difference.

Lemma 5. $\exists z(z \in x \& z \notin y) \Rightarrow x \neq y.$

Proof If some $a \in x \& a \notin y$, then $\neg(a \in x \rightarrow a \in y)$. Then $\exists z \neg(z \in x \rightarrow z \in y)$. By the contraposed axiom of extensionality, $x \neq y$.   □

From this lemma, we have, by cases,

$$x \subset y \vee y \subset x \Rightarrow x \neq y. \tag{5.4}$$

Along these lines, proper subsets form a *strict partial order* – irreflexive, asymmetric, and transitive:

$$x \not\subset x$$
$$x \subset y \Rightarrow y \not\subset x$$
$$(x \subset y) \& (y \subset z) \Rightarrow x \subset z.$$

(For $x \not\subset y$ to hold, either $x \not\subseteq y$ or $\neg\exists z(z \in y \& z \notin x)$ needs to hold.) For irreflexivity, $\forall z(z \in x \vee z \notin x)$ by excluded middle, so by dualities, $\neg\exists z(z \in x \& z \notin x)$. For asymmetry, if $x \subset y$, then something in y is not in x, so $y \not\subseteq x$, so $y \not\subset x$.

If any a were such that $a \subset a$, then a would not be identical to itself. Apropos, the next theorem was originally presented by Restall in the context of naive set theory in the logic LP.

Theorem 6 (Restall, 1992). $\exists x \exists y(x \neq y).$

Proof By naive abstraction, there is an r, the *Russell set*, such that

$$x \in r \leftrightarrow x \notin x.$$

Then here are two proofs (by cases), given in parallel Kantian-antinomy style:

1	$r \in r \rightarrow r \notin r$		1	$r \notin r \rightarrow r \in r$
2	$r \notin r \rightarrow r \notin r$		2	$r \in r \rightarrow r \in r$
3	$r \in r \vee r \notin r$		3	$r \in r \vee r \notin r$
4	$r \notin r$		4	$r \in r$

Therefore,

$$r \in r \mathbin{\&} r \notin r.$$

By Lemma 5, since $\exists x (x \in r \mathbin{\&} x \notin r)$,

$$r \neq r$$

and so $\exists x (x \neq x)$, "which is more than enough for the intended result" [Restall, 1992, theorem 7]. □

This is our first official *non-self-identical* object. The Russell set will still be self-identical, $r = r$, because everything is self-identical (as in Section 5.1.2). It has exactly the same members as itself. It also differs from itself with respect to membership. As a way in to grasping this, one might imagine two Russell sets, one containing itself and the other not – and then realize that these "two" sets are really identical.

An important fact about any non-self-identical object is that nothing is identical to it. (If something isn't even itself, it surely isn't anything else!) This lemma is classically true, although here not vacuously so:

Lemma 7 (Non-self-identity lemma). $x \neq x \Rightarrow \forall y (y \neq x)$.

Proof Suppose $x \neq x$. For any y, either $y = x$ or $y \neq x$. If $y = x$, then by substitution, $y \neq x$. □

If $\exists y (x = y)$ is read "x exists," as tradition following Russell and Quine suggests (contra [Routley, 1980b; Priest, 2005]), then the non-self-identity lemma says that non-self-identical objects do not exist. So the Russell set is like this, $\forall x (r \neq x)$. On the other hand, because $x = x$ for any x as before, we have too that

$$\forall x \exists y (x = y).$$

All objects are self-identical and everything exists. And also some are not and don't. Or, as Wittgenstein puts it [Wittgenstein, 1922, 5.5352],

People have wanted to express 'there are no things' by writing '$\sim (\exists x).x = x$'. But even if this were a proposition, would it not be equally true if indeed 'there were things' but they were not identical with themselves?

Note too that if $x \neq x$, then $\{x\} \neq \{x\}$, and then $\forall y (y \neq \{x\})$.

5.1.4 Unions and Intersections

Define

$$X \cup Y := \{x : x \in X \vee X \in Y\}$$
$$X \cap Y := \{x : x \in X \,\&\, x \in Y\}$$

for finite union and finite intersection, respectively. These are associative and commutative,

$$X \cap (Y \cap Z) = (X \cap Y) \cap Z$$
$$X \cap Y = Y \cap X$$
$$X \cup (Y \cup Z) = (X \cup Y) \cup Z$$
$$X \cup Y = Y \cup X$$

by logic, because & and $\vee$ are, e.g., $\forall u(u \in X \,\&\, (u \in Y \,\&\, u \in Z) \leftrightarrow (u \in X \,\&\, u \in Y) \,\&\, u \in Z)$. Distribution holds,

$$X \cap (Y \cup Z) = (X \cap Y) \cup (X \cap Z)$$
$$X \cup (Y \cap Z) = (X \cup Y) \cap (X \cup Z),$$

by logic too.

Here we gather the basic "lattice" properties of sets that will be used often. It is true that

$$X \cap Y \subseteq X \qquad X \cap Y \subseteq Y \tag{5.5}$$
$$X \subseteq X \cup Y \qquad X \subseteq Y \cup X \tag{5.6}$$

by conjunction elimination and disjunction introduction. (And anything that holds intensionally holds for $\sqsubseteq$ too.) For subset relations,

$$X \sqsubseteq Y \Leftrightarrow X \cup Y \equiv Y. \tag{5.7}$$

This is because our disjunction is extensional. Proving left to right, if $X \sqsubseteq Y$, then $z \in X \vee z \in Y \Rightarrow z \in Y$ using argument by cases; and $z \in Y \to z \in X \vee z \in Y$ by disjunction introduction (not repeating any assumption); so $X \cup Y \equiv Y$. Right to left, the logical form of the argument is:[6] supposing $p \vee q \to r$, then since $p \to p \vee q$, by transitivity, $p \to r$. In particular, then, $X \cup X \equiv X$.

For intersection, our multiplicative conjunction means the orthodox $X \sqsubseteq Y \Leftrightarrow X \cap Y \equiv X$ needs to be broken up into two parts:

$$X \sqsubseteq Y \Rightarrow X \cap X \sqsubseteq X \cap Y \tag{5.8}$$
$$X \cap Y \equiv X \Rightarrow X \sqsubseteq Y. \tag{5.9}$$

The logical form of (5.8) is $(p \Rightarrow q) \Rightarrow (p \,\&\, p \Rightarrow p \,\&\, q)$, as opposed to the contractive $(p \Rightarrow q) \Rightarrow (p \Rightarrow p \,\&\, q)$. So the relation between intersection and subsethood is not exactly as classical, which will come up a lot in point set topology (see Chapter 9).

---

[6] Note that this isn't the principle $(p \vee q \to r) \Rightarrow (p \to r) \,\&\, (q \to r)$, any proof of which appears to reuse the antecedent in some way (try it). Without contraction, one ends up proving $((p \vee q) \,\&\, (p \vee q) \to r) \Rightarrow (p \to r) \,\&\, (q \to r)$.

There are generalizations, the usual

$$\bigcup X := \{y : \exists x (x \in X \ \& \ y \in x)\}$$
$$\bigcap X := \{y : \forall x (x \in X \implies y \in x)\},$$

but these tend to be much less useful, because they don't dualize, because the conditional is not defined in terms of other connectives. Stay with finite unions and intersections when possible, since these are dual, as we now check.

5.1.5 Complements

Sets are metaphors for predicates, so their relations reflect logical connectives – so far, implication, conjunction, and disjunction, and now also negation: every set X has a *complement*:

$$\overline{X} := \{x : x \notin X\}.$$

In classical set theory, this unrestricted complement no more exists than does the universal set, since $\overline{X}$ is the universal set with only X deleted. Here, the existence of complements is immediate from comprehension; the extent to which complementation absolutely "deletes" X, on the other hand, is something we will study in depth.

From negation properties of the logic, "double negation" is immediate, as are exhaustion, exclusion, and de Morgan laws:

$$\overline{\overline{X}} = X \tag{5.10}$$

$$\forall x (x \in X \cup \overline{X}) \tag{5.11}$$

$$\neg \exists x (x \in X \cap \overline{X}) \tag{5.12}$$

$$\overline{X \cap Y} = \overline{X} \cup \overline{Y} \tag{5.13}$$

$$\overline{X \cup Y} = \overline{X} \cap \overline{Y}. \tag{5.14}$$

The exclusion property (5.12), derived from the law of noncontradiction $\neg(p \ \& \ \neg p)$ thus carries the increasingly familiar caveat that there may be exceptions. For example, the Russell set

$$r \in r \cap \overline{r}$$

by Lemma 7, although it is still the case that $r \notin r \cap \overline{r}$ by (5.12). (So $r \cap \overline{r} \neq r \cap \overline{r}$, by Lemma 5.)

From (5.10), then (5.13) has as a special case

$$\overline{X \cap \overline{X}} = X \cup \overline{X}$$

and similarly for (5.14). It is also true that

$$X \subseteq Y \Leftrightarrow \overline{Y} \subseteq \overline{X} \tag{5.15}$$

because $\rightarrow$ contraposes. Since $\Rightarrow$ does not contrapose, (5.15) is not true for $\sqsubseteq$.

Given a particle representing "always false" such as $\bot$ (introduced officially in Section 5.2.1), then an *exclusive* complement can be defined:

$$\tilde{X} := \{x : x \in X \Rightarrow \bot\}.$$

This will have properties similar to intuitionistic negation, like $\forall x (x \in X \cap \tilde{X} \rightarrow \bot)$, double negation introduction $X \sqsubseteq \tilde{\tilde{X}}$, and contraposition $X \sqsubseteq Y \Rightarrow \tilde{Y} \sqsubseteq \tilde{X}$ (note only one direction for these last two). The union $X \cup \tilde{X}$ is not exhaustive (i.e., corresponding property (5.11) is not true, assuming $p \vee (p \Rightarrow \bot)$ is not true). In general, this notion of complement, while more hygienic than complement in terms of $\neg$, will turn out to have little use in what follows. A *relevantly exclusive* complement, $\{x : x \in X \rightarrow \bot\}$ (i.e., as before but with the $\rightarrow$ in place of $\Rightarrow$) will be of even less use, since it is comprised of just the things that X "intrinsically" excludes, those for which being in X is absolutely sufficient for $\bot$. So unless specifically stated, "complement" throughout means $\overline{X}$, in terms of (paraconsistent) negation $\neg$.

The *relative complement* of Y in X is X without Y:

$$X \backslash Y := X \cap \overline{Y}.$$

Note again, though, that disjunctive syllogism is invalid: it may be that $x \in A \cup B = X$, and $x \in X \backslash A$, without it necessarily following that $x \in B$. Maybe $x \in A$ and $x \notin A$, while x is not in B at all. Relative complements are more sensitive to work with than their classical counterparts. This extra "sticky-ness" with complements is rewarded by complementation being completely general, and having an intrinsic expression, in *self-complements*.

What is a self-complement? A nonparaconsistent theory has it that $X \cap \overline{X}$ is always empty; a nonempty self-complement is absurd. The idea of paraconsistency is to be more circumspect about what is absurd and what is not (since, e.g., we already know that $r \cap \overline{r}$ is not empty). I take the notation $X \backslash X$ as invitation for a nonabsurd object satisfying it, and therefore introduce a name:

Definition 1. *The* restricted self-complement *of X, written*

$$\varnothing_X := X \backslash X,$$

is the intersection of X with its own unrestricted complement.

For example, the Russell set again has a nonempty self-complement $r \backslash r = \varnothing_r$; in particular, $r \in \varnothing_r$ (and not, so $\varnothing_r \neq \varnothing_r$). What is $\varnothing_X$? It is the distinctive "shape" left when X is gone, the inconsistent residue (if there is any) of X. Perhaps it is the edge or boundary of X.

For every x,

$$\varnothing_x \subseteq x \tag{5.16}$$

because $z \in x \cap y \rightarrow z \in x$, as in (5.5) of Section 5.1.4; so to get (5.16), let $y = \overline{x}$. And then

$$x \subseteq y \Rightarrow \emptyset_x \subseteq y \tag{5.17}$$

by antecedent strengthening. (If $x \subseteq y$, then $x \cap z \subseteq y$.) But in general, unlike *the* empty set, which we will meet later, $\emptyset_x$ is not a subset of any arbitrary set. It is not a globally bottom element in the subset partial order; assuming $x \cap \overline{x}$ is included in any y, $x \cap \overline{x} \subseteq y$, would be to assume ex contradictione quodlibet. The restricted self-complement of x is the bottom element particular to some set x, each x's own "zero." Different sets may or may not have different zeros, since it wouldn't even classically be the case that $x \cap \overline{x} = y \cap \overline{y}$ is any reason to think $x = y$.[7]

Relative self-complements can be used to keep track of possible inconsistency; also, keeping track of contractions gives general forms of otherwise familiar laws:

Lemma 8. *Let $a \subseteq x$. The following hold:*

(i) "Double negation elimination": $x \backslash (x \backslash a) \sqsubseteq a \cup \emptyset_x$
(ii) "Double negation introduction": $(a \cap a) \sqsubseteq x \backslash (x \backslash a)$
(iii) $x \backslash \emptyset_x \subseteq x$, and $x \cap x \sqsubseteq x \backslash \emptyset_x$
(iv) $x \equiv x \cup \emptyset_x$
(v) $x \equiv a \cup x \backslash a$

Proof To prove (i), calculate the following:

$$x \backslash (x \backslash a) = x \cap \overline{(x \cap \overline{a})}$$
$$= x \cap (\overline{x} \cup \overline{\overline{a}})$$
$$= (x \cap \overline{x}) \cup (x \cap a))$$
$$= (\emptyset_x) \cup (x \cap a)$$
$$\sqsubseteq \emptyset_x \cup a$$

using de Morgan rules from (5.10).

For (ii), $a \subseteq x$ is given, and $a \subseteq a \cup \overline{x}$ is always true. So $a \cap a \sqsubseteq x \cap (\overline{x} \cup a)$ by factor (4.19). We can actually get a stronger result,

$$(a \cap a) \cup \emptyset_x \sqsubseteq x \backslash (x \backslash a)$$

by adding that $\emptyset_x \sqsubseteq x \cap (\overline{x} \cup a)$, which is true because $u \in x \rightarrow u \in x$ and $u \notin x \rightarrow u \notin x \lor u \in a$, so $(u \in x \& u \notin x) \Rightarrow (u \in x \& (u \notin x \lor u \in a))$.

For (iii), the first bit is just &-elimination. And the other bit is

$$(x \subseteq x) \& (x \subseteq x \cup \overline{x}) \Rightarrow x \cap x \sqsubseteq x \cap \overline{x \cap \overline{x}}$$

which comes from identity, $\lor$-introduction, factor, distribution, and de Morgan laws.

---

[7] If Sartre were here, he might suggest that the absence of Pierre is not the same as the absence of Simone – "Pierre absent haunts this café ..." [Sartre, 1958, p. 10].

Prove (iv) by $\vee$-elimination from right to left, and $\vee$-introduction (indeed, the stronger $\sqsubseteq$ fact) from left to right.

And (v) is better behavior than ususal for noncontractive sets. Right to left, since $\forall z (z \in a \vee z \notin a)$, then $x \sqsubseteq (a \cup \overline{a}) \cap x \sqsubseteq (x \cap a) \cup (x \cap \overline{a})$, and therefore

$$x \sqsubseteq a \cup (x \cap \overline{a}).$$

The other direction, $a \cup (x \backslash a) \sqsubseteq x$, is argument by cases. $\square$

From (v), we have $x \equiv x \cup \varnothing_x$ as a special property of the background set x, but *not* in general $a \equiv a \cup \varnothing_x$.

5.1.6 Powersets

The *powerset* of X is the collection of all subsets of X. Since there are two notions of "subset," there are two notions of powerset, one properly included within the other:

$$\{x : x \subseteq X\} \quad \text{and} \quad \{x : x \sqsubseteq X\}$$

These two notions of powerset are importantly different. For example, $\forall x (x \sqsubseteq \{z : z = z\})$, because $\Rightarrow$ weakens; but one does not expect $\forall x (x \subseteq \{z : z = z\})$, since $z \in x \rightarrow z = z$ is *irrelevant* (Sections 4.2.3.4 and 5.1.2). In many instances, it seems that the more "extensional" $\sqsubseteq$ is what is meant or needed to formalize more "casual" arguments about subsets,[8] just as $\Rightarrow$ is often the more apt expression of "if/then." Our fallback definition of powerset, then, will be the more inclusive

$$\mathscr{P}(X) := \{x : x \sqsubseteq X\}$$

unless explicitly mentioned otherwise.[9]

Let us tie up this sequence of the chapter with two unorthodox observations about the inconsistent substructure of subsets.

5.1.6.1 Inconsistent Structure

Inspired by a result of Arruda (cf. [Arruda and Batens, 1982; da Costa, 2000] in [Batens et al., 2000]), recall that the Russell set r is a proper subset of itself, $r \sqsubset r$. Moreover,

$$\mathscr{P}(r) \sqsubset r.$$

For suppose $x \sqsubseteq r$. If $x \notin x$, then $x \in r$; but if $x \in x$, then $x \in r$ on supposition.[10] On the other hand, $r \in r$ but $\neg(r \sqsubseteq r)$, so r has a member that is *not* in its own powerset, and thus the inclusion is *proper*.

---

[8] See Mares [2004a, 11§9].
[9] A classically equivalent way to say that Y is a subset of X is for $Y = Y \cap X$. A "powerset" of X could be $\{Y : Y \cap X = Y\}$. But without contraction, there is no reason to think that it has any members. The more properties one demands a notion to satisfy, the fewer things satisfy that notion.
[10] Note that as that proof used argument by cases, which is only valid over $\Rightarrow$, it will not work for showing that $\mathscr{P}(r) \subseteq r$.

Then iterating the powerset operation takes us ever deeper:

$$\mathscr{P}(\mathscr{P}(r)) \sqsubset \mathscr{P}(r) \sqsubset r.$$

For if $x \in \mathscr{P}(\mathscr{P}(r))$, then $x \sqsubseteq \mathscr{P}(r)$, and then $x \in \mathscr{P}(r)$. Again these inclusion relations are proper. Repeating the argument as much as one likes, if we write $\mathscr{P}^0(x) :=$ x, $\mathscr{P}^{n+1}(x) := \mathscr{P}(\mathscr{P}^n(x))$, then in general

$$\mathscr{P}^{n+1}(r) \sqsubset \mathscr{P}^n(r),$$

which I think is rather lovely, in a haunting sort of way. It shows that the powerset of a non-well-founded inconsistent set is "imploded," a theme we will return to. This "structure in the cracks" is brought out by inconsistency; other features of our substructural logic will manifest in other ways too, as we now see.

5.1.6.2 Noncontractive Structure

There is inner structure to noncontractive sets. Since contraction is invalid, it is not the case that $p \Rightarrow p \,\&\, p$. This has the effect on sets A that $A \sqsubseteq A \cap A$ is not in general true. On the other hand, by the validity of $p \,\&\, p \to p$, it is always true that $A \cap A \sqsubseteq A$. So it will be the case that A properly includes $A \cap A$, which properly includes $A \cap A \cap A$, which properly includes $A \cap A \cap A \cap A \ldots$. A picture of nested self-intersections emerges.

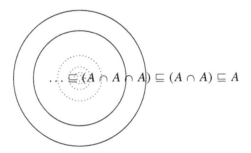

$$\ldots \sqsubseteq (A \cap A \cap A) \sqsubseteq (A \cap A) \sqsubseteq A$$

Classically, there is nothing to see here. Nonclassically, this suggests there is some nuance to how "close" an element of A is to A. Along similar lines, take the set of all φs, for some property φ. Again in general,

$$\{x : \varphi(x) \,\&\, \varphi(x)\} \subseteq \{x : \varphi(x)\}$$

but not vice versa. Something that is $\varphi \,\&\, \varphi$ is more "intensely" φ than something that is only φ. A notable exception are singletons: since identity contracts (5.2), $\{a\} \equiv \{a\} \cap \{a\}$. Elements of a singleton of a are already as close to a as they can be. This *might* provide some insight into why contraction is invalid – that repeating a (bad) assumption is getting "closer" to some (bad) state of affairs – but I leave this for now as just a hint at the possibilities inherent in noncontractive metaphysics.

5.2 A Sketch of the Universe

5.2.1 V and $\varnothing$

The *universe* exists:

$$V := \{x : \exists y(x \in y)\}.$$

It is the collection of all the things that have a property. It is the universe, since

$$\forall x(x \in V). \tag{5.18}$$

Proof: By reductio, if $x \notin V$, then $\neg\exists y(x \in y)$, so by quantifier duality, $\forall y(x \notin y)$. Selecting something for y, for instance the complement of the universe, we have $x \notin \overline{V}$. By definition of complement, and double negation elimination, $x \in \overline{\overline{V}} = V$. If something has no properties ($x \notin V$), it has the property of having no properties, and so has a property after all ($x \in V$), q.e.d.

V is even the universe under $\rightarrow$, because everything "means" that it is in V:

$$\forall x(x \subseteq V). \tag{5.19}$$

This follows by existential introduction: $z \in x \rightarrow \exists yz \in y$.

The *empty set* exists:

$$\varnothing := \{x : \forall y(x \in y)\}.$$

It is the collection of things that have *every* property. The empty set has the dual properties of the universe, such as

$$\forall x(x \notin \varnothing), \tag{5.20}$$

because if $x \in \varnothing$, then x has every property, for example the property of not being in the empty set. More generally, if $x \in \varnothing$, then $x \in \{z : \varphi\}$ for any φ at all, so the empty set is included, invisibly, in everything:

$$\forall x(\varnothing \subseteq x). \tag{5.21}$$

This is the *global* bottom element.

The universe and the empty set relate in the right way:

$$\overline{V} = \varnothing \tag{5.22}$$

$$\overline{\varnothing} = V. \tag{5.23}$$

To prove (5.22), right to left is (5.21). Left to right, if $x \notin V$, then $\forall y(x \notin y)$ and in particular, $x \notin \overline{\varnothing}$, so $x \in \overline{\overline{\varnothing}}$, so by double negation (5.10), $x \in \varnothing$. Then (5.23) can be proved from (5.22) by substitution, or by using (5.19) and ((5.21)) to argue afresh. If something is not in the universe, it is nowhere. If an object isn't nowhere, it is somewhere.

The *absurdity* symbol can now be defined:

$$\bot := \varnothing \in \varnothing.$$

(Any $t \in \emptyset$ would do, but Restall said once in conversation that a self-membered empty set seems more appropriate). It is true that

$$\bot \rightarrow \varphi \tag{5.24}$$

for all φ, since

$$\emptyset \in \emptyset \rightarrow \forall y(\emptyset \in y)$$
$$\rightarrow \emptyset \in \{x : \varphi\}$$
$$\rightarrow \varphi.$$

With

$$\top := \neg\bot,$$

it is also true that

$$\varphi \rightarrow \top \tag{5.25}$$

for any φ, by contrapositive reasoning. Clearly, then,

$$\top \tag{5.26}$$

is a theorem. Then the universe and the empty set may be rewritten:

$$V = \{x : \top\} \tag{5.27}$$
$$\emptyset = \{x : \bot\} \tag{5.28}$$

showing that not being in V is not merely false but absurd.

There are other sets that are classically equivalent to the universe and the empty set (ignoring that, classically, the universe does not even exist). These are the doppelgängers we previewed in Section 4.2.3.4. For example, Whitehead and Russell [Whitehead and Russell, 1910, p. 216] pick

$$\mathcal{U} = \{x : x = x\}$$

for the universe, which makes sense since identity is the relation that everything has to itself and to nothing else. This set is universal, in the sense that $\forall x(x \in \mathcal{U})$. But we have seen that some objects are not self-identical (as well as being self-identical), such as r, and so $r \notin \mathcal{U}$. And then, by the axiom of extensionality, since $r \in \mathcal{U}$, too, then $\mathcal{U} \neq \mathcal{U}$ and then

$$\mathcal{U} \notin \mathcal{U}.$$

A fortiori, $\mathcal{U}$ is not identical (at all) with V, since then otherwise they would be intersubstitutable, from which it would follow that $\mathcal{U} \notin V$, an absurdity. The complement of $\mathcal{U}$ is similarly an "empty" set $\{x : x \neq x\}$, which also has members, and so which is a member of itself and not.

In practice, when you want to say that something is empty, such as the intersection of A and B, it tends not to be apt to try to establish $A \cap B = \emptyset$, or even $A \cap B \equiv \emptyset$, because this would mean that occupying the intersection is *absurd*. It is very hard to prove that $x \in A \cap B$ implies $\bot$ for most ordinary cases. Intensionality is good for positive claims, and extensionality is good for negative ones. For "*A* is empty," it is better to write "$\neg \exists x (x \in A)$." There are empty places in the mathematical universe for which their occupancy would not be apocalyptic.

For any X, we have better lattice behavior with the special top and bottom elements than in the general case Section 5.1.4 (these all being extensional equivalences, not identity):

$$X \cap V \equiv X \tag{5.29}$$

$$X \cup V \equiv V \tag{5.30}$$

$$X \cup \varnothing \equiv X \tag{5.31}$$

$$X \cap \varnothing \equiv \varnothing \tag{5.32}$$

$$X \cup \overline{X} \equiv V. \tag{5.33}$$

To prove $X \cap V \equiv X$, for example, left to right is &-elimination, while right to left,

1	$\top \rightarrow x \in V$	(5.27)
2	$x \in X \rightarrow x \in X$	theorem
3	$x \in X \ \& \ \top \Rightarrow x \in X \ \& \ x \in V$	1, 2, factor
4	$\top \Rightarrow (x \in X \Rightarrow x \in X \ \& \ x \in V)$	3, residuation
5	$\top$	(5.26)
6	$X \sqsubseteq X \cap V$	4, 5, modus ponens

This is a special case of a property of $\top$, that theorems may be added "for free" to the consequent of an $\Rightarrow$ implication; cf. Section 4.3.2.2. Other derivations are left as an exercise.

In the case of the universe, it would be natural to expect relative complements $V \backslash X$ to coincide with unrestricted complements $\overline{X}$. And from (5.29), they do, at least extensionally:

$$V \cap \overline{X} \equiv \overline{X}.$$

The limiting case is

$$V \backslash V = V \cap \overline{V} = \varnothing$$

which we can even write with an $=$ sign, because $\bot$ is so explosive it intrinsically "means" everything.

With respect to relative complements, the empty set has the special property

$$\overline{\varnothing}^{\varnothing} = \varnothing \backslash \varnothing$$
$$= \varnothing \cap V$$
$$= \varnothing.$$

The empty set is a fixed point on relative complementation. It is not unique in this regard – just use comprehension to make some $A = A \backslash A$; but $x \in A$ is not explosive and so A is not $\varnothing$. But this relative complementation property is still unusual.

Before pressing on, enough language has now been defined to vindicate the basis of one of the main paradoxes, the Cantor–Frege paradox of the universal set from Section 1.1.2.2, although there is nothing as yet paradoxical about it here.

Theorem 9. *The universe is its own powerset:*

$$\mathscr{P}V \equiv V.$$

Proof There are two directions to establishing an (extensional) identity, but one of them is instantaneous, since $X \subseteq V$ for all X. In the other direction, $x \subseteq V$ for all x; so by weakening $x \in V \Rightarrow x \in \mathscr{P}V$. $\qquad\square$

5.2.2 The Routley Set, Zermelo's Axioms, and Mappings

5.2.2.1 Routley's Set

Now we officially introduce a set we encountered back in Section 3.1.3.2, perhaps the most distinctive object for the dialetheic paraconsistent approach to foundational mathematics.

Using full comprehension, behold the *Routley set*:[11]

$$\mathcal{Z} = \{x : x \notin \mathcal{Z}\}.$$

(The notation "$\mathcal{Z}$" is original from [Routley, 1977, p. 915].) This set has the beautiful property that

$$\mathcal{Z} = \overline{\mathcal{Z}}.$$

By the law of excluded middle, for any x, either $x \in \mathcal{Z}$ or not. But if it is, then it isn't, so it isn't; if it isn't, then it is, so it is. Thus

$$\forall x(x \in \mathcal{Z})$$

and

$$\neg\exists x(x \in \mathcal{Z}),$$

matching properties of both V and $\emptyset$. Luckily, $\mathcal{Z}$ isn't either of these, by dint of intensional identity, but it is a bit like a shadow cast behind (or between) them. Routley calls it

a completely bizarre set, everything belonging to it iff it does not, whereas very many sets do not belong to [the Russell set] without also belonging, namely all those sets that are straightforwardly non-self-membered such as the set of all integers, the set of all purple items, all concrete objects, etc. If the dialectician is going to tolerate some inconsistent sets isolated further inconsistent sets might as well be admitted as well, especially if there are reasons and advantages in doing so. In this case there are [Routley, 1977, p. 915].

---

[11] If we prefer to generate it without unrestricted comprehension but the fixed point Theorem 13 instead, we will only have the extensional $\mathcal{Z} \equiv \{x : x \notin \mathcal{Z}\}$, not an identity; but this makes no difference in what follows.

Some of these advantages, we will be considering. Note immediately the Routley set's "elasticity" – in terms of relative complements, $\varnothing_{\mathcal{Z}} = \mathcal{Z} \backslash \mathcal{Z}$ may be empty, but it is far from empty. It can be "countenanced" as having no members, and as many members as the universe.[12] As Routley puts it memorably (if a bit dubiously, in terms of Meinong scholarship), $\mathcal{Z}$ is out there, bizarre or not, "if sets are all out there, in *Aussersein*, as they seem to be" [Routley, 1977, p. 916].

The Routley set is terribly inconsistent. For any X, the intersection

$$\mathcal{Z}(X) := X \cap \mathcal{Z} = \{x : x \in X \ \& \ x \in \mathcal{Z}\}$$

gives a similar local effect, since one of the conjuncts is both true and false: for every $x \in X$, then

$$x \in \mathcal{Z}(X) \ \& \ x \notin \mathcal{Z}(X).$$

This is a dramatic way to achieve what can also be done by parameterizing any set by a true contradiction, e.g., for any set X, the subset $X \cap \{z : r \in r\}$ will also contain and not contain every member of X, with r the Russell set. And this means, as foreshadowed, that every single object has at least one inconsistent property, that of being in $\mathcal{Z}$ and not.

Routley set in view, with respect to our overarching question, "why are there paradoxes?," I suggest that the following simple fact is a profoundly important part of the answer.

Theorem 10. *Every nonempty set X has a(n inconsistent) proper subset $Y \subset X$ such that $Y \equiv X$.*

Proof Let $Y = \{x : x \in X \ \& \ x \in \mathcal{Z}\}$. Then $Y \subseteq X$ and $X \subseteq Y$. Since X is presumed nonempty, take $a \in X$. But $a \notin Y$ because $a \notin \mathcal{Z}$, so the inclusion is proper. $\qquad\square$

For every nonempty X, there is an inconsistent object $\mathcal{Z}(X)$, and $X \equiv \mathcal{Z}(X)$.

Now, note well that for arbitrary X, while $X \equiv \mathcal{Z}(X)$, and $\mathcal{Z}(X)$ is inconsistent, this does *not* mean that X itself is inconsistent, because $\equiv$ is not identity. The noncontraposable nature of $\Leftrightarrow$ keeps the contradictions from spreading immediately. Saying what Theorem 10 *does* mean will be the task ahead.

5.2.2.2 Zermelo's Axioms

Zermelo explicitly stated that his selection of axioms in 1908 was an attempt to pick instances of naive comprehension that seem safe [Zermelo, 1967, p. 200], until a more principled approach could be found. (Let that be a warning to any of us when we institute what we think will be "temporary" measures.) The fact that Zermelo's axioms are therefore theorems of naive set theory is not a surprise. The following are true, as instances of comprehension.

---

[12] As opposed to counting, in terms of cardinality. See [Varzi, 2014].

Pairs: $\forall x \forall y \exists z (z = \{x, y\})$
Separation: $\forall z \exists y (y = \{x : \varphi \; \& \; x \in z\})$
Union: $\forall x \exists y (y = \{z : \exists u (u \in x \; \& \; z \in u)\})$
Powerset: $\forall x \exists y (y = \{z : z \subseteq x\})$
Infinity: Two versions:

> (a) $\exists X (\exists y (y \in X) \; \& \; \forall z (z \in X \Rightarrow \{z\} \in X))$
> (b) $\exists X \exists Y (Y \subset X \; \& \; X = Y)$　　　　　(Dedekind infinite)

Version (a) of the axiom of infinity is proved by the existence of a set, call it i, such that

(i) $\varnothing \in i$ and
(ii) $\forall x (x \in i \Rightarrow \{x\} \in i)$,

that is, a set closed under zermelodic successor, $\{\varnothing, \{\varnothing\}, \{\{\varnothing\}\}, \ldots\}$. Theorem 9 shows that the universe V satisfies these requirements. For a smaller set, take the circularly defined

$$i = \{x : x \subseteq i\}.$$

(This exists either by "impredicative" comprehension or using the fixed point Theorem 13.) We have $\varnothing \in i$ by (5.21), and $x \in i \Rightarrow \{x\} \in i$ by (5.3) and the definition of i.

So we have several witnesses for $\exists X \forall x (x \in X \leftrightarrow x \subseteq X)$, which suffices for proving infinity version (a). (For yet another, see Section 6.1.) This is an instance of the classical *reflection principle*, that "it always seems a plausible step ... to take a property of the whole universe and postulate that it already holds at some [lower] level" [Drake, 1974, p. 124]; this is a theme of [Kanamori, 1994]. The universe is infinite if anything is; then some set reflected inside the universe is, too. Compare this proof to Dedekind's infamous theorem 66 from 1888 [Dedekind, 1901] – that an infinite set exists, because the set of all thoughts contains the thought of itself, and the thought of the thought of itself, and ... [13]

Classically, a set is (Dedekind) infinite iff it has a proper subset of the same size as itself (mentioned in the Introduction). Version (b) of the axiom of infinity is (more than) enough to show that some set is Dedekind infinite. The Russell set is like this: $r = r \; \& \; r \subset r$. In a sense, though, we've already seen that *every* nonempty set has the "Dedekind" property, because of Routley's set, Theorem 10.

To be clear, I don't think – and the theory in this book does not say – that every set is infinite. In light of Routley's set, Dedekind's definition of the infinite is not appropriate here.[14] But every set has, perhaps, an infinite shadow cast over it, or casts an infinite shadow. The fact that either of the two standard glosses on infinity – either being isomorphic to the natural numbers, or being Dedekind infinite – can both be almost trivially (but nontrivially!) obtained from inconsistency is striking. It suggests (and I will suggest again later) that perhaps the Cantorian transfinite is a consistentizing euphemism for more

---

[13] Priest points out that this much maligned argument can be reconstructed as a reasonable proof that the set of all singletons is equipollent to, yet properly included in, the set of all sets. See Priest [2006b, p. 142].
[14] Even classically, whether or not the Dedekind definition of infinity links up with other definitions (e.g., whether a set that is not Dedekind infinite is finite) will vary depending on the presence of the axiom of choice [Jech, 1978, 1§3, p. 28].

fundamental inconsistency; or put the other way, perhaps contradictions touch the infinite. But rather than gaze too long out into the *paradoxien des unendlichen*, into the *Abgrund*, there are more ordinary things for us to attend to here on the ground.

5.2.2.3 Relations and Mappings

Ordered pairs are defined, following Kuratowski, as

$$\langle x, y\rangle := \{\{x\}, \{x, y\}\}.$$

Note the special case $\langle x, x\rangle = \{u : u = \{x\} \vee u \equiv \{x\}\}$ (cf. Section 7.3.3). The purpose of the definition is to deliver the following:

Proposition 11.   *[Law of ordered pairs]* $\langle x, y\rangle = \langle u, v\rangle \Leftrightarrow (x = u)\ \&\ (y = v)$.

Proof   The proof is unproblematic because identity contracts. Begin by observing that

$$\{x\} = \{y\} \Leftrightarrow x = y.$$

Right to left is substitution: if $x = y$, then since $\{x\} = \{x\}$, also $\{x\} = \{y\}$. Left to right is also substitution: if $\{x\} = \{y\}$, then since $x \in \{x\}$, also $x \in \{y\}$. Similarly,

$$\{x, y\} = \{u, v\} \Leftrightarrow (x = u\ \&\ y = v) \vee (x = v\ \&\ y = u).$$

Then using the definition of ordered pairs,

$$\langle x, y\rangle = \langle u, v\rangle \Leftrightarrow \{\{x\}, \{x, y\}\} = \{\{u\}, \{u, v\}\} \Leftrightarrow x = u\ \&\ y = v$$

by a lengthy but straightforward exercise in expanding definitions, just as in [Cantini, 2003, p. 356]; cf. Weber [2010d, p. 81]. $\square$

Instances of comprehension on ordered tuples, of the form

$$\langle x, y\rangle \in \{z : \varphi(z)\} \leftrightarrow \varphi(\langle x, y\rangle),$$

are abbreviations for $u \in \{z : \varphi(z)\} \leftrightarrow \exists x \exists y (u = \langle x, y\rangle\ \&\ \varphi(u))$.[15]
    A *relation* is a set of ordered tuples. The *Cartesian product*

$$X \times Y := \{\langle x, y\rangle : x \in X\ \&\ y \in Y\}$$

is the largest binary relation over itself; but any $Z \sqsubseteq X \times Y$ is a relation, too, relating $x \in X$ to $y \in Y$. When Z is such a relation, this can be expressed in the notation

$$Z : X \longrightarrow Y,$$

read "Z is a relation from X to Y." In such cases, X is called the *domain* of Z and Y is called the *range*.
    When $Z : X \longrightarrow Y$, then the *restriction* of Z to $A \sqsubseteq X$ is

$$Z|A :=: Z_{\restriction A} := Z \cap (A \times Y).$$

---

[15]   Thereby not following Brady in assuming such instances as primitive [Brady, 2006, pp.178, 251].

The *image* of A under Z is

$$Z''(A) := \{y : \exists x(\langle x, y \rangle \in Z \ \& \ x \in A)\}$$

using notation due to Whitehead. The two apostrophes are sometimes omitted, just writing $Z(A)$ for the image (or, later, preimage), when it is clear from context.

Images respect the subset order: where $Z : X \longrightarrow Y$ is a relation, with A, B subsets of X,

$$A \sqsubseteq B \Rightarrow Z''(A) \sqsubseteq Z''(B). \tag{5.34}$$

For let $\langle a, c \rangle \in Z$ for $a \in A$ and $c \in Y$. Suppose $c \in Z''(A)$. Supposing $A \sqsubseteq B$, then $a \in B$, too, and $c \in Z''(B)$. This produces the almost-substitution-like fact

$$A \equiv B \Rightarrow Z''(A) \equiv Z''(B)$$

using factor.

If a relation is also univocal, where any input has exactly one output, it is a *function*. More precisely, $f : X \longrightarrow Y$ is a (total) *function* from X to Y iff

- $f \sqsubseteq X \times Y$
- $\forall x(x \in X \Rightarrow \exists y(y \in Y \ \& \ \langle x, y \rangle \in f))$
- $\forall u \forall v \forall x \forall y(\langle u, x \rangle \in f \ \& \ \langle v, y \rangle \in f \Rightarrow (u = v \Rightarrow x = y))$

This phrasing of "function" does not exhaust the possibilities. There is a weaker sense of function where $=$ is replaced by $\equiv$, meaning the output is unique, *up to deductive equivalence*. And there is the notion of a *relevant* function, with $\rightarrow$ in the place of $\Rightarrow$, for which many fewer relations would turn out to be functions (cf. Restall [1994, p. 210]).

Functions are everywhere in classical mathematics. For applications in paraconsistent mathematics, though, functions are often not the right tool. For example, thinking about the treatment of definite descriptions in Chapter 1, there are expected cases where "*the* first n such that ..." may be multiply realized. Comprehension gives us so much existence, so to speak, that there is less uniqueness. To be a bit impressionistic about it, paraconsistent mathematics – with its attention on inconsistent objects – is open to ways that things like the first n can be *two places at the same time*. If that is what operations are going to be flexible enough to express, then the third requirement on functions stated earlier, uniqueness, is in some cases too much. It would force the "two" places at the same time also to be the *same* place.[16]

We wish to strike a balance. On the one hand, it is in the spirit of the enterprise to allow the possibility of "bad" functional behavior, where some "functions" can take an input to two distinct outputs – the paradigm case being an evaluation on a proposition that says the proposition *is* true and *is* false, and that either of these values is *the* value of the proposition

---

[16] Along these lines, *if* function symbols are added to the language via skolemization, and f is a function, one can write $\langle x, y \rangle \in f$ in the infix way, $f(x) = y$; but the issue is fraught [Øgaard, 2017]. "Handling functions in a logic that admits contradictions is a sensitive matter" [Priest, 2006b, p. 288].

(Section 3.3.2). On the other hand, "being in two places at the same time" cannot entail that the two places are numerically identical; if (distinct) outputs such as 0 and 1 are identified, the result would be absurdity. We would like to pursue the idea that relations can in many ways *approximate* functions as closely as we would like, and see how far we get.

To meet all these competing demands, here and throughout, unless explicitly stated otherwise, the term *mapping* will be used for this concept:

Definition 2.   *A relation f is a* total relational mapping, *or* mapping, *from X to Y iff*

- *f is a relation on $X \times Y$.*
- $\forall x(x \in X \Rightarrow \exists y(y \in Y \ \& \ \langle x, y \rangle \in f))$.
- $\langle x, y \rangle \in f \Rightarrow (z \neq y \Rightarrow \langle x, z \rangle \notin f)$.

We write

$$f(a) := \{b : \langle a, b \rangle \in f\}$$

for the image of $\{a\}$ under f, or $f''\{a\} = \{x : \exists y(y \in \{a\} \ \& \ \langle y, x \rangle \in f))\}$, the *set* of outputs delivered by f upon input a. And so we obtain

$$b \in f(a) \leftrightarrow \langle a, b \rangle \in f.$$

And $b \notin f(a)$ iff $\langle a, b \rangle \notin f$. Sometimes we use the "$\ni$" notation from Dedekind, $f(a) \ni b$ and $f(a) \not\ni b$. Note well, though, that even if f relates a to only one object, this only means that $f(a) = \{b\}$, *not* the "naked" $f(a) = b$ that would require function symbols.

A heuristic suggestion: $f(a)$ is a collection of objects, but sometimes we will want to think about relations more like "functions that have more than one output," rather than having a static set of outputs. So when possible we try to think of the notation $f(a)$ as indefinite, representing *an* output of a. To see how to do this, notice that the third clause in the definition of mapping enforces some "exclusiveness" without too much. For example, given two objects 0 and 1 such that $0 \neq 1$, and a mapping $v : X \longrightarrow \{0, 1\}$ from some set of objects X to $\{0, 1\}$, then

$$1 \in v(p) \Leftrightarrow 0 \notin v(p), \qquad 0 \in v(p) \Leftrightarrow 1 \notin v(p)$$

for every $p \in X$, even if also

$$1 \in v(q) \ni 0$$

for some $q \in X$. (In that case, $1 \in v(q)$ and $1 \notin v(p)$ and $0 \in v(p)$ and $0 \notin v(p)$.) We may say, truthfully, that even when a mapping takes objects to more than one place, once an object is mapped to one place, it does not go anywhere else.[17] It follows that if f is a mapping, then

---

[17] So if any element x of the domain of a mapping f so defined is mapped to an object y such that $y \neq y$, then f itself will be an inconsistent set: $\langle x, y \rangle \in f$ and $\langle x, y \rangle \notin f$. We could very crudely read this as saying that if a map takes points to nowhere, then the map does not exist, or more conservatively (but misleadingly) say that there are no maps to nowhere.

$$\langle x, y \rangle \in f \Rightarrow (z = y \vee \langle x, z \rangle \notin f)$$
$$\Leftrightarrow (\langle x, z \rangle \in f \supset z = y),$$

so the definition is stronger than, but close to, that with a material conditional. In any case, it is worth emphasizing that our definition of a mapping is *classically equivalent to the classical definition* of function.

Define the *composition* of relations f, g as usual:

$$f \circ g := \{\langle x, y \rangle : \exists z (\langle x, z \rangle \in g \,\&\, \langle z, y \rangle \in f)\}.$$

The composition of f on g is empty unless the range of g is a subset of the domain of f. For a relation, the notation $f(g(a))$ stands for an f output of an g output of a.

Two specific types of functions get names that will be used occasionally: using function symbols, for $f : X \longrightarrow Y$,

- f is an *injection* or *into* iff $f(x) = f(y) \Rightarrow x = y$ for all $x, y \in X$.
- f is a *surjection* or *onto* iff for all $y \in Y$ there is some $x \in X$ where $f(x) = y$.

A surjective relation would be the same as the preceding except $\forall x \exists y y \in f(x)$.

5.2.3 Fixed Point Theorems

The *generalized* notion of a fixed point on a mapping f will be any z such that

$$z \in f(z).$$

Input z into f, and z is one resulting output; a relational fixed point is *an* output of itself.

5.2.3.1 Knaster's Theorem

To begin on a conventional note, here is a standard fixed point theorem that in some ways prefigures later investigations into continuous mappings. As this is a more conventional theorem, the fixed point shown is not in the general sense (in terms of $\in$), but in terms of the extensional equivalence $\equiv$.

A relation f from X to Y is *monotone* iff it preserves order:[18]

$$u \sqsubseteq v \Rightarrow f(u) \sqsubseteq f(v).$$

Note that in the proof the assumption that f is monotone is used twice, so it is assumed twice.

Theorem 12 (Knaster 1928). *Let $f : \mathscr{P}(X) \longrightarrow \mathscr{P}(X)$ be a mapping that is monotone, and monotone. Then*

$$f(z) \equiv z$$

for some $z \sqsubseteq X$.

---

[18] This notion can be generalized to any partial order, not just $\sqsubseteq$. For fixed points in order theory, see [Schröder, 2003].

Proof They key to this, as in many similar situations, is finding the right object to be the fixed point; then the argument is simple. Let

$$\Delta := \bigcup \{x : x \sqsubseteq X \ \& \ x \sqsubseteq f(x)\}$$
$$= \{y : \exists x (x \sqsubseteq X \ \& \ x \sqsubseteq f(x) \ \& \ y \in x)\}$$

be the union of all subsets of X that are included in their own mapping. The proof proceeds by making three observations.[19]

Observation one: if $x \sqsubseteq X$ and $x \sqsubseteq f(x)$, then $x \sqsubseteq \Delta$. That's because, if $y \in x$, then along with our other suppositions, all the conditions are satisfied for $y \in \Delta$. Observation two: $\Delta \sqsubseteq X$. That's because, if $y \in \Delta$, then $y \in x \ \& \ x \sqsubseteq X$ for some x, which in turn implies that $y \in X$. Observation three: $\Delta \sqsubseteq f(\Delta)$. That's because, if $y \in \Delta$, then $y \in f(x)$ for some $x \sqsubseteq \Delta$; but if $x \sqsubseteq \Delta$ then by observation two we can apply monotonicity and get $f(x) \sqsubseteq f(\Delta)$. So $y \in f(\Delta)$.

By definition, $f(\Delta) \sqsubseteq X$, and by observation three and monotonicity (again),

$$f(\Delta) \sqsubseteq f(f(\Delta)).$$

Therefore, from observation one,

$$f(\Delta) \sqsubseteq \Delta.$$

This with observation three (which we proved, so can use again) delivers

$$f(\Delta) \equiv \Delta$$

as required.   □

5.2.3.2 The Fixed Point Theorem

This next result is very important. It is also very basic, the set theoretic expression of the fixed point theorem for the untyped lambda calculus from the Introduction. It arrives this far into the chapter only because it is more powerfully stated using ordered pairs and mappings; otherwise, it could have been stated almost immediately after the axiom of comprehension (even without assuming "impredicative" instances of comprehension, cf. Section 1.1.2).

Theorem 13 (Fixed Point Theorem). *For any formula φ, there is a $\mathfrak{t}$ such that*

$$x \in \mathfrak{t} \Leftrightarrow \varphi(\mathfrak{t}).$$

Proof We just need to find the right object. Let

$$t(z) := \{x : \langle x, z \rangle \in z\}.$$

This exists by comprehension, one for each z. Where z is a mapping, then $t(z)$ is the part of the universe that z takes to z itself. Now, again by comprehension there is a set

$$s := \{\langle y, z \rangle : \varphi(\{x : \langle x, z \rangle \in z\})\}.$$

---

[19] Following [Smullyan and Fitting, 1996, p. 109].

This s is a map that takes anything in the universe to z exactly when $t(z)$ has property φ, that is,

$$\langle y, z \rangle \in s \leftrightarrow \varphi(t(z)).$$

Writing this without abbreviations,

$$u \in s \leftrightarrow \exists y \exists z (u = \langle y, z \rangle \,\&\, \varphi(t(z))).$$

It is now a matter of considering $t(s)$, and showing it is a fixed point.[20] There are two directions.

Since $x \in t(s) \rightarrow \langle x, s \rangle \in s$ by definition of $t(s)$, then by definition of s and transitivity,

$$x \in t(s) \rightarrow \exists y \exists z (\langle x, s \rangle = \langle y, z \rangle \,\&\, \varphi(t(z))).$$

If $\exists y \exists z (\langle x, s \rangle = \langle y, z \rangle \,\&\, \varphi(t(z)))$, then by existential instantiation and the law of ordered pairs, $s = z \,\&\, \varphi(t(z))$, whence by substitution, $\varphi(t(s))$, so

$$x \in t(s) \Rightarrow \varphi(t(s)).$$

In the other direction,

$$\varphi(t(s)) \Rightarrow \langle x, s \rangle = \langle x, s \rangle \,\&\, \varphi(t(s))$$

because $\top \rightarrow \langle x, s \rangle = \langle x, s \rangle$, and $\varphi(t(s)) \rightarrow \varphi(t(s))$, so use factor and the truth of $\top$. Then by existential generalization,

$$\varphi(t(s)) \Rightarrow \exists y \exists z (\langle x, s \rangle = \langle y, z \rangle \,\&\, \varphi(t(z))).$$

So $\varphi(t(s)) \Rightarrow \langle x, s \rangle \in s$, and since $\langle x, s \rangle \in s \rightarrow x \in t(s)$, we have

$$\varphi(t(s)) \Rightarrow x \in t(s)).$$

All up,

$$x \in t(s) \Leftrightarrow \varphi(t(s)),$$

so $t(s) =: \mathfrak{t}$ is what was sought, which completes the proof. $\square$

Corollary 14 (Diagonal Lemma). *For any formula $\varphi(x)$ with one free variable, there is a sentence g such that*

$$g \Leftrightarrow \varphi(\ulcorner g \urcorner).$$

Proof This is as forecast in Section 1.1.2.1. By the fixed point theorem, $b \equiv \{x : \varphi(\{z : a \in b\})\}$ exists, for some set a. Then $a \in b \Leftrightarrow \varphi(\{z : a \in b\})$. Availing ourselves of the naming device $\ulcorner \psi \urcorner := \{x : \psi\}$ for any ψ, then $\ulcorner a \in b \urcorner$ is the term $\{x : a \in b\}$. Then if g is the sentence "$a \in b$," this is the required fixed point. $\square$

---

[20] Closely following [Shirahata, 1999]. See also [Petersen, 2000, p. 382; Cantini, 2003, p. 357; Terui, 2004, p. 15]. The fixed point theorem for linear set theory comes from untyped lambda calculus [Girard, 1998, p. 30].

One interpretation of this theorem is straightforward: naive set theory includes a truth predicate $T(x)$, and so a fixed point for the predicate $\neg T(x)$, a sentence that says of itself that it is a lie. But matters are not so simple (!). This theorem is at the edge of an abyss. Since Theorem 13 applies to every formula, including those defining relations (as in Section 5.2.2.3), there would seem to be a striking consequence: for any function, there is a fixed point. Prima facie this looks perilously close to proving absurdity, as with the "blip" function in Section 0.3. Take an operation that, for two putatively distinct objects t and f, is intended to take anything that is f to t, and anything that is *not* f to f. There is a fixed point on this operation. How is t = f, implosion, averted?

We have some good *reassurance* that Theorem 13 is not absurd. In light affine set theory in BCK, this theorem is *consistent* (Section 4.2.3.1). In these and other noncontractive approaches, excluded middle (effectively) fails, so "the fixed point cannot be shown to be empty, although nothing can be established to fall under it either" [Petersen, 2000, p. 383]. But in the paraconsistent tradition, we have excluded middle, no gaps. And yet the fixed point theorem is an easy result in Routley–Brady set theory; they even purposely *impose* fixed points [Brady and Routley, 1989, p. 419]. Brady's constructions since the late 1970s have shown repeatedly that this set theory is not trivial, in the sense that there is a model of set theory in DKQ in which at least one sentence (e.g., $\forall x \forall y (x \in y)$) is not satisfied. We are assured (up to the consistency of classical metatheory!) that the fixed point theorem is not catastrophic for such theories. But consistency/nontriviality or no, how could it be that there is a fixed point on every mapping? Priest calls the fixed point theorem "the hardest problem" [Priest, 2005, ch. 5]. This theorem is a *question*. There are at least two prongs to the question: how is coherence maintained, and then, what is the (coherent) situation the fixed point theorem is describing?

For the worry that any two objects are collapsed together, we have indirectly already prepared for this in the long discussion of extensionality in Chapter 4. The fixed point in Theorem 13 is an extensional equivalence, proved in terms of $\Leftrightarrow$ or $\equiv$, not identity. Extensionality is not good for sameness. Specifically, relating t to $\varphi(t)$ goes through an existential quantifier elimination, which (by design) is not an intensional ($\rightarrow$) principle of subDLQ. If the fixed point entities in question here aren't identical, then worries about how, e.g., a number might be "identical" to its own successor are premature.[21]

---

[21] In the proof that $x \in t(s) \Rightarrow \varphi(t(s))$, substitution can be avoided, but not the existential [Istre, 2017, p. 81]:

1	$\varphi(t(s)) \rightarrow \varphi(t(s))$	
2	$\langle x,s \rangle = \langle x,s \rangle \ \& \ \varphi(t(s)) \rightarrow \varphi(t(s))$	(1, antecedent strengthening)
3	$\exists y \exists z (\langle x,s \rangle = \langle y,z \rangle \ \& \ \varphi(t(z))) \Rightarrow \varphi(t(s))$	(2, $\exists$)
4	$\langle x,s \rangle \in s \Rightarrow \varphi(t(s))$	(def of s)
5	$x \in t(s) \Rightarrow \varphi(t(s))$	(def of $t(s)$)

See Øgaard [2016, appendix B]. Following Petersen, Priest relates this problem to a paradox of Hilbert and Bernays from *Grundlagen der Mathematik* 1934 [Priest, 1997b]. Priest's diagnosis is that fixed point terms have *more than one* denotation (dual to a more conservative diagnosis of denotation *failure*); as elsewhere, he recommends giving up the transitivity (and substitutivity) of identity.

Now, if we bypass the derivation of Theorem 13 and *assume* Routley–Brady's unrestricted comprehension axiom, where for any f at all there is *instantaneously* a set t such that

$$z \in t \leftrightarrow z \in f(t),$$

in terms of intensional $\leftrightarrow$, then $t = f(t)$ appears to be an identical fixed point for arbitrary f. It looks like, if f takes values from $\{0, 1\}$, that $t = 0$ or $t = 1$. But look again. As flagged earlier, when we talk about "functions," we are working with the images of singletons, $f(x) := \{y : \langle x, y \rangle \in f\}$, so the fixed point given by comprehension, once we write it out, is just $t = \{y : \langle t, y \rangle \in f\}$. And this set may simply be *empty*; nothing suggests that $t \in f''\{t\}$, and nothing impels us to say that $\langle t, t \rangle \in f$. Even if we write out the extension of, e.g., "negation,"

$$f_\neg = \{\langle 0, 1 \rangle, \langle 1, 0 \rangle\}$$

and show that each input has only one output, and accept that $f_\neg$ has a "fixed point" $t = f_\neg(t)$, still we are *not* forced to say that either $f_\neg(t) = 0$ or $f_\neg(t) = 1$.

The problem here is, abstractly, the original problem: the attempt to sort the universe into two categories as classically understood, e.g., as with the aforementioned blip function, the thing that is f, and everything else. The operation has a fixed point, showing it *cannot* be an exclusive-and-exhaustive-on-pain-of-triviality function. The orthodox response is incompleteness, giving up exhaustion; e.g., in computability theory, the fixed point theorem – in the form of Kleene's second recursion theorem – is deemed not a problem because there are *partial* computable functions, which are "undefined" on some (problematic) inputs.[22] Contrawise, the response here is to maintain exhaustion, but acknowledge that fixed points are a failure of exclusivity. The blip function, then, like all functions, is a *relation*, as are negation and the other connectives (Section 3.3.2). Such a relation may even be univocal (for each x, there is exactly one y such that $f(x) = \{y\}$) but it does not work as cleanly as classically imagined.[23] And this, as is becoming increasingly familiar (Section 1.2.3), is expressed in a failure of *uniqueness*, where extensionally equivalent terms are not identical.

As for the second prong of the question: *why* is there always an overlap in an otherwise well-intentioned division of the universe? This is the fundamental driving question of the whole project, but here is the start of an answer.

Set-defining formulae draw from the full universe of sets. The fixed point theorem says that *something* is a "diagonal" object on the formula defining a relation. But as we just intimated, fixed point t *is not necessarily anything from the domain or range of a given function*. For example, t is clearly not a natural number. By (close) comparison, in the untyped lambda calculus, numbers can be encoded, and a successor function is well behaved on those numbers – but not on objects that are not numbers; such fixed points

---

[22] But see [Sylvan and Copeland, 2000].
[23] Cf. 'boolean negation' does not exist (as classically imagined) [Priest, 1990]; see Priest [2006b, p. 288].

are considered to be syntactic debris.[24] As Dedekind worried about "alien intruders" in arithmetic,[25] so the fixed point may be some (inert) object from elsewhere in the universe.[26] "That effective formal systems are unable to unmask such impostors," write Meyer and Mortensen, "is at once their shame and glory" [Meyer and Mortensen, 1987, p. 6].

How do operations "pick up" objects that are not in either their domain or range? Naive set theory, and naive closed theories more broadly, aim for *universality*. They are designed to be general. Naive set theory in paraconsistent logic is expressive as no other theory can be: when we say $\forall x \varphi(s)$, we really mean *for all x*, quantifying over *everything*. And this level of generality makes *non*generality a problem (Section 4.3.2.3). According to Theorem 13, the naive set universe V has an extreme version of the "fixed point property" – *every map from V to V has a fixed point*. But *in an untyped universe, every map whatsoever is, in some sense, a map from the universe to itself.* For restricted domains of V, while every map from, say, the real numbers $\mathbb{R}$ to $\mathbb{R}$ does have a fixed point z, it need not be that $z \in \mathbb{R}$. It may be something else. So the source of fixed points, and the source of paradox, may be this: every relation is a relation over the whole universe. (See one of the last results in the book, Theorem 66 in Chapter 9.)

Putting these two lines of thought together, the fixed point result shows that adding function symbols ("skolemization") cannot be done uniformly for all univocal relations. Classical functions are too classical. I try as far as possible to work with relations rather than functions – including relations that have a unique output. When results do rely on special function symbols (particularly, addition and multiplication functions on numbers), we think of these as unusual instances where a specific function symbol has been tentatively inserted "by hand," and where the results are taken as conditional on the coherence of adding these individual symbols.

That's still not a whole answer, of course. This is the hard problem for a glutty would-be universal theory. This is a question that will recur. My ultimate response is encapsulated in the final proofs (Lemma 69 and Theorem 70) and in Chapter 10. Sticking with the mathematics for now, let's see one of the fixed point theorem's most famous and transfixing applications.

5.2.3.3 Cantor's Theorem

You know the story: on December 7, 1873, Cantor discovered set theory when he saw that there are more real numbers than natural numbers, opening the way to the realm of transfinite numbers, a place Hilbert called a paradise. But there was a serpent in the garden . . .

---

[24] As in Scott's model, exposited in, e.g., Hindley and Seldin [2008, ch. 16]. Terui shows that "t is a number" is provable if and only if t is a numeral [Terui, 2004, lemma 2.12].

[25] In his "Letter to Keferstein" [van Heijenoort, 1967, p. 100].

[26] Istre puts it insightfully: "A useful way to think of this [fixed point] theorem is as finding sets that are stable after the application of some defined operation.... There is some set in the universe that remains unchanged after applying [the operation].... Note that this does not necessarily say that the operation has no effect on the elements of the fixpoint term. It only implies that we get back the same set after applying the operation. So, for example, if we took a successor operation, there would be some subset of the universe which was not altered, but it is not necessarily implied that it means that there must be some element which is its own successor, i.e. a number not 'affected' by the successor operation, nor does it imply those elements which behave oddly with regard to the successor function are actually interpretable as natural numbers" [Istre, 2017, p. 83].

Here we show how the serpent turns back on itself, ouroboros. This study is not concerned with a theory of cardinality (partly for reasons indicated later in the Excursus), but look how Cantor's celebrated theorem – that there is no way to pair off every element of a set one-to-one with a member of its powerset – is proved using a fixed point. (For effect, I will use the Routley set, but any number of paradoxical sets would do.)

Theorem 15 (Cantor). *There is no surjective $f : X \longrightarrow \mathscr{P}(X)$.*

Proof Note that the theorem is true automatically if X has no members, so we assume X is not empty. We show two proofs, depending on whether surjective f is understood to be a full-fledged function as is traditional, or whether it is softened to a mapping. Either way the result carries through.

Proof 1: This is supposing the hypothetical surjection in question is a function, with a symbol added to the language such that $f(x) = y$ iff $\langle x, y \rangle \in f$. We show in fact that for *every* function $f : X \longrightarrow \mathscr{P}(X)$ that there is some $y \subseteq X$ with $f(x) \neq y$, for every $x \in X$.

For any X, the set $\mathcal{Z}(X) = \{x : x \in X \,\&\, x \notin \mathcal{Z}(X)\}$ is a subset of X. (It is a subset in the sense of $\subseteq$ (with the $\rightarrow$ arrow), and therefore also in the sense of $\subseteq$.) For every $x \in X$, both $x \in \mathcal{Z}(X)$ and $x \notin \mathcal{Z}(X)$ (Section 5.2.2.1). Thus if X is not empty, then

$$\mathcal{Z}(X) \neq \mathcal{Z}(X)$$

Then everything is nonidentical to $\mathcal{Z}(X)$ (Lemma 7). In particular, for any $f : X \longrightarrow \mathscr{P}(X)$,

$$f(x) \neq \mathcal{Z}(X)$$

for every $x \in X$. A fortiori, for all f, and so all surjective f, there is some $y \in \mathscr{P}(X)$ such that for no x is it the case that $f(x) = y$.

Proof 2: Let f be a mapping, as per Definition 2. Then argue just as in Proof 1, so that $\mathcal{Z}(X) \in \mathscr{P}(X)$ and $\mathcal{Z}(X) \neq \mathcal{Z}(X)$. Suppose that $x \in X$ and $\mathcal{Z}(X) \in f(x)$. But then, since mappings are "exclusive," anything that is *not* $\mathcal{Z}(X)$ is *not* in the image of $\{x\}$; and since $\mathcal{Z}(X)$ is not even itself, it is not in the image of x under f. As x was arbitrary, there will always be something left out by any mapping from a set to its powerset.[27]

□

For every pairing between X and $\mathscr{P}(X)$, there is something out of range. Juxtapose this with Theorem 9, which says that, for the universe V, extensional *identity* would be a pairing-off with $\mathscr{P}(V)$ that has nothing out of range – Cantor's paradox. Welcome to our inconsistent universe.

---

[27] Another way to argue the same point – that Cantor's theorem would still hold without function symbols (so if a relation f has a unique output, $f(x) = \{y\}$ for some y) – would be again to note that $\mathcal{Z}(X) \neq \mathcal{Z}(X)$, and so we still have $f(x) \neq \{\mathcal{Z}(X)\}$, since non-self-identical objects have non-self-identical singletons (Section 5.1.2).

In a paraconsistent set theory, the familiar fable of Cantorian paradise lost has a new moral. In the orthodox proof of the existence of transfinite cardinals, and especially the more general version of the proof that we just saw in abstract, were the ingredients for a paradox, as Cantor knew[28] and Russell announced. Interpreting the theorem as an indicator of a vast transfinite tower of different infinities is strongly reminiscent of Tarski's infinite hierarchy of metalanguages, trying to outrun the liar. It is the hierarchical example par excellence of "latterday Kantian attempts to retain a certain control over conceptual production" [Priest, 2006b, p. 48]. In the naive setting, the theorem turns out to be an almost trivial observation about subsets. The standard interpretation is not wrong – all the refracted transfinite cardinalities *are* there, and they are all different from one another – but that is an incomplete account of the situation; cf. [Priest, 2017a].

An inconsistency is classically perceived as a never-ending chase for the horizon. And it *is* that, but not *only* that; it is also just an inconsistency.[29] Transfinite cardinals may be an expression of consistency strength in the classical setting [Kanamori, 1994]; but since the defining property of the Routley set $\mathcal{Z}$ applies to *any* set, however small, the paradoxes really have nothing to do with size.

A General Policy

Going forward, we will try to find fixed points "organically" rather than grafting in synthetic ones using Theorem 13. That is, while unrestricted set comprehension proves the existence of fixed points rather easily, the resulting objects are often artificial or otherwise unilluminating – conjuries of syntax. It is preferable to do as we did earlier in Knaster's theorem, to show the existence of a fixed point through a meaningful argument based on the subject matter, not imported deus ex machina style, through a portal from another dimension – though when necessary, we do that, too. This is rather in the way that many mathematicians prefer a constructive or direct proof to an indirect one, but accept unreservedly an indirect one.

5.3 Order

We've covered a lot of ground: enough now to carve out a more substantial mathematical theory within set theory, and with it, a substantial paradox.

The mathematical universe has a spine of *ordinal numbers*. These form a transfinite index set, to keep track of everything else. So the ordinals are a practical essential for mathematical developments. And the ordinals are inherently beautiful. Once set theory has provided a language for expressing mathematical concepts, it needs to confirm the existence of the ordinals. It was the purpose of the previous sections to do the former; it is the purpose of this section to do the latter, and with it, prove the Burali-Forti paradox Section 5.1.2.5: that the set of all ordinals is itself an ordinal.

---

[28] Arguably, given his use of "inconsistent multiplicities" in 1899 [Cantor, 1967]; see Myhill [1984, p. 130]; Lavine [1994].

[29] "Is it you, / or is that a fault-line / on the horizon / a second sky?" (Matthew Francis, *Ocean*).

5.3.1 Well-foundedness

A *well-founded set* is one where all nonempty parts have ∈-least members. The collection of all well-founded sets is

$$X \in \mathfrak{M} \leftrightarrow \forall Y (Y \sqsubseteq X \ \& \ \exists z (z \in Y) \Rightarrow \exists z (z \in Y \ \& \ \neg \exists u (u \in z \cap Y))).$$

We say that z is ∈-least in Y if $\neg \exists u (u \in z \cap Y)$ rather than $z \cap y = \varnothing$, for reasons discussed earlier and in Chapter 1.

Proposition 16. *If Y is any nonempty subset of a well-founded set X, then Y is well founded.*

Proof Let $Y \sqsubseteq X \ \& \ \exists z (z \in Y)$. Any $U \sqsubseteq Y$ is also a subset of well-founded X by transitivity of ⊑s, so if U has a member, it has a least member. □

Lemma 17. *(i) A well-founded set X is not self-membered, or else not self-identical:*

$$X \in \mathfrak{M} \Rightarrow X \notin X \vee X \neq X.$$

(ii) If a set is self-membered, then it is not well founded:

$$X \in X \ \& \ X \in X \Rightarrow X \notin \mathfrak{M}.$$

Proof In this lemma, (i) is equivalent to the classical statement modulo the material conditional; (ii) is equivalent modulo contraction.

For (i), let $X \in \mathfrak{M}$. If $X \notin X$, then $X \notin X \vee X \neq X$. If on the other hand $X \in X$, then $\{X\} \sqsubseteq X$ is a nonempty subset of X (5.3). Since X is well founded,

$$\exists z (z \in \{X\} \ \& \ \forall u (u \notin z \vee u \notin \{X\})).$$

That is, there is some z such that $z = X$ and every u is either not in z or not identical to X. In particular, $X \notin X \vee X \neq X$.

For (ii), suppose $X \in X$. Then $X \in X \ \& \ X \in \{X\}$, and then $\exists u (u \in X \ \& \ u = X)$ by existential generalization. So $\neg \forall u (u \notin X \vee u \neq X)$ by duality. But if $X \in X$ (again), then $\{X\} \sqsubseteq X$ is a nonempty part of X. Hence ex hypothesi, we have a counterexample to the well-foundedness of $\{X\}$,

$$\exists z (z \in \{X\} \ \& \ \{X\} \sqsubseteq X) \text{ but } \neg \exists z (z \in \{X\} \ \& \ \forall u (u \notin z \vee u \notin \{X\}))$$

and so

$$\neg \forall Y (\exists z (z \in Y \ \& \ Y \sqsubseteq X) \Rightarrow \exists z (z \in Y \ \& \ \forall u (u \notin z \vee u \notin Y)))$$

by the logical counterexample axiom, as required. □

From Lemma 17(ii), by reductio (Section 4.3.2.1), we have half of Mirimanoff's paradox (Section 1.1.2.4):

Proposition 18 (Mirimanoff). $\mathfrak{M} \notin \mathfrak{M}.$

5.3.2 Ordinals

Ordinals are the order-types of well-ordered sets. They form a (transfinite) set of indices that starts at zero and proceeds in a straight line absolutely forever. Cantor held two generating principles: that for any ordinal, there is a next one; and for any sequence of ordinals (like $0, 1, 2, \ldots$), there is always a least ordinal greater than all of them (e.g., ω is next). Any transfinite process can be *tracked* by the ordinals because any nonempty subset of ordinals has a *least* member. There is always a place to begin (or fall back to, depending on which direction you are coming from). Intuitively, a paradox comes about just because taking the *entire* sequence of ordinals, there must be a least ordinal greater than them all, and yet this ordinal would itself be in the sequence, and so not the last. The ordinals On are, *per impossibile*, the *longest line*.[30]

The work in producing a theory of ordinal numbers, then, is giving the right definition of them – one suitable for the logic – that will make $\in$ a transitive well-ordering. The idea, due to von Neumann and others, is to define the *inner* structure of an ordinal α as ordered by membership so as to be the following:

- Strict (no loops!): for all $x \in \alpha$, x is not self-membered; and for all $y \in \alpha$, if x is a member of y, then y is not a member of x
- Transitive: members of members of α are members of α.
- Linear (no branches!): any other ordinal is either a part of α or has α as a part.
- Well-founded: any nonempty subset of α has a least member.

A set is *well-ordered* when it is in a well-founded strict linear order. What we discover is that the *collection* of all well-ordered ordinals is itself well-ordered, self-similarity upward, so that the inner structure becomes outer structure too. This may be achieved for paraconsistent set theory through the following (circular) definition.

An *ordinal* is a transitive, well-ordered set of ordinals. By comprehension, the set of all ordinals On exists:

$$\alpha \in On \leftrightarrow \alpha \subseteq On$$
$$\&\ \forall \beta (\beta \in \alpha \rightarrow \beta \subseteq \alpha)$$
$$\&\ \forall \beta (\beta \in \alpha \Rightarrow \beta \notin \beta)$$
$$\&\ \forall \beta \forall \gamma (\beta \in \alpha\ \&\ \gamma \in \alpha \Rightarrow (\beta \in \gamma \Rightarrow \gamma \notin \beta))$$
$$\&\ \forall \beta \forall \gamma (\beta \in \alpha\ \&\ \gamma \in \alpha \Rightarrow (\beta \in \gamma \rightarrow \beta \subseteq \gamma))$$
$$\&\ \forall \beta (\beta \in On \Rightarrow (\alpha \neq \beta \Rightarrow \alpha \subseteq \beta \vee \beta \subseteq \alpha))$$
$$\&\ \forall y (y \subseteq \alpha\ \&\ \exists z(z \in y) \Rightarrow \exists z(z \in y\ \&\ \forall u(u \notin y \vee u \notin z))).$$

The ultimate recursive perversity is that even the classical definition of ordinals is *designed* to prove that On itself is a transitive, well-ordered set of ordinals. That is, showing that the

---

[30] Aristotle took lines to be the edges of physical objects. He took the universe – the largest physical object – to be a finite sphere. Thus the circumference of that sphere is that than which no line can be longer (*Physics* Book 3 [Hellman and Shapiro, 2018, p. 8]).

ordinals can provide the central load-bearing column, or foundation, of mathematics is at the same time to prove a paradox.

Cantor thought of the set of all ordinals as *the absolute*: "Actually conceived, the absolute bounds the entire ordinal sequence, whereas potentially conceived, the ordinal sequence is absolutely unbounded" [Jané, 1995, p. 383]. Actual and potential infinity come together in the Burali-Forti paradox.[31]

Theorem 19 (Burali-Forti). $On \in On$.

Proof As a warm-up, since ordinals are transitive, the contrapositive is $\beta \nsubseteq \alpha \rightarrow \beta \notin \alpha$. Then it also follows that if $\gamma \in \beta$ & $\gamma \notin \alpha$, then $\beta \nsubseteq \alpha$ (by logical counterexample) and hence $\beta \notin \alpha$.

So we set out to prove that On has all the properties of being an ordinal. Take the clauses in the definition of ordinals in order. Well, $On \subseteq On$, of course. And On is transitive: $\alpha \in On \rightarrow \alpha \subseteq On$, by definition. To show how On is ordered, let $\alpha, \beta, \gamma \in On$. We have the following:

- $\alpha \notin \alpha$. Proof: Either $\alpha \in \alpha \vee \alpha \notin \alpha$. But $\alpha \in \alpha \Rightarrow \alpha \notin \alpha$, by the third clause of the definition of On.
- $\alpha \in \beta \Rightarrow \beta \notin \alpha$. Proof: Let $\alpha \in \beta$. Then $\alpha \subseteq \beta$. Suppose that $\beta \in \alpha$. Then $\beta \subseteq \alpha$. Then $\beta = \alpha$. But $\alpha \notin \alpha$, so by substitution, $\beta \notin \alpha$.
- By definition, On is linearly ordered, both in the conditional sense given in the definition, and the weaker

$$\alpha = \beta \vee \alpha \subseteq \beta \vee \beta \subseteq \alpha.$$

It is then simple to show that ordinals themselves are linearly ordered: for any $\beta, \gamma \in \alpha$, we have $\beta, \gamma \in On$, whereby $\beta \neq \gamma \Rightarrow \beta \subseteq \gamma \vee \gamma \subseteq \beta$. And observe that On is itself in the linear order, since

$$\forall \beta (\beta \in On \Rightarrow (On \neq \beta \Rightarrow \beta \in On \vee On \in \beta))$$

by reflexivity, weakening of $\Rightarrow$, $\vee$-introduction, and then apply transitivity of ordinals.

Therefore, On is itself a strict, transitive, linearly ordered set of ordinals.

This leaves us to show that On is well-founded – any nonempty set of ordinals has a least member. Thus suppose that $Y \subseteq On$, with $\beta \in Y$. Either β is least in Y or not. If $\forall u(u \notin \beta \vee u \notin Y)$, then β is least in Y and we can stop. Otherwise, $\exists u(u \in \beta$ & $u \in Y)$, that is, $\exists u(u \in \beta \cap Y)$. But since $\beta \cap Y \subseteq \beta$ and β is well-founded, then $\beta \cap Y$ is well-founded (Proposition 16); so if it is nonempty, then it has a least member,

$$\gamma \in \beta \text{ & } \gamma \in Y \text{ & } \forall u(u \notin \gamma \vee u \notin \beta \cap Y).$$

---

[31] According to Jané, from the 1883 *Grundlagen* until 1896, Cantor believed in the actual existence of the absolutely infinite; in his late work, on the other hand (with the 1895–1897 *Beitrage* silent on the matter), Cantor took the absolute to be "irreducably potential" – closer to the Aristotelian notion of potential infinity, the $\alpha\pi\epsilon\iota\rho o\nu$ [Jané, 1995]. Cf. [Lavine, 1994; Dauben, 1979].

Looking at the second disjunct, if $u \notin \beta \cap Y$, then $u \notin \beta \vee u \notin Y$. But if $u \notin \beta$, then $u \notin \gamma$, since $\gamma \in \beta$ and so $\gamma \subseteq \beta$. Thus we can simplify down to

$$\gamma \in Y \ \& \ \forall u(u \notin \gamma \vee u \notin Y),$$

which shows that γ is the least element of Y as required.   □

On is not empty, then, since it is a member of itself. By irreflexivity of membership for ordinals:

Corollary 20. *$On \notin On$ and so $On \neq On$.*

$\varnothing \in On$, too, because all the clauses are conditionals of the form: if $\beta \in \varnothing$, then $\ldots$; since being a member of $\varnothing$ entails anything, these are all trivially satisfied. Similarly, nonempty subsets of $\varnothing$, were there any, all vacuously have least members. The only tricky part is the clause for trichotomy. Let $\beta \in On$. If $\beta = \varnothing$, then $\varnothing \in On$ by substitution. If $\beta \neq \varnothing$, then we know anyway that $\varnothing \subseteq \beta$, so weakening gives the result.

With ordinals around, there is way to keep track of processes:

Theorem 21 (Transfinite Recursion). *Let $h : V \longrightarrow V$ be any mapping. Then there is a mapping $f : On \longrightarrow V$ such that, for $\alpha \in On$,*

$$f(\alpha) = h(f|\alpha),$$

where $f|\alpha$ is the restriction of f to α.

Proof From naive comprehension, the set $f = \{\langle x, y \rangle : y = h(f|x)\}$ exists, and is a mapping because h is.   □

One such process would be the defining and naming of numbers. We will turn to the Peano–Dedekind postulates for arithmetic in the next chapter.

5.3.3 Well-Ordering the Universe

As a coda to this chapter, and in keeping with our overall interest in "implosion" of a fixed-pointed universe, we will note how the naive conception of set allows one to perform a kind of injection of the universe into a point.

Let a *soft injection* be a map $f : X \longrightarrow Y$ such that $x \neq y \Rightarrow f(x) \neq f(y)$. This is the contrapositive statement of what it means for f to be an injective function, one–one. The softer version gives us this:

Theorem 22. *There is a soft injection of the universe V into the ordinals.*

Proof Define the map

$$\Omega : V \longrightarrow \{On\}$$

to be the only one it could be:

$$\Omega(x) = \{On\}$$

for every $x \in V$. Since $On \neq On$, then $\Omega(x) \neq \Omega(y)$ for any x, y. Then by weakening,

$$x \neq y \Rightarrow \Omega(x) \neq \Omega(y).$$

Thus Ω is a soft injection. □

When there is a one–one injection of a set into the ordinals, then since the ordinals are well-ordered, so is that set. Since $\{On\}$ is a subset of the ordinals, it is well-ordered. If a set is *softly well-ordered* iff there is a soft injection of it into a subset of the ordinals, then the Ω map induces a soft well-order on V.[32] This proves the last result for this chapter.

Corollary 23 (Zermelo 1904). V *is softly well-ordered.*

Since nothing about the proof is essential to V, every set can be softly well-ordered[33] – a point of much conjecture and research in the post-Cantorian era. What say you, Hausdorff in 1914 [Hausdorff, 1957, p. 66]?

The method by which a well ordering necessarily results … is basically very simple, although it places something of a burden on the abstract thinking of the reader.

The existence of a well-order on any set is a classical equivalent of the axiom of choice. This thereby goes some way toward answering a conjecture of Routley, that the axiom of choice is a consequence of the naive set concept as given by comprehension.[34] The idea would be, as with Cantor's theorem, that inconsistency gives ordinary finite sentences the expressive power of an infinitary language. From there, all things are possible.

Set theory is marvelous, and perhaps paraconsistency sets it free. But we will leave this, and other tantalizing directions in inconsistent foundational mathematics[35] here for today, and turn to humbler things: the counting numbers.

Excursus: Partitions, Equivalence Classes, and Cardinality

To what extent can an arbitrary set be *partitioned* in to two parts?

A set $\mathcal{D} = \{U_0, U_1, \dots, U_n\}$ of subsets of X is a *partition* of X iff:

---

[32] Cantor: "It is a basic law of thought [*Denkgesetz*] … that it is always possible to bring any *well-defined* set into the form of a *well-ordered* set" [Hallett, 1984, p. 155, emphasis in the original text].

[33] But not necessarily well-ordered. Subsets of a well-ordered set are well-ordered (Proposition 16), but the proof does not carry over for soft well-orders. By the same token, Theorem 22 does not show that all sets are *finite* (or, a fortiori, of cardinality 1). Contrapositive injections tell us nothing about how many members X has; see [Weber, 2012].

[34] See Priest and Routley [1989a, p. 374]; Routley [1977, p. 925]. See also Priest [2006b, p. 142]; Weber [2012]. Cf. [Rubin and Rubin, 1963].

[35] Like a reconstruction of Cantor's proof that every cardinal is an aleph, via a "projection" of On into V [Cantor, 1967, p. 117]. Or a demonstration that all *large cardinals* exist (cf. [Drake, 1974]) – basically because On is the *largest* cardinal and therefore reflects all 'lower' large cardinal properties. Cf. Routley [1977, p. 926].

(1) $\exists x(x \in U_0)$ and ... and $\exists x(x \in U_n)$.
(2) If $\exists x(x \in U_i \, \& \, x \notin U_j)$ or $\exists x(x \in U_j \, \& \, x \notin U_i)$, then $\neg \exists x(x \in U_i \cap U_j)$, for each i, j.
(3) $X \equiv \{U_0\} \cup \{U_1\} \cup \ldots \cup \{U_n\}$.

Each member of $\mathcal{D}$ is nonempty and pairwise disjoint, and the union of $\mathcal{D}$ is X.[36] A binary relation is an *equivalence* iff it is reflexive, symmetric, and transitive. Classically, partitions and equivalence relations are two sides of the same coin. A partition gives rise to an equivalence relation, and an equivalence relation can be used to form a partition. How do the two notions relate here?

Partitions Do Not Support Transitive Equivalence Relations. In the paraconsistent setting, a partition may have some overlapping cells (as long as they also do not overlap). Elements of a partition $\mathcal{D}$ are pairwise disjoint, and can also be not always disjoint, as in $\neg \exists x(x \in U_i \cap U_j)$, for each i, j but also $\exists x(x \in U_i \cap U_j)$, for some i, j. (For an extreme example, take the partition between the Routley set $\mathcal{Z}$ and *itself*.) But any overlap will collapse two cells in to one equivalence class, because an equivalence relation is transitive. Now, we could tighten the definition of a partition so that $U_i \cap U_j = \emptyset$ for any pair of cells. But then it would be impossible in general to satisfy the condition that the partition covers the space X, because the condition is too demanding: "all the cells that do not overlap on pain of absurdity" are not enough to catch each $x \in X$.

So a partition is not sufficient to pin down equivalence classes. In detail, given a set X, let $\mathcal{D}$ be a partition. Then for $x, y \in X$, let $x E y$ be the relation that x and y are in the same member of $\mathcal{D}$:

$$x E y := \exists C (C \in \mathcal{D} \, \& \, x \in C \, \& \, y \in C).$$

Since $\mathcal{D}$ is a partition, every $x \in X$ is a member of some $C \in \mathcal{D}$. It follows that E has the properties:

- Reflexive, up to contraction: $x E x$, if $x \in C \, \& \, x \in C$ for some C.
- Symmetric: $x E y$ implies $y E x$ just because conjunction commutes.
- Not transitive: suppose $x E y$ and $y E z$. So $x, y \in C$ and $y, z \in C'$, meaning that $C \cap C'$ is not empty. Classically, that means $C = C'$ (by contraposing (2) in the definition of a partition), in which case $x, z \in C$ and $x E z$, but those last steps don't follow here. Even with the contrapositive of (2), we would just have that there is no extensional difference between C and C', which isn't even enough to say that they are extensionally equivalent, let alone identical.

There may be better ways to phrase (2) in the definition of partition, but I suspect that a "strict" partition that forbids any overlap of cells will fail to cover the space, and vice versa, as was the message of the incompleteness vs. inconsistency harangue in the first half of this book.

---

[36] This is the finite case; more generally, $X \equiv \{x : \exists U (U \in \mathcal{D} \, \& \, x \in U\}$.

Equivalence Relations Generate Partitions – Up to Inconsistency. On the other hand, let E be an equivalence relation over X and let

$$[x] = \{y \in X : yEx\},$$

which is the equivalence class of x under E. Quotient X into equivalence classes, as in

$$\mathcal{D} = \{[x] : x \in X\}.$$

To what extent is this a partition?

- For (1), each $x \in [x]$, so the cells are not empty.
- For (2), suppose $\exists z(z \in [x]\ \&\ z \notin [y])$ or $\exists z(z \notin [x]\ \&\ z \in [y])$. Does this show $[x]$ is completely disjoint from $[y]$? If also $\exists z(z \in [x]\ \&\ z \in [y])$, then $zEx\ \&\ zEy$, so xEy because equivalence relations are transitive, and then $[x] \sqsubseteq [y]$. Repeat the argument to show that $[y] \sqsubseteq [x]$. But ex hypothesis $\neg([x] \sqsubseteq [y])$ and $\neg([y] \sqsubseteq [x])$, by the counterexample principle. So we can roughly see that, up to contraction, if two members of $\mathcal{D}$ differ at all, then either they are disjoint, *or* they are both identical and not identical (and so not even self-identical [but also self-identical]).
- For (3), we have by construction that $X \equiv \bigcup \mathcal{D}$.

The division of a set based on equivalence classes approximates a partition as defined earlier, modulo paraconsistent exceptions and substructural stuttering. In general, though, these notions are not interchangeable.

Example: Cantor–Bernstein Theorem. This plays out in concrete ways. For example, the celebrated *Cantor–Bernstein theorem*[37] asserts that, if two sets inject into one another, then they are the same size. (You may have seen this as $|A| \leqslant |B|\ \&\ |B| \leqslant |A| \Rightarrow |A| = |B|$, where $|X|$ is the cardinality of X.) The classical argument considers injective functions $f : A \longrightarrow B$ and $g : B \longrightarrow A$. The idea is to use a back-and-forth mapping h on subsets of A,

$$h(X) = A \backslash g(B \backslash f(X)),$$

which is meant to separate A into two classes. The function reaches a fixed point

$$\mathcal{E} = \bigcup \{X \subseteq A : X \subseteq h(X)\} = h(\mathcal{E}).$$

This in hand, we want to define $H : A \longrightarrow B$ as

$$H(x) = \begin{cases} f(x) & \text{if } x \in \mathcal{E} \\ g^{-1}(x) & \text{otherwise.} \end{cases}$$

---

[37] For lovingly detailed history of the theorem, see [Hinkis, 2013], which incidentally argues that the sometimes-used name "Cantor–Schröder–Bernstein theorem" is not appropriate.

where g^{-1} is the inverse of g. But this is the crucially problematic point. To complete the proof, H needs to be a bijective function. There is every possibility, though, that membership on $\mathcal{E}$ is inconsistent. So H may not even be a function, let alone a bijection.[38]

For the remainder, then, we will essentially leave the theory of cardinality alone (give or take finite arithmetic in the next chapter).[39] The questions we are addressing about the paradoxes are mainly "qualitative," and so we can leave the cardinality problem here.

---

[38] Note that this paragraph corrects an error in my [Weber, 2012]. Thanks to Badia here.
[39] Compare that mereology is not sufficient to express counting [Lewis, 1991; Hamkins and Kikuchi, 2016].

6

Arithmetic

This chapter follows the axiomatic development of elementary arithmetic, through the fundamental properties of natural numbers (suitably substructurally expressed), to the original limitive theorem: the Pythagorean revelation that $\sqrt{2}$ is an irrational magnitude in the diagonal of the unit square.

Set theory in the last chapter was overtly inconsistent, even while providing some necessary foundations. The goals for this chapter are more staid – to develop a more "classical" arithmetic, providing some *reassurance*. No contradictions about the natural numbers are proved, and the attitude is neutral as to whether arithmetic may ultimately be inconsistent or not; the appeal of the paraconsistent approach is that it wouldn't matter either way.[1] As the book of Ecclesiastes does not say, one paradox rises and another passes away, but the numbers abideth forever.[2]

6.1 Thither Paraconsistent Arithmetic!

6.1.1 $\mathbb{N}$

Arithmetic is about the natural numbers. What are those? One approach is to take them to be the initial segment of the ordinals.[3] Another approach, following Frege,[4] is to take the natural numbers to be the intersection of all inductive structures. From comprehension,

$$\mathbb{N} = \{z : \forall y(0 \in y \ \& \ \forall x(x \in y \Rightarrow \mathbf{s}x \in y) \Rightarrow z \in y)\}$$

---

[1] "We do not think of such negated formulas as $2 + 2 \neq 4$ as *true*. But the point is, even if we did, it *still* wouldn't follow that $2 + 2 = 5$. That's the idea, anyway" [Meyer and Urbas, 1986, p. 49]. On the inconsistency of arithmetic (for Gödelian reasons), see [Priest, 1994a; Shapiro, 2002].

[2] For this chapter, I have benefited from studying two (unpublished) sources in particular – notes by Slaney ([Slaney, 1982], cf. [Slaney et al., 1996]), for the "Quixotic purpose" of proving $\sqrt{2}$ irrational, using the relevant logic TW; and Restall's Ph.D. thesis, [Restall, 1994, ch. 11]. Thanks to Dave Ripley for detailed comments on earlier versions of this chapter.

[3] The fact that the ordinals are well-founded (Theorem 19) would be sufficient using *classical* logic to prove that the ordinals satisfy a (transfinite) induction scheme. Not so here. To illustrate just for the initial segment, one supposes $\varphi(0)$ and $\varphi(n) \rightarrow \varphi(n + 1)$, but then for reductio supposes that $\neg\varphi(k)$ for some k; then by well-foundedness, there must be a least such, which is both φ and $\neg\varphi$. But the reductio fails because there may just be ordinals that are φ and not φ.

[4] Here I am following [Terui, 2004, p. 18], following Girard in linear logic.

is the set of things that are in every set containing *zero*

$$0 := \varnothing$$

and its *successors*

$$s(x) := \langle \varnothing, x \rangle.$$

This serves as a pretty nice basis for the standard axioms of arithmetic. Reading set membership as property instantiation, $\mathbb{N}$ is the set of exactly those objects n such that, for any φ, if $\varphi(0)$ and $\varphi(x) \Rightarrow \varphi(s(x))$, then $\varphi(n)$.

Let us check the *Peano postulates*[5] for $\mathbb{N}$ so defined.

(1) Zero is a number: $0 \in \mathbb{N}$.

(2) and is not the successor of any number: $s(x) = 0 \Rightarrow \bot$.

(3) Every number has a successor: (up to contraction): $x \in \mathbb{N} \Rightarrow s(x) \in \mathbb{N}$.

(4) which is unique: $s(x) = s(y) \Rightarrow x = y$.

(5) These are all the numbers: $\varphi(0)$ & $\forall x(\varphi(x) \Rightarrow \varphi(sx)) \Rightarrow \forall n(n \in \mathbb{N} \Rightarrow \varphi(n))$.

The proofs are all simple, except for (3).

For (1), 0 is a number, because $0 \in y \Rightarrow 0 \in y$, for all y; so by antecedent strengthening,

$$\forall y(0 \in y \ \& \ \forall x(x \in y \Rightarrow sx \in y) \Rightarrow 0 \in y))$$

and therefore $0 \in \mathbb{N}$.

For (2), 0 is not the successor of any number, because $\{\varnothing\} \in s(x)$ for all x, by the definition of ordered pairs (Proposition 11). So if $s(x) = 0$, then $\{\varnothing\} \in \varnothing$, which by Section 5.2.1 is absurd.

For (3), the successor of a number is a number – almost. By modus ponens,

$$\forall x(\varphi(x) \Rightarrow \varphi(sx)) \ \& \ \varphi(n) \Rightarrow \varphi(sn). \tag{6.1}$$

Antecedent strengthening, we add $\varphi(0)$,

$$\varphi(0) \ \& \ \forall x(\varphi(x) \Rightarrow \varphi(sx)) \ \& \ \varphi(n) \Rightarrow \varphi(sn).$$

Now let $\mathcal{A}$ stand for $\varphi(0)$ & $\forall x(\varphi(x) \Rightarrow \varphi(sx))$. Then

$$\mathcal{A} \ \& \ \varphi(n) \Rightarrow \varphi(sn)$$

and therefore, by transitivity of the conditional

$$(\mathcal{A} \Rightarrow \varphi(n)) \Rightarrow (\mathcal{A} \ \& \ \mathcal{A} \Rightarrow \varphi(sn))$$

since if $p \ \& \ q \Rightarrow r$, then $s \Rightarrow q$ implies $p \ \& \ s \Rightarrow r$. Then we are a contraction away[6] from $n \in \mathbb{N} \Rightarrow s(n) \in \mathbb{N}$.

---

[5] Or the Dedekind–Peano Postulates, since they originate in Dedekind's 1888 essay [Dedekind, 1901]. See Peano's 1889 "Principles of Arithmetic" and Dedekind's 1890 "Letter to Keferstein" in [van Heijenoort, 1967].

[6] In [Shirahata, 1999; Terui, 2004], this is dealt with because the logic includes a modality that allows some contraction; but as per Section 4.3.2.2, that is not part of the logic here. So we have this Peano axiom holding "up to contraction." One interpretation of the breakdown of this proof is that $\mathbb{N}$ is *stratified* – we could define (with $\mathcal{A}$ still as before) $\mathbb{N}_0 := \{z : \mathcal{A} \Rightarrow z \in y\}, \mathbb{N}_1 := \{z : \mathcal{A} \ \& \ \mathcal{A} \Rightarrow z \in y\}, \mathbb{N}_2 := \{z : \mathcal{A} \ \& \ \mathcal{A} \ \& \ \mathcal{A} \Rightarrow z \in y\}$, and so forth. There is room to explore here, but we leave this idea for future and in Section 6.1.2 assume a "flat" domain of natural numbers.

For (4), the successor of a number is unique, because

$$\langle\varnothing,x\rangle = \langle\varnothing,y\rangle \Rightarrow x = y$$

by the law of ordered pairs (Proposition 11).

For (5), the numbers satisfy mathematical induction, because if $0 \in y$ and $x \in y \Rightarrow sx \in y$, then $z \in \mathbb{N}$ implies $z \in y$ by permuting the definition of $\mathbb{N}$. That's especially good, since induction without contraction is a sensitive issue.[7]

Numbers have names. The first few look like this:

$$1 := s0 = \langle 0,0\rangle$$
$$2 := s1 = \langle 0,\langle 0,0\rangle\rangle = \langle 0,1\rangle$$
$$3 := s2 = \langle 0,\langle 0,\langle 0,0\rangle\rangle\rangle = \langle 0,1,2\rangle$$

$$\vdots$$

Addition $+$ and multiplication $\times$ may be (very) recursively defined [Terui, 2004, p. 20]:

$$\langle x,y,z\rangle \in + \leftrightarrow (y = 0 \,\&\, x = z) \vee \exists u \exists v (y = s(u) \,\&\, z = s(v) \,\&\, \langle x,u,v\rangle \in +)$$
$$\langle x,y,z\rangle \in \times \leftrightarrow (y = 0 \,\&\, z = 0) \vee \exists u \exists v (y = s(u) \,\&\, \langle v,x,u\rangle \in + \,\&\, \langle x,u,v\rangle \in \times).$$

These exist by full comprehension.

6.1.2 Axioms

Set theory provides a model (mostly) of the numbers, as we've just seen. Nevertheless, arithmetic in this chapter isn't officially reducible to set theory. The approach is purely axiomatic.[8] The following formulation of the Peano postulates will be our official starting point for this chapter, stipulating recursion clauses defining addition and multiplication.

$$\text{III} \quad x + 0 = x$$
$$\text{I} \quad 0 = sx \Rightarrow \bot \qquad\qquad x + sy = s(x + y)$$
$$\text{II} \quad sx = sy \Rightarrow x = y \qquad \text{IV} \quad x \times 0 = 0$$
$$x \times sy = (x \times y) + x$$

$$\text{V} \quad \varphi 0 \,\&\, \forall x(\varphi x \Rightarrow \varphi sx) \Rightarrow \forall x \varphi x.$$

From axiom V, mathematical induction, note that if $\vdash \varphi(0)$, and $\vdash \forall x(\varphi x \Rightarrow \varphi sx)$, then $\vdash \forall x \varphi x$.[9] The axiom is used to show that all numbers have property φ by proving the following: a *base case*, that zero is φ; and an *induction step*, that assuming a number is φ,

---

[7] Restall observes that induction "looks too much like an extended case of *pseudo modus ponens* ... to be comfortable" [Restall, 1994, p. 194] (cf. Section 4.1.1.3). If one thinks of the universal quantifier in the consequent in the induction scheme as an infinitary conjunction, then it does appear that the antecedent would be used many times. In linear set theory, the induction form that we use is called "light induction."

[8] Following precedent in paraconsistent arithmetic, e.g., [Meyer and Mortensen, 1984]; see Priest [2006b, ch. 17]. On the one hand, this approach may "isolate" it from dangerous fixed points (Theorem 13), but on the other hand is obviously undesirable from a foundationalist standpoint.

[9] This is the form of the axiom given by Meyer and Mortensen [1984, p. 918].

then so is its immediate successor. Other classically equivalent formulations of induction are considered in the excursus on number theory later in the chapter.

Let's start by confirming that, as one would hope, the numbers are zero and its successors:[10]

Proposition 24. $\forall x (x = 0 \vee \exists y(x = sy))$.

Proof Show this using induction. Base case: $0 = 0$, so

$$0 = 0 \vee \exists y(x = sy)$$

by disjunction introduction. Induction step: $sx = sx$ is always true, so $\exists y(sx = sy)$ by existential generalization, and then $sx = 0 \vee \exists y(sx = sy)$ by disjunction introduction. Then by weakening,

$$x = 0 \vee \exists y(x = sy) \implies sx = 0 \vee \exists y(sx = sy).$$

Therefore, by axiom V, Proposition 24 is true of all numbers.   □

6.1.3 Consistency at Zero

Axiom I, the assumption that zero can have no predecessor, is predicated on pain of absurdity (or "shrieked," in Beall's idiom). Any predecessors of 0 are "annihilated." With 0 defined as $\emptyset = \{x : \bot\}$, this will simply follow on most definitions of successor, as we saw earlier – but a more common phrasing of Peano's first axiom is the simple negation

$$\neg(sx = 0)$$

as in relevant arithmetic [Meyer and Mortensen, 1984]. (Of course, the distinction here is erased under classical logic.) This phrasing, or the even weaker

$$s0 = 0 \rightarrow 0 = 1$$

from [Friedman and Meyer, 1992, p. 827], which says only that $0 = 1 \rightarrow 0 = 1$, allows a lot more flexibility in how the axioms may be interpreted – even allowing finite models, and with that, decidability. Why not use simple negation for the axiom instead (and with it a different empty set, like $\{x : x \neq x\}$)?

Some might argue that any "arithmetic" that has a finite interpretation is simply not *arithmetic*.[11] This is an appealing, and also completely question-begging, line (as arguments based on scare quotes and italics tend to be). In any case, I have no such principle to stand on. Rather, the decision is practical: in my trial-and-error experience, many standard proofs in arithmetic simply appear to track back to the *absolute* fact that zero is the *first* number, or else. Using only (paraconsistent) negation in this axiom, or no negation at all,

---

[10] [Restall, 1994, lemma 11.18]. One could *define* the numbers this way, using a circular/fixed-point definition [Terui, 2004; Weber, 2010d].
[11] See [Meadows and Weber, 2016].

is so friendly to inconsistency that it simply does not rule enough out to prove much. Working through standard proofs, the recursive structure of the numbers all but demands some unassailable consistent base point.[12] On the other end of the spectrum, Dunn [Dunn, 1980] has suggested the "induction" axiom for relevant Robinson arithmetic Q be given as the very strong:

$$x \neq 0 \rightarrow \exists y(x = \mathbf{s}y)$$

(cf. Proposition 24). This builds a (lawlike!) disjunctive syllogism into the conception of number, and is much stronger – so strong, in fact, that relevant Q collapses back into classical arithmetic.[13]

The arithmetic here tries to find a happy medium. Axioms I–V are true when interpreted in classical PA (with ⇒s read as a material conditional); everything in this chapter is sound with respect to classical arithmetic as a model. A classical argument shows that the theory has only infinite models.[14] And yet there are still prospects for controlled inconsistencies, should they arise. Full classical negation is not required, but a little annihilation negation helps. A paraconsistent negation (and additive disjunction) allows us to reason by cases, exhaustively, about a structure that then has some inherent, localized explosiveness. The outcome is proofs that may have a "classical" flavor – which should be welcome, given the otherwise dim prospects for a theorem-for-theorem classical recapture (Section 3.2). On this middle-way approach, we aren't trying only to mimic PA, and neither are we rebuking it. As with ZFC and set theory, we're simply trying to do *arithmetic*, using *logic*. This chapter is not about a *different* "paraconsistent arithmetic," It is offered as a finer-grained way to reason about a logic-invariant core of the natural numbers.[15]

6.2 Addition, Multiplication, and Order

6.2.1 Addition

Some reassuring platitudes: it is true that

$$1 + 1 = 2$$

because in general,

$$x + 1 = \mathbf{s}x.$$

Proof: $x + 1 = x + \mathbf{s}0 = \mathbf{s}(x + 0) = \mathbf{s}x$, using Axiom III.

Also reassuring is that, given the consistency of 0, many other claims are simply false (on pain of absurdity). For instance, $2 = 4 \Rightarrow \bot$, and $4 = 1 \Rightarrow \bot$. Both of these reduce

---

[12] Badia makes a good case that $\bot$ is needed to fix mathematically interesting structures [Badia, 2017]. Compare this with the more general use of $\bot$ in Section 10.1.6.

[13] It has the bad consequence that $0 \neq 0 \rightarrow \bot$ [Dunn, 1980, p. 412]. Dunn goes on to show that dropping 0 (starting at 1) does not so collapse, but this seems more to do with Q than anything about logic.

[14] See [Weber, 2016a], with thanks to Badia.

[15] See [Meyer, 1975, 1976] for extended sermons, and results, on this topic.

to something having zero as its successor, which is $\perp$. (Proofs: $2 = 4$ is absurd because $ss0 = sss1 \Rightarrow 0 = s1$ by repeated applications of Axiom II, and $4 = 1$ is absurd since $s3 = s0 \Rightarrow s2 = 0$.) This shows that some "ground-level" sums work as expected.

The next sequence of proofs is based on [Restall, 1994, ch. 11].

Lemma 25. $x + sy = sx + y$.

Proof By induction (on y). The first five lines that follow prove the base case, and lines 6–8 argue the induction step:

1	$s(x + 0) = x + s0$	$\parallel$	Axiom III
2	$x + 0 = x$	$\parallel$	Axiom III
3	$s(x + 0) = sx$	$\parallel$	2, substitution
4	$sx + 0 = sx$	$\parallel$	Axiom III
5	$x + s0 = sx + 0$	$\parallel$	1, 3, 4, =
6	$x + sy = sx + y$	$\parallel$	assume
7	$s(x + sy) = s(sx + y)$	$\parallel$	6, substitution
8	$x + ssy = sx + sy$	$\parallel$	7, Axiom III, =	$\square$

Proposition 26. *Addition is commutative, $x + y = y + x$, and associative, $x + (y + z) = (x + y) + z$.*

Proof For commutativity, the base case is to show $x + 0 = 0 + x$. First we show by induction that

$$0 + x = x \tag{6.2}$$

for all x. Trivially, from Axiom III, $0 + 0 = 0$. And if $0 + x = x$, then $s(0 + x) = sx = 0 + sx$ by substitution and Axiom III. That proves (6.2). And now, since

$$x = x + 0$$

the base case for the theorem is proved by transitivity of identity.

For the induction step,

1	$x + y = y + x$	$\parallel$	assume
2	$sx + y = x + sy$	$\parallel$	lem. 25, =
3	$x + sy = s(x + y)$	$\parallel$	Axiom III
4	$s(x + y) = s(y + x)$	$\parallel$	1, substitution
5	$s(y + x) = y + sx$	$\parallel$	Axiom III
6	$sx + y = y + sx$	$\parallel$	2–5, =

The base case of associativity is left as an exercise. For the induction step,

1	$x + (y + z) = (x + y) + z$	$\parallel$	assume
2	$(x + y) + sz = s((x + y) + z)$	$\parallel$	Axiom III
3	$s((x + y) + z) = s(x + (y + z))$	$\parallel$	1, substitution
4	$s(x + (y + z)) = x + s(y + z)$	$\parallel$	Axiom III
5	$x + s(y + z) = x + (y + sz)$	$\parallel$	Axiom III, substitution
6	$(x + y) + sz = x + (y + sz)$	$\parallel$	2–5, =	$\square$

The next proposition, cancellation, is extremely important (and frankly, from an "intensional" perspective, a bit suspicious). It allows extensionality principles (Proposition 33), which in turn allow arguments that would otherwise require contraction, such as Lemma A(ii). We will have a lot more to say about cancellation when we get to group theory and subtraction in the next two chapters.

Proposition 27. $y = z \Leftrightarrow y + x = z + x$.

Proof For the right to left direction, the base case is

1	$y + 0 = z + 0$	assume
2	$y + 0 = y$	Axiom III
3	$z + 0 = z$	Axiom III
4	$y = z$	$1, 2, 3, =$

Therefore, $y + 0 = z + 0 \Rightarrow y = z$. Here is the induction step:

5	$y + x = z + x \Rightarrow y = z$	induction hyp.
6	$y + \mathsf{s}x = z + \mathsf{s}x$	assume
7	$y + \mathsf{s}x = \mathsf{s}(y + x)$	Axiom III
8	$z + \mathsf{s}x = \mathsf{s}(z + x)$	Axiom III
9	$\mathsf{s}(y + x) = \mathsf{s}(z + x)$	$6, 7, 8, =$
10	$y + x = z + x$	9, Axiom II
11	$y = z$	5, 10, modus ponens

Therefore, $(y + x = z + x \Rightarrow y = z) \Rightarrow (y + \mathsf{s}x = z + \mathsf{s}x \Rightarrow y = z)$. This proves the theorem from right to left.

Left to right can be proved without need of substitution; with lines 1–4 the base case and 5–8 the induction:

1	$y = z$	assume
2	$y + 0 = y$	Axiom III
3	$z + 0 = z$	Axiom III
4	$y + 0 = z + 0$	$1, 2, 3, =$
5	$y = z \Rightarrow y + x = z + x$	assume
6	$y + x = z + x \Rightarrow \mathsf{s}(y + x) = \mathsf{s}(z + x)$	Axiom II
7	$\mathsf{s}(y + x) = \mathsf{s}(z + x) \Rightarrow y + \mathsf{s}x = z + \mathsf{s}x$	from Axiom III, $=$
8	$y = z \Rightarrow y + \mathsf{s}x = z + \mathsf{s}x$	5, 6, 7, transitivity □

Then if $x = y$ and $u = v$, then $x + u = y + v$. Less generally,

Corollary 28. $x + n = x \Leftrightarrow n = 0$.

6.2.2 Multiplication

Notation: $nm := n \times m$ and $n^2 := nn$, the latter just being a special case of the former and all the "exponentiation" we will need. Assume multiplication binds harder than addition, $x \times y + z = (x \times y) + z$.

Proposition 29. $0x = x0$, *and* $x1 = x$.

Proof   To show that multiplication commutes with zero, we actually prove that

$$0x = 0$$

for all x. Base case is $0 \times 0 = 0$ from Axiom IV. For induction,

1	$0x = 0$	assume
2	$0sx = 0x + 0$	Axiom IV
3	$0x + 0 = 0 + 0$	1, substitution
4	$0 + 0 = 0$	Axiom III
5	$0x + 0 = 0$	3, 4, =
6	$0sx = 0$	2, 5, =

Thus $\forall x (0x = 0)$. Then since $\forall x (x0 = 0)$ by Axiom IV, transitivity gives $0x = x0$.

To show that 1 is identity for multiplication: since $s0 = 1$ by definition, substitution gives $x \times 1 = x \times s0$. Then Axiom IV gives $x \times s0 = (x \times 0) + x$. Since $x \times 0 = 0$ by Axiom IV, substitution gives $(x \times 0) + x = 0 + x$, and we know from Axiom III and Proposition 26 that $0 + x = x$. So the result holds by transitivity of $=$.   □

Lemma 30. $sx \times y = (x \times y) + y$.

Proof   Base: $sx \times 0 = 0 = 0 + 0 = (x \times 0) + 0$, just by axioms. Induction: suppose $sx \times y = xy + y$, to prove that $sx \times sy = xsy + sy$. Compute, using the induction hypothesis, axioms, and the associativity and commutativity of addition:

$$sx \times sy = sx \times y + sx$$
$$= (xy + y) + sx$$
$$= xy + (y + sx)$$
$$= xy + s(y + x)$$
$$= xy + (x + sy)$$
$$= (xy + x) + sy$$
$$= xsy + sy.$$

□

Proposition 31.   *Multiplication commutes,* $xy = yx$, *distributes over addition,* $x(y+z) = xy + xz$, *and associates,* $x(yz) = (xy)z$.

Proof   For commutativity, induction. Base case is Proposition 29. For the induction step, suppose $xy = yx$. Then $xy + y = yx + y$ by adding y to both sides (Proposition 27). But since Axioms III and IV give $sxy = xy + y$ and $yx + y = ysx$, respectively, transitivity gives $sxy = ysx$, as required.

The proof of (right) distribution, $(y + z)x = yx + zx$, is a straightforward induction. (The solution is in [Restall, 1994, p. 199].) Full distribution follows by commutativity of multiplication: since $(y + z)x = x(y + z)$, by right distribution and transitivity, $x(y + z) = yx + zx$. Since $yx + zx = xy + xz$, transitivity again yields the result.

For associativity, base case: since $y0 = 0$, then $x(y0) = x0 = 0 = (xy)0$, by Axiom IV and substitution. Here is the induction step, which uses distribution. First, since $ysz = yz + y$ by Axiom IV, multiplying both sides by x and using distribution gives $x(ysz) = x(yz + y) = x(yz) + xy$. Now suppose $x(yz) = (xy)z$ for the induction hypothesis. Then $x(yz) + xy = (xy)z + xy$ by adding xy to both sides. Axiom IV says that $(xy)sz = (xy)z + xy$, so by transitivity of identity, $x(ysz) = (xy)sz$. □

With commutativity, then, we have

$$2x = x + x$$

because $2x = \mathsf{ss}0 \times x = (x \times 1) + x = x + x$. In fact, from Propositions 29 and 31, we can simplify notation in standard ways:

$$\underbrace{x + \cdots + x}_{n \text{ times}} = nx,$$

e.g., in the case of $n = 3$, we have $x + x + x = 1x + 1x + 1x = x(1 + 1) + 1x = x(1 + 1 + 1) = x3$. This procedure can in principle be repeated by hand for any finite n.

6.2.3 Order

The next definition introduces a notational trick (from Slaney's notes) of working explicitly with successors. In various places, we will want to presume that $n > 0$, but it is cumbersome to have this as an additional hypothesis. E.g., in a reductio we want to show something of the form $\neg\varphi \Rightarrow \bot$, but would instead show $(\neg\varphi$ and $n > 0) \Rightarrow \bot$, which isn't enough to establish φ. Working with successors directly allows us to argue, in effect, about all numbers greater than 0 without qualifications.

Define

$$x \leqslant y := \exists n(x + n = y)$$
$$x < y := \exists n(x + \mathsf{s}n = y).$$

Immediately, $x < y \Rightarrow x \leqslant y$. Less trivially [Restall, 1994, p. 203],

Proposition 32. $x \leqslant y \Rightarrow x = y \vee x < y$

Proof Suppose $x + n = y$ for some n. We want to reason through the two cases from Proposition 24, that n is either zero or a successor. But we can't use the supposition more than once. To avoid contraction, first conjoin the supposition with Proposition 24, and distribute over the disjunction, to get

$$(x + n = y \ \& \ n = 0) \vee (x + n = y \ \& \ \exists m(n = \mathsf{s}m))$$

Now eliminate the cases. In the first, $x = y$ by Corollary 28 and substitution. In the second, $x < y$ by substitution and definition, as discussed in Section 6.2.3. □

We record further that

$$\forall x (0 \leqslant x) \qquad\qquad \forall x \forall y (x \leqslant x + y)$$
$$\forall x (0 < sx) \qquad\qquad \forall x \forall y (x < x + sy)$$
$$\forall x (x < 0 \Rightarrow \bot) \qquad\quad \forall x (x < sx)$$

with proofs an exercise in applying definitions. For instance, if $x < 0$, then $x + sn = 0$ for some n. Then conjoin this with Proposition 24, so either ($x = 0$ and $x + sn = 0$), whereby $sn = 0$; or else ($\exists m (x = sm)$ and $x + sn = 0$), whereby $s(sn + m) = 0$; either case is $\bot$.

Proposition 33. $\leqslant$ *is a partial order on* $\mathbb{N}$:

$$x \leqslant x$$
$$x \leqslant y \,\&\, y \leqslant x \Leftrightarrow x = y$$
$$x \leqslant y \,\&\, y \leqslant z \Rightarrow x \leqslant z.$$

Proof   Reflexivity is just because $x + 0 = x$.

Antisymmetry left to right (the other direction is substitution): by cases on $\mathbb{N}$. If $x + 0 = y$ or $y + 0 = x$, then $x = y$ by Corollary 28. If $x + sn = y$ and $y + sm = x$, then by substitution, $x + sn + sm = x$, which by Corollary 28 implies $sn + sm = 0$. Then $ss(n + m) = 0$, which implies absurdity, which implies $x = y$.

Transitivity is straightforward; let's just ensure there is no contraction.

1	$x + n = y$	assume
2	$y + m = z$	assume
3	$y + m = y + m$	=
4	$y + m = (x + n) + m$	1, 3, substitution
5	$(x + n) + m = x + (n + m)$	associativity (prop. 26)
6	$y + m = x + (n + m)$	4, 5, =
7	$z = x + (n + m)$	2, 6, sub
8	$x \leqslant z$	7, def Section 6.2.3

$\square$

Since $<$ implies $\leqslant$, antisymmetry is also available in the "impossible" case, $x < y \,\&\, y < x \Rightarrow x = y$.

Proposition 34. $<$ *is a strict partial order:*

$$x < x \Rightarrow \bot$$
$$x < y \Rightarrow (y < x \Rightarrow \bot)$$
$$x < y \,\&\, y < z \Rightarrow x < z.$$

Proof   For strong irreflexivity, argue by induction. The case $0 < 0$ is absurd. Suppose $x < x$. Then by definition Section 6.2.3, $x + sy = x$, which by Corollary 28 implies that $sy = 0$, which is also absurd. A corollary: $\neg(x < x)$.

For strong asymmetry, let $x < y$. Suppose $y < x$. Then $x + sn = y$ and $y + sm = x$, so $y = (y + sm) + sn = y + ss(m + n)$. By Corollary 28, $ss(m + n) = 0$, which is absurd. Corollary: $x < y \Rightarrow \neg(y < x)$. To prove the corollary directly,

1	$x < y \Rightarrow (y < x \Rightarrow x = y)$	Proposition 33
2	$x = y \Rightarrow (y < x \Rightarrow x < x)$	substitution law
3	$x < x \Rightarrow \perp$	fact
4	$x < y \Rightarrow (y < x \Rightarrow (y < x \Rightarrow y \nless x))$	1-3, $\Rightarrow$, $\perp$
5	$x < y \Rightarrow (y < x \Rightarrow y \nless x)$	4, $\Rightarrow$, reductio
6	$x < y \Rightarrow y \nless x$	5, $\Rightarrow$, reductio

For transitivity, let $x < y$ and $y < z$. Then $x + \mathsf{s}m = y$ and $y + \mathsf{s}n = z$ for some m, n. Then substituting and applying Axiom III,

$$z = (x + \mathsf{s}m) + \mathsf{s}n = x + (\mathsf{s}m + \mathsf{s}n) = x + \mathsf{s}\mathsf{s}(m + n)$$

as required. $\qquad\qquad\qquad\qquad\qquad\qquad\qquad\qquad\qquad\qquad\qquad\qquad\qquad\qquad\qquad$ □

It follows immediately that $x \leqslant y$ & $y < z \Rightarrow x < z$. More dramatically, because every number is strictly less than its own successor, and $<$ is irreflexive on pain of absurdity, the chain of successors never ends:

Corollary 35. $x = \mathsf{s}x \Rightarrow \perp$ *for every number x.*

This places a hard limit on what sort of inconsistency, if any, might turn up in arithmetic. Speculations about inconsistent arithmetic in [Priest, 1994a] and [van Bendegem, 2003] tend to anticipate some number that is its own successor. We've now seen that, essentially because of the strong from of Axiom I, this can't happen on the current arrangement. It also means, with respect to the fixed point Theorem 13 in the ambient set theory, that any fixed point on s cannot be a number. (Note that strong irreflexivity of $<$ is proved by induction; it is only shown to be true of those things that are members of every inductive structure – numbers, not alien intruders.) In terms of what inconsistencies are still possible, nothing rules out inconsistency in identity, some $n \neq n$.

The definition of strict order as $x \leqslant y$ & $x \neq y$ would not do as well in reasoning through arguments, both due to the weaker negation (the information that $x \neq y$ is more or less useless in a proof) and the failure of contraction. If $x < y$, then $x \leqslant y$ & $x \neq y$, but not vice versa. This is analogous to the way a proper subset was defined in the previous chapter.

The Peano properties interact with the order properties as follows:

- $x \leqslant y \Leftrightarrow x + z \leqslant y + z$

- $x \leqslant y \Leftrightarrow x\mathsf{s}z \leqslant y\mathsf{s}z$

- $x < y \Leftrightarrow x + z < y + z$

- $x < y \Leftrightarrow x\mathsf{s}z < y\mathsf{s}z$

- $n > 0 \Rightarrow x \leqslant xn$ $\qquad$ (Archimedes' axiom)

The proofs are parasitic on those for identity. The first $\leqslant$-invariance property is by commutativity and additive cancellation:

$$x \leqslant y \Leftrightarrow x + n = y$$
$$\Leftrightarrow (x + n) + z = y + z$$
$$\Leftrightarrow (x + z) + n = y + z$$
$$\Leftrightarrow x + z \leqslant y + z.$$

The strict version holds for $x + \mathsf{s}n = y$.

To show multiplicative cancellation over $<$, the base case $x1 < y1 \Rightarrow x < y$ is true by substitution on Proposition 29. Now assume $x\mathsf{s}z < y\mathsf{s}z \Rightarrow x < y$ for induction. What if $x\mathsf{ss}z < y\mathsf{ss}z$? Then $x\mathsf{ss}z + \mathsf{s}n = y\mathsf{ss}z$ for some n, so with Axiom IV and commutativity, $(x\mathsf{s}z + \mathsf{s}n) + \mathsf{s}z = y\mathsf{s}z + \mathsf{s}z$, so by additive cancellation $x\mathsf{s}z + \mathsf{s}n = y\mathsf{s}z$, which implies $x < y$ on hypothesis.

For the Archimedean property, $0 > 0$ implies everything, so it implies $x \leqslant x0$. So suppose $n = \mathsf{s}m$ for some m. By Axiom IV and definition of order, $x \leqslant (x + xm) = x\mathsf{s}m$.

Immediately, then, $x \leqslant y \Rightarrow \mathsf{s}x \leqslant \mathsf{s}y$ and $x < y \Rightarrow \mathsf{s}x < \mathsf{s}y$, and

$$x < y \mathbin{\&} u < v \Rightarrow x + u < y + v. \tag{6.3}$$

To check, observe that $x < y \Rightarrow x + u < y + u$, and $u < v \Rightarrow y + u < y + v$. Then factor and transitivity gives the result.

Proposition 36.   *The numbers are totally ordered:*

$$x \leqslant y \vee y \leqslant x. \qquad\qquad\qquad\qquad\qquad\qquad\qquad \textit{(linearity)}$$
$$x < y \vee x = y \vee y < x. \qquad\qquad\qquad\qquad\qquad\qquad \textit{(trichotomy)}$$

Proof   Linearity.[16] For the duration of this proof, let $\varphi(x, y) := x \leqslant y \vee y \leqslant x$.
Base: we know that $0 \leqslant y$; so

$$\varphi(0, y).$$

Induction: suppose $\varphi(x, y)$, that is, $x \leqslant y \vee y \leqslant x$. In the first case, for some n, $x + n = y$. Either $n = 0$ or there is some m that immediately precedes n, whence $x + \mathsf{s}m = y$. (As is now familiar, distribute the induction hypothesis over the two disjuncts to avoid contraction.) If $n = 0$, then $x = y$, and so $\mathsf{s}x \geqslant y$. If n is the successor of m, then $\mathsf{s}(x + m) = \mathsf{s}x + m = y$, so $\mathsf{s}(x) \leqslant y$. In the second case, if $y \leqslant x$, then immediately $y \leqslant \mathsf{s}(x)$. Thus either way

$$\varphi(x, y) \Rightarrow \varphi(\mathsf{s}x, y)$$

as required.

---

[16]   Proved by double induction in [Restall, 1994, p. 204] (and similar in [Slaney, 1982]). Cf. Meyer [1975, p. 112].

Trichotomy now follows: from linearity, use Proposition 32 to get

$$(x < y \vee x = y) \vee (y < x \vee x = y).$$

Associate, and then use the fact that $x = y \vee x = y \Rightarrow x = y$ to finish the proof.   □

A corollary:

$$xz > 0 \Rightarrow z > 0.$$

To check, suppose $xz > 0$. By trichotomy, either $z > 0$ or $z = 0$, or $z < 0$. In the first case, there is nothing to prove. If $z = 0$, then $xz = 0$ by Axiom IV. Then $0 < 0$ by substitution on the supposition (again), but this implies $z > 0$ because it implies everything. The third case, $z < 0$, implies $\bot$ immediately.

The full Archimedean property is that all numbers may be surpassed,

$$\forall x \forall y(x > 0 \Rightarrow \exists z(y < xz)). \tag{6.4}$$

Proof: Let $0 < x$. From Archimedes' axiom, $y \leqslant yx$. Then $y + n = yx$ for some n. To prove the theorem, a witness for z is needed. If $n = 0$, then $x = 1$; then $z = sy$ suffices. If $n > 0$, then $z = y$ suffices.

Excursus: Number Theory

This optional section is a detour. The concern is the recovery of some cornerstones in the theory of prime numbers, most saliently the unique prime factorization theorem – the fundamental theorem of arithmetic, and with it Euclid's theorem about the infinitude of primes. The results are conditional. This section (and only this one) is contingent on assuming further forms of induction that are not equivalent to the version we've been using:

VI *Infinite descent:*
$$\forall x(\neg \varphi x \Rightarrow \exists y(y < x \,\&\, \neg \varphi y)) \Rightarrow \forall x \varphi x.$$

Least number(s):
$$\exists x \exists y \varphi(x, y) \Rightarrow \exists x \exists y(\varphi(x, y) \,\&\, \forall u \forall v(\varphi(u, v) \Rightarrow x \leqslant u \,\&\, y \leqslant v)).$$

Complete induction:
$$\forall x(\forall y(y < x \Rightarrow \varphi y) \Rightarrow \varphi x) \Rightarrow \forall x \varphi x.$$

These various forms of induction are independent, mainly due to the lack of contraposition. The second form is given explicitly in terms of two variables to avoid contraction.

The least number principle and antisymmetry of $\leqslant$ entail that least numbers are *unique* – if x and y are both least φ, then $x = y$. Proof: if $\forall u(\varphi u \Rightarrow x \leqslant u)$, then $\varphi y \Rightarrow x \leqslant y$; so if φy, then $x \leqslant y$; mutatis mutandis $y \leqslant x$. For this reason, Axiom VI, although very useful (cf. Theorem 40), seems too strong, in light of the philosophical picture concerning uniqueness from earlier in the book.

Since the results in this section are under the extra induction hypotheses, we will suspend our main numbering of theorems and label the ones in this section with letters rather than numbers.

Division. Notation:

$$\mathsf{div}(x, y) := \exists n(xn = y)$$

means that x *divides* y.

Two miscellaneous results:

- $2sx = sy \Rightarrow sx < sy$
- $sx < 2sx$

These are again about all $x, y > 0$, but notation relieves the need for restricting arrows. To prove the first, if $2sx = sy$, then $sx + sx = sy$ so $\exists n(sx + n = sy)$. And for proving the second, $sx + sx = sx + sx$, so $\exists n(sx + sn = 2sx)$.

Lemma A.   *The following are true:*

(i) $\mathsf{div}(x, y) \,\&\, \mathsf{div}(x, z) \Rightarrow \mathsf{div}(x, yn + zm)$
(ii) $\mathsf{div}(x, y) \,\&\, y > 0 \Rightarrow x \leqslant y$

Proof   Proof of (i) is calculation. If $xr = y$ and $xt = z$, then $yn + zm = xrn + xtm = x(rn + tm)$, using substitution and Proposition 31.

The argument for (ii) could have avoided making $y > 0$ explicit by writing $\mathsf{div}(x, sy) \Rightarrow x \leqslant sy$, but as it turns out the extra conjunct isn't in the way. On the other hand, proof of (ii) requires the information contained in $xn = y$ to be extracted twice. Even though $=$ contracts, we avoid any appeal to contraction by breaking this premise up into separate pieces, via extensionality. Suppose $xn \geqslant y$, and $y > 0$. Then $xn > 0$ by Proposition 34, and $n > 0$ from the corollary to Proposition 36. Then $x \leqslant xn$ by Archimedes (6.4). Now, supposing that $xn \leqslant y$, then $x \leqslant y$ by transitivity of $\leqslant$. That shows

$$(xn \leqslant y) \,\&\, (y \leqslant xn) \,\&\, y > 0 \Rightarrow x \leqslant y.$$

As $xn = y \Rightarrow (xn \leqslant y) \,\&\, (y \leqslant xn)$, then transitivity of $\Rightarrow$ and existential generalization completes the proof.   □

The "divides by" relation is an equivalence:

- $\mathsf{div}(x, x)$ because $x1 = x$ (by Proposition 29).
- $\mathsf{div}(x, y) \,\&\, \mathsf{div}(y, x) \Rightarrow x = y$ by two applications of Lemma A(ii) (if $y = 0$, then $x = 0$ too) and antisymmetry of $\leqslant$.
- $\mathsf{div}(x, y) \,\&\, \mathsf{div}(y, z) \Rightarrow \mathsf{div}(x, z)$ because if $xn = y$ and $ym = z$, then $xnm = z$ by substitution.

This brings us to a substantial number theoretic result.

Theorem B (Division Algorithm).   *Let* $0 < y \leqslant x$. *Then* $\exists q \exists r(x = yq + r)$, *with* $r < y$ *and* r, q *both unique.*

Proof   Let $0 < y \leqslant x$.

Existence: there is at least one u such that $\exists q(x = yq + u)$, namely x itself: $x = y0 + x$. So there must be a least such u, called r. That shows existence – and as a theorem, it can be reused.

Now we show that $r \leqslant y$, using linearity (Proposition 36) – twice: $r \leqslant y \vee y \leqslant r$ is a theorem, so conjoin it to itself, distribute, and drop off any unwanted extra conjuncts, so that

$$r \leqslant y \vee (y \leqslant r \,\&\, y \leqslant r)$$

is a theorem, too. We argue that the second disjunct implies absurdity, using the minimality of r. If $y \leqslant r$, then $y + n = r$ and

$$x = yq + (y + n) = (yq + y) + n = y(q + 1) + n$$

using Proposition 31. But then $\exists v(x = yv + n)$, and so $r \leqslant n$ because r is the least. But, again using the assumption that $y + n = r$, it must be that $r > n$, since $y > 0$. Then $r < r$, which is absurd; so $r \leqslant y$ as desired. Again this is proven, and reusable. So to show that the relation is *strict*, consider $r + m = y$. If m is a successor, then stop; if $m = 0$, then $r = y$; then $x = yq + y = y(q + 1)$, and then 0 is the least remainder, so $r = 0 < y$.

Uniqueness: let q, r be the *least* such that $x = yq + r$. Suppose $x = yu + v$. Then $q \leqslant u$ and $r \leqslant v$ by minimality. If $q = u$ or $r = v$, then they are unique (stop), so suppose $q < u$ and $r < v$. Then by definition there is an n such that $q + sn = u$ and an m such that $r + sm = v$. By substitution and (6.3),

$$x = yq + r < y(q + sn) + (r + sm) = x,$$

whence $x < x$. So $q < u \,\&\, r < v$ imply $\bot$, which in turn implies $q = u \,\&\, r = v$; since trichotomy exhausts the cases, and all the inequalities are absurd, q, r are unique after all.   □

Read the following as: t is a *greatest common divisor* of x and y:

$$\mathsf{gcd}(x, y, t) := \mathsf{div}(t, x) \& \mathsf{div}(t, y) \,\&\, \forall u(\mathsf{div}(u, x) \& \mathsf{div}(u, y) \Rightarrow u \leqslant t).$$

By uniqueness of least number(s), $\mathsf{gcd}(x, y, s) \,\&\, \mathsf{gcd}(x, y, t) \Rightarrow s = t$, making t *the* greatest common divisor. Using a function symbol, we write $\mathsf{gcd}(x, y) = t$. It has properties:

- $\mathsf{gcd}(0, 0) = 0$
- $\mathsf{gcd}(x, y) = \mathsf{gcd}(y, x)$
- $\mathsf{gcd}(x, y) = \mathsf{gcd}(y, r)$ when $x = yq + r$

To show the third property, compute. Since a greatest common divisor is a common divisor, $\mathsf{gcd}(x, y)$ divides x and $\mathsf{gcd}(x, y)$ divides y. Then $\mathsf{div}(\mathsf{gcd}(x, y), r)$. Then

$$\mathsf{gcd}(x, y) \leqslant \mathsf{gcd}(y, r)$$

because a greatest common divisor is the greatest. (To make sure this is proved without contraction, don't *start* with the conjunctive claim that $\mathsf{gcd}(x, y)$ is the greatest common divisor; then you would have to repeat the assumption every time you wanted to get at one of the conjuncts. Instead, assume it is a common divisor, and assume it is a greatest divisor; then put these together at the end.) Similarly, $\mathsf{div}(\mathsf{gcd}(y,r), y)$ and $\mathsf{div}(\mathsf{gcd}(y,r),r)$ implies $\mathsf{div}(\mathsf{gcd}(y,r),x)$, so

$$\mathsf{gcd}(y,r) \leqslant \mathsf{gcd}(x, y).$$

Antisymmetry gives the result.

Theorem C. $\forall x \forall y \exists t\,(\mathsf{gcd}(x, y) = t)$.

Proof Consider x, y. Let g be the least number such that $\exists u \exists v (g = xu + yv)$. From Theorem B, there are unique q, r such that $x = gq + r$. By minimality, $g \leqslant r$; but if $r > 0$, then $r < g$, which by Proposition 34 implies $g < g$. So $r = 0$ and therefore g divides x. Similarly, g divides y. So g is a common divisor. To show it is greatest, just observe that for any g' that divides x and y, then g' divides g, and therefore by Lemma A, $g' \leqslant g$.   □

Corollary D. $\exists u \exists v (\mathsf{gcd}(x, y) = xu + yv)$.

Prime Numbers. The aim now is proving the fundamental theorem of arithmetic and the infinitude of primes. A theme recurs, of theorems *up-to-inconsistency*, e.g., claims of the form "$\varphi(x)$ or else $x \neq x$," which are classically equivalent to their classical statements but are not.[17]

A number $p > 1$ is *prime* iff

$$\forall x (\neg \mathsf{div}(\mathsf{ss}x, p) \lor \mathsf{ss}x = p).$$

A number $n > 1$ is *composite* iff $\exists x (\mathsf{div}(\mathsf{ss}x, n) \mathbin{\&} \mathsf{ss}x \neq n)$.

Proposition E.   *Every number greater than 1 is either prime or composite.*

Proof Duality of conjunction/disjunction and existential/universal quantifiers.   □

The definition of being prime is given materially, rather than with an implication (e.g., "if $\mathsf{s}x$ divides p, then it equals p"); working without disjunctive syllogism, the material phrasing is not entirely satisfactory – but it lets us show Proposition E. Looking at the proof of Proposition F, an alternative definition might be: $p > 1$ is prime iff $\mathsf{div}(x, p) \mathbin{\&} \mathsf{div}(x, p) \Rightarrow x = p$. Then the proof in Proposition F goes through – but only moves a bump in the carpet.

Proposition F.   *The smallest divisor of n is prime.*

Proof Let q be the smallest divisor of n. That is,

$$\mathsf{div}(q, n) \mathbin{\&} \forall x (\mathsf{div}(x, n) \Rightarrow q \leqslant x),$$

---

[17] This idea (and terminology) was developed in [McKubre-Jordens and Weber, 2016].

which exists by the least number principle. By transitivity of div, if x divides q, then x divides n; if x divides n, then $q \leqslant x$. And if $\text{div}(x,q)$, then, by Lemma A, $x \leqslant q$. We have used $\text{div}(x,q)$ twice, to show that $x = q$ by extensionality. But $\text{div}(x,q)$ & $\text{div}(x,q)$ $\Rightarrow$ $x = q$ implies $\neg\text{div}(x,q) \vee x = q$, by the instance of excluded middle $(\varphi \& \varphi) \vee \neg(\varphi \& \varphi)$. Therefore, by definition of prime numbers, q is prime. $\qquad\square$

Lemma G. *If p is prime and $\text{div}(p, xy)$, then $\text{div}(p,x)$ or $\text{div}(p,y)$, or $p \neq p$.*

Proof Assume p divides xy. By trichotomy, either $\gcd(p,x) > 1$ or $\gcd(p,x) = 1$, since the third case $\gcd(p,x) = 0$ is absurd. But if $\gcd(p,x) > 1$, then by the definition of p being prime, then either p does not divide p (and so $p1 \neq p$) or else $\gcd(p,x) = p$ – whence p divides x. The last case, then, is $\gcd(p,x) = 1$. Then there exist u, v such that

$$1 = pu + xv$$

by Corollary D. Therefore,

$$y = puy + xvy$$

by multiplying y to both sides. But since p divides pu, and p divides xy on assumption, p must divide both puy and xvy. Then p also divides the sum $puy + xvy = y$. $\qquad\square$

The lemma generalizes (with the standard notation, $\prod_{i=0}^{n} x_i := x_0 \times \cdots \times x_n$ and $\bigvee_{i=0}^{n} \varphi_i := \varphi_0 \vee \ldots \vee \varphi_n$).

Corollary H. $\text{div}\left(p, \prod_{i=0}^{n} x_i\right) \Rightarrow \bigvee_{i=0}^{n} \text{div}(p, x_i)$, *or $p \neq p$, for p prime.*

Usually, "Euclid's lemma": (Lemma G) is used in the following proof, but the extensional phrasing makes it inadequate. Induction is used instead; even so, we end up with something weaker than what might be expected.

Theorem I (Fundamental Theorem of Arithmetic). *Let $n > 1$. Then there are primes $p_0, \ldots, p_m$ such that*

$$n = \prod_{i=0}^{m} p_i$$

unique up to inconsistency: for any other such $q_0, \ldots, q_\ell$, either each p_i is identical to exactly one q_j, or some $p_i \neq p_i$.

Proof Existence: by complete induction (Axiom VI). Let $n \geqslant 2$. Suppose there is a prime factorization for all $m < n$. Now, either n is prime, or else $n = ab$ for some $a,b < n$ (pro. E); then by hypothesis $a = \prod_{i=0}^{k} p_i$ and $b = \prod_{i=0}^{\ell} p_i$, so

$$n = \prod_{i=0}^{k} p_i \times \prod_{i=0}^{\ell} p_i$$

as needed.

Uniqueness: consider n with prime factorizations:

$$\prod_{i=1}^{k} p_i = n = \prod_{i=1}^{\ell} q_i.$$

Assume the primes appear in ascending order. Then the plan is to show that $p_i = q_i$ for each i, or else $p_i \neq p_i$ for some i, by complete induction. For suppose that there is a unique prime factorization for all $m < n$, with $1 < m$; then if $p_0 = q_j$ for any of the j, cancel p_0 from both sides (assuming [which we probably should not] multiplicative cancellation, $x = y \Leftrightarrow xsz = ysz$), and the result is, on hypothesis, uniquely the product of primes.

So, using trichotomy, either $p_0 = q_0$ (stop), or else suppose without loss of generality that $p_0 < q_0$. Then $p_0 + a = q_0$ for some $a > 0$. Notably, $a < n$, so on hypothesis a has a unique prime factorization, $a = \prod_{i=0}^{t} r_i$. Then take $q_0 \times \cdots \times q_\ell$, but replace q_0 with a, giving some number

$$m = a \times \prod_{i=1}^{\ell} q_i = \prod_{i=0}^{t} r_i \times \prod_{i=1}^{\ell} q_i.$$

Then $m + (p_0 \times \prod_{i=1}^{\ell} q_i) = n$, and $m < n$. By induction hypothesis, m has a unique factorization. Since p_0 divides n and also divides $p_0 \times \prod_{i=1}^{\ell} q_i$, then p_0 divides m (by Lemma A(i)). Because m has a unique factorization including p_0, then either $p_0 = r_i$, for some one of the rs, or $p_0 = q_i$ for some one of the qs. In the second case, stop. In the first case, p_0 divides a. Then p_0 divides q_0. Since q_0 is prime, then either $p_0 = q_0$, or p_0 both divides q_0 and does not, whence $p_0 \neq p_0$.   □

Corollary J.   *If* $p \neq p \Rightarrow \bot$ *for prime* p, *then prime factorization is unique.*

Uniqueness would be important in applications, e.g., if primes were being used as Gödel codes.

Theorem K (Infinitude of Primes). $\forall p(p \text{ is prime} \Rightarrow \exists q(q > p \ \& \ q \text{ is prime})).$

Proof   For prime p_n, consider all the primes p_i such that $p_i \leqslant p_n$. Take

$$\mathfrak{p} = \left(\prod_{i=0}^{n} p_i \right) + 1.$$

Since p_n divides $\prod_{i=0}^{n} p_i$, by Lemma A, $p_n \leqslant \prod_{i=0}^{n} p_i$ and so $p_n < \mathfrak{p}$. If $\mathfrak{p}$ is a prime, we are done. Otherwise, $\mathfrak{p}$ has a prime factorization $\prod_{i=0}^{\ell} q_i$ by the "existence" part of the fundamental theorem of arithmetic (Theorem I). One of these prime factors of $\mathfrak{p}$ is greater than p_n, because if not, using trichotomy it must be that $p_n \geqslant q_i$ for every $i < \ell$. But in that case $\mathfrak{p} = \prod_{i=0}^{\ell} q_i \leqslant \prod_{i=0}^{n} p_i < (\prod_{i=0}^{n} p_i) + 1 = \mathfrak{p}$, i.e.,

$$\mathfrak{p} < \mathfrak{p},$$

which is absurd (Proposition 34). This exhausts the cases given from trichotomy.   □

6.3 Descent: Inconsistency and Irrationality

Resuming our program, this final section shows two things. First, with respect to the goals announced at the end of Chapter 3, we see that the sorites paradox can be expressed, and the target contradiction vindicated, in paraconsistent arithmetic. Second, with respect to the goal announced at the start of the chapter, we see that the existence of irrational numbers can be established – or at least, it can if there are no infinite descents.

As we are returning from the Excursus of the last section, the induction Axiom VI is suspended and not used again, and we resume the numbering of theorems left off at the end of Section 6.2.3.

6.3.1 Recovering the Discrete Sorites

The discrete sorites is driven by two conflicting assumptions: an empirical premise, TOLERANCE for vague φ,

$$\forall n(\neg\varphi(n) \vee \varphi(n+1))$$

and the least number principle, that if there is a number k that is φ, then there is a *least* such. Without using Axiom VI of the previous Excursus, how can the least number be invoked to recapture the paradox?

Sorites sequences tend to be *finite*.[18] Altitudes, heaps of sand, and the age of a baby are bounded by finite constraints. The least number principle for finite sets can be established by replacing the existential quantifier with (finite) disjunctions.

Let $\theta \in \mathbb{N}$, so $\{n \in \mathbb{N} : n \leqslant \theta\}$ is a finite initial segment of the natural numbers. If there is a $k \leqslant \theta$ such that $\varphi(k)$, then there is a least such:

$$\varphi(0) \vee \cdots \vee \varphi(\theta) \Rightarrow \quad \varphi(0)$$
$$\vee \quad (\neg\varphi(0) \,\&\, \varphi(1))$$
$$\vee \quad (\neg\varphi(0) \,\&\, \neg\varphi(1) \,\&\, \varphi(2))$$
$$\vdots$$
$$\vee \quad (\neg\varphi(0) \,\&\, \cdots \,\&\, \neg\varphi(\theta-1) \,\&\, \varphi(\theta)).$$

The proof simply uses application of (as many copies as needed of) the law of excluded middle to each number. If one of the numbers less than θ is φ, then either 0 is the least; or it's not, but 1 is; or neither 0 nor 1 are, but 2 is; or ... or none of the numbers before θ is, but θ is φ. This is entirely manual "proof by inspection."

---

[18] Thanks to Marcus Rossberg for discussion on this topic.

We may now claim recapture of all finite cases of the sorites. Assuming φ is vague, then it is given that something is φ and something is not. TOLERANCE is the conjunction of disjunctive pairs,

$$(\neg\varphi(0) \vee \varphi(1)) \,\&\, \ldots \,\&\, (\neg\varphi(\theta) \vee \varphi(\theta+1)).$$

Establishing the contradiction is now a matter of putting TOLERANCE and the consequent of the finite least number principle together. Conjoining the (big) TOLERANCE conjunction with the (big) "some k is least" disjunction, and distributing, we have

$$(\varphi(0) \,\&\, \neg\varphi(0)) \vee \ldots \vee (\varphi(\theta+1) \,\&\, \neg\varphi(\theta+1)).$$

One of $k \leqslant \theta$ is the first non-φ, and hence there is some k both φ and $\neg\varphi$. A contradiction, as required.

Thus one can check, given enough time, Wang's version of the sorites in a paraconsistent setting. If 1 is a small number, but a googolplex (1 followed by a googol zeros) is not a small number, then some number (either 1 or ... or $10^{10^{100}}$) is both small and not small. That's a lot of "or's," but so it goes. As for infinitary sorites, where this finite-check procedure no longer works (even discounting "merely medical" limitations), these are possible, but will be covered in the case of continuity and topology, to be handled in the next chapters. The finite argument shows just how low tech the sorites paradox really is.

6.3.2 Even, Odd, and Irrational Numbers

We near completing our task for this chapter, proving $\sqrt{2}$ irrational, with some elementary observations about even and odd numbers.[19]

A number x is *even* iff $\exists n(2n = x)$, and *odd* iff $\exists n(2n+1 = x)$. Odd numbers are the successors of even numbers: if a number is even, its successor is odd. And only slightly less trivially, if n is odd, then its successor $sn = 2x + 2 = 2(sx)$ is even.

Lemma 37. *Every number is either even or odd. No number is both even and odd – on pain of absurdity.*

Proof First, exhaustion. Base case: 0 is even, so it is even or odd. Induction step: if n is even, then sn is odd; and if n is odd, then sn is even. So if n is even or odd, then sn is even or odd, and so too for all the numbers.

Next, exclusion, on pain of absurdity. The case 0 is not both even and odd, because if it were odd, then $0 = 2y + 1$ for some y (no). Suppose that

$$\exists y \exists z(2y = n = 2z + 1) \Rightarrow \bot.$$

That is, n being both even and odd is absurd. But then, if $sn = 2y + 1 = 2z + 2$ is both even and odd, then $n = 2y = 2z + 1$, which is, on assumption, absurd. $\square$

---

[19] Abstractly, the target theorem is implied by the fundamental theorem of arithmetic: the equation $a^2 = 2b^2$ contradicts unique prime factorization [Stillwell, 1998, p. 19]. But, not having quite the uniqueness of prime factors that the classical argument would require, and to avoid reliance on Axiom VI, we take a more ground-level approach.

Corollary: if a number is even, then it is not odd; if odd, then not even. A fortiori,

$$(n \text{ is even} \Rightarrow n \text{ is odd}) \Rightarrow n \text{ is odd}$$

and *mutatis mutandis* for "odd implies even."

Lemma 38. *If x is odd, then x^2 is odd.*

Proof If $x = 2y + 1$ for some y, then using distribution of multiplication over addition,

$$\begin{aligned}
x^2 &= (2y + 1)^2 \\
&= (2y + 1)(2y + 1) \\
&= 2y(2y + 1) + (2y + 1) \\
&= 2y2y + 2y + 2y + 1 \\
&= 2(2y^2 + 2y) + 1
\end{aligned}$$

so $\exists z(x^2 = 2z + 1)$. □

To prove the contrapositive of Lemma 38 requires Lemma 37:

Proposition 39. *If x^2 is even, then x is even.*

Proof Suppose that x^2 is even. If x is even, then stop. If x is odd, then x^2 is odd (Lemma 38), so it is both even and odd. Then by Lemma 37, x is even. □

This brings us to the final theorem of the chapter, showing $\sqrt{2}$ is irrational "and no funny business" [Slaney, 1982].

Pythagoras knew that $a^2 + b^2 = c^2$, when a, b are the legs of a right triangle and c the hypotenuse. A unit square with sides of length 1 then would have a diagonal $\sqrt{2}$. The story goes that he hoped there would be rational numbers to represent this magnitude. But there are dreams that cannot be. How does the standard proof go?

We assume that $\sqrt{2}$ is rational, in order to derive the negation of the assumption by reductio. The main problem to face is that, on close inspection, the proof requires using the assumption more than once. The informal argument is as follows. Either $\sqrt{2}$ is expressible as a rational number, or not. Suppose there are natural numbers x, y such that $\sqrt{2} = \frac{x}{y}$. Then the rational expression can be put in lowest terms. So (on assumption) there exist x, y such that, rearranging, $\sqrt{2} = x/y$ iff $2 = (x/y)^2 = x^2/y^2$ iff

$$2y^2 = x^2,$$

where either x or y is odd. But since x^2 is even (on assumption, for a second time), then x is even: there is some z such that $2z = x$. Then on assumption (again!) by substitution, $2y^2 = 4z^2$, so dividing out, $y^2 = 2z^2$. But then both x and y are even, not in lowest terms at all – contradiction.

A geometric illustration of this proof makes the idea vivid.[20] Consider a square that is twice another:

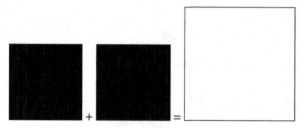

Then move the two smaller squares into the bigger one, allowing them to overlap,

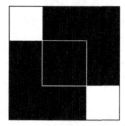

and the result is three new squares, in exactly the same proportion as the originals.

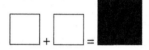

Repeating the argument again and again will result in successively smaller and smaller squares. If infinite descents are impossible in $\mathbb{N}$, the result follows.

There are many problems with the argument, if it is to be captured in subDLQ. What exactly is the formulation of the contradiction? Is it grounds for rejecting the assumption? (Paraconsistently, this sort of question does not have a uniform answer.) If it is, can this assumption be singled out for removal without using disjunctive syllogism? And most damningly, the initial assumption was used (at least) three times – dangerous practice when Curry sentences are out there in the aether.[21] And *most* most damningly, what if there are infinite descents after all?

To the proof. We skip the step of showing that $2y^2 = x^2$ is equivalent to $\sqrt{2} = \frac{x}{y}$, which requires foray into multiplicative inverses, beyond our goals here. So, following Slaney, we consider whether "at least a close relative of the irrationality of $\sqrt{2}$ is in a good sense free of contraction," and also follow his council of "postponing the decision as to whether to laugh or cry until more evidence is in" [Slaney, 1982].

---

[20] Attributed to John Conway.

[21] Since it is an *equation* that is being contracted on, perhaps the proof is not so damning, given that identity contracts (5.2). In any case, the proof that follows steers around the issue.

Theorem 40. *For any x, y, if either x or y is odd, then*

$$2(\mathbf{s}y)^2 \neq x^2.$$

Proof Suppose $\exists x \exists y(2(\mathbf{s}y)^2 = x^2)$. Then x^2 is even, so x is even, using Corollary 39. But then $\exists n(x = 2n)$, so by substituting into the supposition,

$$2(\mathbf{s}y)^2 = (2n)^2 \Rightarrow 2(\mathbf{s}y)^2 = 4n^2$$
$$\Rightarrow (\mathbf{s}y)^2 = 2n^2$$
$$\Rightarrow (\mathbf{s}y)^2 \text{ is even.}$$

So both x^2 and $(\mathbf{s}y)^2$ are even, using the supposition twice.

Now, we are presuming that one them is odd. Then on the supposition three times, one of them is both even and odd, which by Lemma 37 means $\bot$: by reductio, the supposition that $\sqrt{2}$ is rational is false, as was to be shown. □

What if both x and y are always even? It would mean the fraction x/y has no lowest terms. If we assume the least number principle (Axiom VI) from our Excursus midchapter, then the following becomes available:

Lowest Terms: For all z, if $\exists x \exists y(x = zy)$, then $\exists x \exists y((x = zy)$, and either x or y is odd).

Proof If there exists x, y, such that $x = zy$, then by least number (Axiom VI), there are such x, y where $\forall u \forall v(u = zv \Rightarrow x \leqslant u \ \& \ y \leqslant v)$. That is, if $n = zm$, then $x \leqslant n$ and $y \leqslant m$ by minimality. Assuming multiplicative cancellation, if $2n = z2m$, then $n = zm$. And recall that $n < 2n$ and $m < 2m$. Putting these together, if x and y are both even, $x = 2n$ and $y = 2m$, then $x \leqslant n < 2n = x$ and $y \leqslant m < 2m = y$; so $x < x$ and $y < y$, which is absurd. Then one of x or y is odd, by the corollary to Lemma 37. □

With this, the main theorem can be sharpened up – one of x^2 and $(\mathbf{s}y)^2$ must be odd. But Axiom VI is a substantial additional assumption that almost certainly cannot be derived from the set theory of Chapter 5. We know that there is no natural number $x < x$, and can establish lowest terms up to any given finite n (as in the sorites recapture). But without lowest terms, infinite descents can't be ruled out.

This chapter has been predominantly a "recapture" of some basic arithmetic, in part as steps toward addressing the Feferman objection (Section 3.1.2.1): we sustained long chains of (fairly) ordinary reasoning, give or take a non-self-identity; we proved that $1 + 1 = 2$. But, as with set theory in the last chapter, there are new things to see in old places. Where we've ended up is that either there are irrational numbers, and so a field of *real* numbers to consider; or there are ratios of natural numbers with no lowest terms. Either way, this means it is time to turn our attention beyond the counting numbers.

7

Algebra

This chapter transitions from the foundational topics of the previous two chapters, to more "working" mathematics. Already some striking phenomena have emerged, especially in set theory, suggesting the prevalence of inconsistent substructure in ordinary places; from here, we will expand our technical and conceptual horizons further, with the goal of explaining the paradoxes by drawing a mathematical picture of them.

If our question is asking for something spatial, geometric – where are the paradoxes, and how are they shaped? – then to answer it, we must develop a more qualitative sense of *space*. That is the purview of real analysis and topology, the subjects of the next two chapters. But one of the great (ironic?) lessons of modern mathematics from Descartes and Galois is that to get a grip on the geometry of space, we will need some *algebra*. We have puzzled over equations with "inconsistent residue," and will increasingly need clear rules for how to work with these objects in a systematic way. The field of real numbers and its generalizations are draped over the frame of abstract algebra.[1]

This chapter builds up intuitions about some simple algebraic concepts – vectors, groups, rings, and fields. The discussion is organized around how to deal with an open problem in inconsistent mathematics from Dunn and Mortensen, of how to allow some nontrivial inconsistency in fields.

7.1 Algebra for Inconsistent Mathematics: A Triviality Problem

How can we *calculate* in an inconsistent environment, balancing equations amid controlled chaos? Here is a short version of a long story.

A longstanding source of intrigue for paraconsistent mathematics is the possibility of returning to the intuitive *infinitesimal calculus* of Leibniz and Newton, which was inconsistent.[2] Early calculus was concerned with rates of change. A change occurs over time, but the genius of calculus is to work with *change at an instant*: the limiting case of "change

---

[1] See [Grey, 2008]. A textbook on Riemannian geometry opens: "If you've just completed an introductory course on differential geometry, you might be wondering where the geometry went" [Lee, 1997, p. 1]. For standard sources on some of the material in this chapter, see, e.g., [Cohn, 2003; Beardon, 2012].

[2] For example, [Priest and Routley, 1989a, p. 374; Mortensen, 1995, chs. 5–7; Brown and Priest, 2004; Colyvan, 2012, 7§1.2]. Along a different paraconsistent calculus track, see [da Costa, 2000].

over time" as time approaches zero. On the one hand, an instant has no temporal extension, but on the other hand it must, because there cannot be "change" over *no time whatsoever*. Calculus embraces this tension and gives a way of measuring how a curve is changing at a point. The method looks to be by giving an approximation, but then instead of "rounding off" the error, at the limit any discrepancy *disappears*.[3] This was done using *infinitesimals*, infinitely small but nonzero quantities. These were numbers (7.1) big enough to divide by, but (7.2) small enough to vanish at the end of a calculation. It was soon pointed out, then, that the original calculus contained "ghosts of departed quantities," in Berkeley's phrase.[4]

Now, if infinitesimals are somehow "there but also not there," then given the availability of paraconsistent logic, and the intuitive beauty of the original Newton–Leibniz idea, it seems appealing to try to rehabilitate the infinitesimal calculus, since some "small" inconsistency is not necessarily a problem. As usual, though, matters turn out not to be so simple as to plunk down standard looking mathematical principles into some nonexplosive logic. There are some immediate impediments. Let us set up the constraints on how field operations can behave, by looking at a problem, the *Dunn–Mortensen triviality result* for fields.

The real numbers $\mathbb{R}$ form a field (Section 7.3.2.2, and Chapter 8), which is widely thought to confer properties on $x, 0, 1 \in \mathbb{R}$ such as

- $x - x = 0$.
- $x \times 0 = 0$.
- If $x \neq 0$, then $\exists y (x \times y = 1)$.
- $x \times 1 = x$.
- $x = y \Rightarrow x - y = 0$.
- $x \neq y \Rightarrow x - y \neq 0$.

However, as has been observed since the early days of paraconsistency, assuming all of this standard algebra makes it very difficult to keep inconsistency from spreading. The binary relations $\leqslant$ and $=$ seem *designed* for properties of numbers to be transmitted around. Suppose, for example, that for some a, b that

$$a = b \tag{7.1}$$

but also

$$a \neq b \tag{7.2}$$

That is, suppose (as we have already found in set theory and will visualize in Section 7.2) that there is a non-self-identical number. Then from (7.2) and the last assumption listed, that nonidentical quantities have a nonzero difference,

$$a - b \neq 0. \tag{7.3}$$

---

[3] See the introductions to [Bell, 2008; Goldblatt, 1998]; see also [McKubre-Jordens and Weber, 2016].
[4] In *The Analyst* 1734; see Boyer [1959, ch. 6].

From (7.3) and the assumption of the existence of multipicative inverses for nonzero quantities (written in the usual way),

$$\frac{(a-b)}{(a-b)} = 1. \tag{7.4}$$

From (7.1) and the assumption that the difference between any quantity and itself is zero,

$$a - b = 0. \tag{7.5}$$

From (7.5), the existence of multiplicative inverses for nonzero numbers, and the assumption that any quantity multiplied by zero is zero,

$$\frac{(a-b)}{(a-b)} = \frac{0}{(a-b)} = 0. \tag{7.6}$$

From (7.4) and (7.6) and the transitivity of identity,

$$0 = 1.$$

This elementary observation is due to Dunn via Mortensen, and is one of the outstanding problems in inconsistent mathematics and, in particular, paraconsistent real analysis: "... the functional structure of fields interacts with inconsistency to produce triviality in even the purely equational part of the theories ..." [Mortensen, 1995, p. 43].[5]

As ever, there are many moving parts. To hone in on what I think is driving the problem, let's see how a simpler argument establishes something similar over inequalities, namely that if any quantity is strictly less (greater) than itself, everything is. (We saw in the last chapter (Proposition 34) that no natural number $n \in \mathbb{N}$ can be self-separated, on pain of triviality, but that does not prohibit deviant behavior elsewhere.) That is, if there is any x such that

$$x < x,$$

then, using some conventional moves, it seems every y is also self-separated:

$$x < x \Rightarrow x + y < x + y$$
$$\Rightarrow (x + y) - x < (x + y) - x$$
$$\Rightarrow y + (x - x) < y + (x - x)$$
$$\Rightarrow y + 0 < y + 0$$
$$\Rightarrow y < y.$$

Add y to both sides of the inequality, and then take x away, and local inconsistency becomes global inconsistency.

---

[5] See Mortensen [1995, ch. 6]; Batens et al. [2000, pp. 203–208]. For the suggestion that this might not be so bad, see [Estrada-González, 2016].

While there may be some metaphysical story one could tell to make sense of a totally self-separated continuum (the dual of an indecomposable continuum?[6]), or more generally a reality in which every object is separate from itself (if only Sartre were here), it seems wholly antithetical to the paraconsistent project: wherein noise should not overwhelm sense, and local inconsistency is meant to be contained. Minimally, $0 = 1$ is out of bounds; we demurred at the prospect of $\forall x(x \neq x)$ back in Chapter 4; so too, I think, for $x < x$ for *all* x.

To solve the problem, we look for how similar problems are handled elsewhere paraconsistently. To this end, observe that in the preceding derivation of $x < x \Rightarrow y < y$, it is the assumption

$$x + (-x) = 0$$

that generates the spread. Squinting, this equation is analogous to an "ex falso" condition, that

$$p \mathbin{\&} \neg p \Leftrightarrow \bot.$$

In the logical case, we don't achieve paraconsistency with the protocol that "all contradictions are $\bot$, but also some are not";[7] rather, we drop ex falso entirely, perhaps still allowing specific (provable) instances thereof. So too with $x - x = 0$. We are *expecting* there to be (or at least not prohibiting) numbers x and infinitesimal quantities $\eta > 0$ for which $x - x \approx \eta$. If there exist nonzero nullities – infinitesimals – the field algebra should be altered the way logic is made nonexplosive: for *some* x, the self-difference $x - x$ is not zero. The universal connection between $x - x$ and 0 is to be not merely negated (say by $\exists x(x - x \neq 0)$, which is compatible with $\forall x(x - x = 0)$), but severed. Local information is not automatically forwarded to the global network.

That's the idea: find the right algebra to fit the space – a method that informs and is informed by the *structure* of space. One way to think of a space is in terms of an ordering on the parts of the space, where, minimally, the part that is the whole space itself is the "top," and the part that isn't any of the space – the empty set – is the "bottom." Already in the set theory in Chapter 5, though, this idea got complicated when we turned up some funny objects: most saliently, relative complements $\varnothing_X = X \cap \overline{X}$ that are not necessarily identical to the absolute empty set $\varnothing = \{x : \bot\}$. For two spaces X and Y, then, the part of X that isn't any of X and the part of Y that isn't any of Y may not be the same object; each space has its own "intrinsic bottom." At the algebraic level, this means that the ordinary group operations such as $x - x = 0$ need to be reconsidered.

At the same time, the intent is that the mathematical core of real analysis go undisturbed; so we don't want an overly radical reconfiguring of the field properties. That means keeping the door open for $x - x = 0$ in *some* cases, and more generally for operations such as

---

[6] [Hellman and Shapiro, 2018, p. 3 and passim]
[7] Ripley floats the idea of *paracoherentism*, which might allow some isolated pools of absurdity, as long as transitivity of consequence is foregone. See his [Ripley, 2015c] in [Caret and Hjortland, 2015].

addition and multiplication to be *functions*, with unique outputs.[8] The solution we pursue is as with $\varnothing_X$, to introduce, for each x, its own "intrinsic" zero,

$$\mathbf{0}_x := x - x.$$

The idea for heading off the Dunn–Mortensen problem, then, is that if $x < x$, then we can add y to both sides, $y + x < y + x$, and we can subtract x from both sides, $y + (x - x) < y + (x - x)$, but in doing so we just get $y + \mathbf{0}_x < y + \mathbf{0}_x$, and that's it. And for the trouble with dividing by the residual of $a - b$ when $a \neq b$, once the assumptions for the "zeros" of a field are suitably adjusted, the argument to $0 = 1$ breaks down, e.g., because $a - b = 0$ does not follow from $a = b$. (Cf. Section 8.1.3.2.) This will, for balance, also require for each x intrinsic units,

$$\mathbf{1}_x := \frac{x}{x}.$$

These can be thought of as inconsistent parts or the "residue" of inconsistent quantities – but before we start telling stories about the equations, let's try it out, first through some informal practice.

7.2 Vectors

Spaces are places you move around in. Calculus describes certain kinds of motion in space, and the sorites paradox indicates that some of this movement is, in some sense, inconsistent: you take a step forward and in doing so both enter and do not enter the desert (to desecrate Lewis's example [Lewis, 1986, p. 212]). What does that look like? We can get some intuition for these spaces by looking at pictures.[9] Wittgenstein will give the benediction [Wittgenstein, 1922, 4.461]:

Tautologies and contradictions lack sense. (Like a point from which two arrows go out in opposite directions to one another.)

7.2.1 Addition

Vectors are basic for thinking about physical movements such as displacement and force. A vector is a *magnitude* with *direction*, that is well-behaved when combined with other vectors.[10] A *free* vector is not attached to any particular point of origin and is pictured as an arrow; let's focus on those. The first thing to get clear on is what "well behaved when combined with other vectors" means. Suppose there are two vectors:

---

[8] While keeping in mind the sensitive nature of working with these functions, as per Chapter 5. It would take us too far afield to investigate "relational" field operations.

[9] Pictures in mathematics can be "windows into Platonic heaven," according to Brown [Brown, 2008].

[10] Despite being ubiquitous and intuitive for physics, *defining* a vector is fraught. Banesh Hoffman, a collaborator with Einstein, late in his career wrote a little book about vectors that is intended "as much to disturb and annoy as to instruct" [Hoffman, 1966, p. v]. Recommended. Inconsistent vector spaces are considered in [Mortensen, 1995, p. 84].

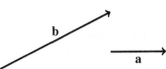

(Vectors are written with boldface letters, **a**, **b**, ...) Then their sum, the result of "doing" one vector and then the other, is given by shifting the tail of one of the vectors to the tip of the other. The direct path from start to finish is drawn with a dashed line:

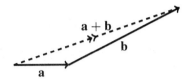

The order doesn't matter, since you finish in the same place if you start in the same place,

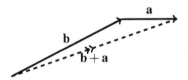

so vector addition is commutative, **a** + **b** = **b** + **a**. This is called the *parallelogram law*.

7.2.2 Subtraction

Every vector **a** has an opposite, −**a**, which has the same magnitude but pointed in the opposite direction:

I am drawing the "deleted" vector – the space it once occupied – as a dotted line. Conventionally, the length of a vector can be scaled by a real number; so in particular, the opposite of a vector can be obtained by scaling it by (the number) −1. Having not yet developed any theory of negative numbers, though, we will be content with saying "a vector pointing in the opposite direction."

Now, what is the result of adding a vector to its opposite? A vector is a magnitude with direction – including the degenerate case of a vector with *zero* magnitude. The vector with magnitude 0 is the *absolute zero* vector **0**. It is the additive identity element:

$$\mathbf{a} + \mathbf{0} = \mathbf{a}$$

for every vector **a**. But since a vector must have some direction, and **0** does not, then according to the classical theory, the absolute zero vector has *every* direction, in good explosive fashion. The absolute zero vector "points in all directions." Compare this to the true empty set, $\varnothing$, which has no members and so "all" of its members are purple unicorns.

Objects such as **0** exist, but from a paraconsistent viewpoint are insufficient for describing all data; there may be objects corresponding to the sum $\mathbf{a} + -\mathbf{a}$ that have some specific direction(s) and not others. Consider thus *relative zero* vectors $\mathbf{0_a}$, which satisfy the equation

$$\mathbf{a} - \mathbf{a} = \mathbf{0_a}$$

(writing $\mathbf{a} + -\mathbf{a}$ as $\mathbf{a} - \mathbf{a}$). The idea is that, when **a** has an inconsistent magnitude, its zero retains a nontrivial sense of direction – in fact, two directions. Some *points*, unextended, are pulling in two directions. Let's explore.

7.2.3 Inconsistency

Recalling Chapter 1, take as a vector the shift generated by walking from the top of Olympus Mons, which is high up, to the first not-high-up point on the mountain. There is more than one first such point; or again, the first such point may be in two different places. So the journey to that first place might be pictured as two (or more) journeys overlaid:

That is, "the journey to the first non-high-up-point" is **a**, and so is "the journey to the first non-high-up point" for some other first point:[11] Now reverse this (single) vector:

Since $\mathbf{a} + -\mathbf{a} = -\mathbf{a} + \mathbf{a}$ (by commutativity), this means that the zero of **a** – the magnitude left when the two journeys **a** and $-\mathbf{a}$ are undertaken consecutively – points in two opposing directions:

Deleting **a** from itself leaves something remaining. On the return journey, you may overshoot the original starting point, or not quite reach it, relative to the nonzero distance

---

[11] That's the intuition, anyway. More carefully, even if x and y are both the first φ, it does not follow that $x = y$ (Section 1.2.3). I am using some ordinary expression here, "identifying" (in some sense) the path(s) to x and to y; but this is offered as evocative motivation, not rigorous theory.

between two of the first-not-high points. So the sum **a** + −**a** may not completely cancel out, at least not as a consistentist would expect it to.

7.2.4 *Adding Inconsistent Vectors*

Let us calculate. Take $\mathbf{0}_a$ and combine it with some **b**:

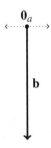

Then the parallelogram rule gives us a whole *range* of destinations – any point "within $\mathbf{0}_a$" of the endpoint of **b**, so to speak:

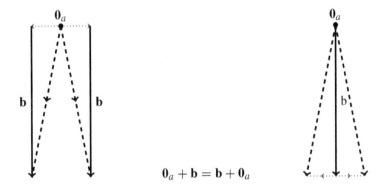

$$\mathbf{0}_a + \mathbf{b} = \mathbf{b} + \mathbf{0}_a$$

To add together two inconsistent zero vectors, then, gives a bigger (and disconnected) range of destinations – but commutativity holds. Given say these two inconsistent zeros,

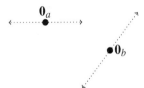

then they add together

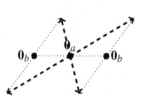

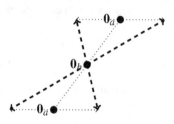

$$0_a + 0_b = 0_b + 0_a$$

So even "bad" vectors remain well behaved under addition:

$$\mathbf{a} + (\mathbf{b} + -\mathbf{a}) = (\mathbf{a} + \mathbf{b}) + -\mathbf{a}$$

This exercise shows what happens when adding inconsistent vectors in general. Given two vectors, their sum is still the result of "doing" one vector and then the other, and the order doesn't matter. There is a range of arrival points – inconsistent (vague?) *direction*.

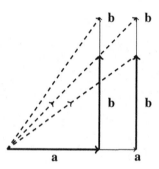

7.2.5 *Diagramming Dialetheias*

I hasten to suggest, though, that whatever their other faults, these pictures of "vague vectors" may be too classical, in that they attempt to depict inconsistency consistently, by parameterizing it (cf. Section 3.3.2). Since what we want to say is that any one point in the range of destinations can correctly be described as *the* point of destination (as in "the first high-up point on the mountain," as in Section 1.2), then even when **a, b** are inconsistent it is perfectly correct to depict their sum as the entirely ordinary and unspectacular

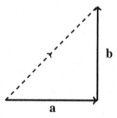

After all, all contradictions are false.

I offer these suggestions with a bigger than usual sized grain of salt. It is time to get more rigorous. Formally, a vector is any element of a *vector space* (defined officially in Section 7.3.2.3), which is a (commutative) group over a field. The next sections build up the simple machinery underneath these terms, and show how our proposed solution to the Dunn–Mortensen triviality problem is implemented.

7.3 Groups, Rings, and Fields

Pictures are good; a language to describe those pictures, and that even makes predictions about those pictures, is good too. Notation may not have the "tactile" quality of a diagram, but good notation repays the loss; as Janich says in praise of the Ricci calculus,"it guides the user through explicit computations . . . [it] *thinks for the user*" [Jänich, 2010, pp. 31, 42]. We have already encountered a slew of binary operations in the last two chapters, but now give the rudiments of a general theory of collections equipped with operations, as abstract algebra.

7.3.1 Groups

A *group* is a structure

$$\langle G, * \rangle,$$

where G is a set and

$$* : G \times G \longrightarrow G$$

is a function such that

(G1) G is closed under the $*$ operation: $\forall a \forall b (a, b \in G \Rightarrow a * b \in G)$.

(G2) For all $a, b, c \in G$, the operation $*$ is associative,

$$a * (b * c) = (a * b) * c$$

and commutative,[12]

$$a * b = b * a.$$

(G3) There is an *absolute unit* $e \in G$ such that, for all $a \in G$,

$$a * e = a.$$

(G4) For every $a \in G$ there is an *inverse* $-a \in G$ and *relative unit* $e_a \in G$ such that

$$a * -a = e_a.$$

---

[12] A group is *abelian* if $*$ commutes, but in this book, all the groups we have cause to consider are abelian.

A "classical" group is one where $e_a = e$ for every $a \in G$. What the preceding definition does, in effect, is swap the standard order of quantifiers: instead of "there is some e for all $a \in G$," axiom (G4) says much more weakly "for all $a \in G$, there is some e_a." These are nonclassical groups.[13]

7.3.1.1 Cancellation

Cancellation for groups, in the presence of relative units, is drastically restricted; that's the aim.

Proposition 41. *For a group,*

(i) $a * c = b * c \Rightarrow a * e_c = b * e_c$.
(ii) $e_c = e \Rightarrow (a * c = b * c \Rightarrow a = b)$.

Proof For (i), by substitution of identicals,

$$a * c = b * c \Rightarrow (a * c) * -c = (b * c) * -c$$
$$\Rightarrow a * (c * -c) = b * (c - c)$$
$$\Rightarrow a * e_c = b * e_c.$$

but the reasoning stops there. In the special case that $e_c = e$, then we have the further fact that $a = b$, but only in this special case. If $a * c = b * c$, then a is identical to b modulo some possible inconsistency of c. They are *identical up to inconsistency*, and, for (ii), behave classically under hypothesis. □

Since there is more visual noise left on the page than the eye is used to seeing in similar contexts elsewhere, there is an option of hiding "leftovers" e_a behind some more notation (arguably, that's what classical mathematics does): introduce an "identity up to inconsistency" relation such as $\simeq$ and say that $a \simeq b$ whenever the only difference between them is some (product of) relative unit elements. An immediate fact about any such $\simeq$ relation is that it would not substitute. An immediate worry about such notation is that it works against the purpose of the program, which after all is to pay attention to, not hide, the possibility of paradox. Whether there would be eventual merit or not in finding ways to clean things up syntactically, I think for now we need to keep everything in sight, so to speak.

---

[13] A *monoid* is a structure satisfying (G1)–(G3) (see Section 7.3.2). One might worry that (G4) is merely notational – it says that for each a there is some object b satisfying $a * b = a * b$, a triviality – and so one could see this as a renaming of monoids as groups. Now, these structures *can* be classical groups, when $e_a = e$, suggesting that this is a generalization of the classical notion of group, not just dropping inverses; but then, the same could be said for monoids. A more direct approach, then, would be to see (G4) as adding inverses as a primitive operation, not just notation, which has the cost of being yet another nontrivial assumption about functionality, but delivers a more robust generalization of groups.

7.3.1.2 The Unit Element(s) and Inverses

Proposition 42.   *The unit element e of a group is unique.*

Proof   Suppose some other e' satisfied (G3), so

$$a * e' = a$$

for any a. Then in particular,

$$e * e' = e. \tag{7.7}$$

But also $a * e = a$ for all a, so in particular

$$e' * e = e'. \tag{7.8}$$

But since commutativity holds by (G2),

$$e * e' = e' * e \tag{7.9}$$

then by transitivity of identity on (7.1)–(7.3),

$$e = e' \qquad\qquad \Box$$

By the same transitivity token, for each a its relative unit e_a is unique, because $e_a = a * -a = e'_a$.

Inverse elements, though, are not unique. Even keeping the relative unit e_a fixed, suppose $a * -a = e_a$ and $a * -a' = e_a$, and try to prove that $-a$ is identical to $-a'$. You know that $-a = -a * e$ and $-a' = -a' * e$, but that's as far as it goes without further assumptions about cancellation. In the special case where there are two inverse elements $-a, -a'$ relative to the *absolute* unit,

$$a * -a = e \quad \text{and} \quad e = a * -a',$$

then using substitution and associativity,

$$\begin{aligned}
-a &= -a * e \\
&= -a * (a * -a') \\
&= (-a * a) * -a' \\
&= e * -a' \\
&= -a'
\end{aligned}$$

so $-a = -a'$. But this requires very special conditions, namely, that e be *the* absolute unit. So there may be more than one thing you can take away from a to get to its bottom e_a.

Accordingly, a group will have solutions in x for the equations

$$a * x = b * e_a \tag{7.10}$$

$$x * x = x, \tag{7.11}$$

but not uniquely. In the first case (7.10), $-a * b$ is one possible value for x, since $a * (-a * b) = (a * -a) * b = e_a * b$, but proving it is the *only* solution calls on cancellation: supposing there is another solution y, we get as far as

$$a * x = b * e_a = a * y \quad \Rightarrow \quad x * e_a = y * e_a$$

and stop, since the whole point of introducing e_as is that they *don't automatically disappear*.

For (7.11), $e * e = e$ is one solution, but not necessarily the only one. Supposing $e' * e' = e'$, then since $e' = e' * e$ by (G3), we get $e' * e' = e' * e$ but cannot cancel beyond

$$e' * e_{e'} = e * e_{e'}.$$

When there is inconsistency around (and there is), these initial calculations show that we are more likely to succeed if we look for solutions in x parameterized by *relative* units. Looking for more is grasping to consistency, about which there is already several hundred years of theory developed.

7.3.1.3 Further Requirements on Inverses?

The double negation property

(G5) $- - a = a$

is usually proved using cancellation. Here is the standard proof for absolute units: if $a * -a = e$ for all a, then $-a * - - a = e$ too, so adding a to both sides, $a * (-a * - - a) = a * e$. Associating, $(a * -a) * - - a = a * e$, so by the definition of (absolute) inverse, $e * - - a = a * e$, and then $- - a = a$ by the definition of unit element. Running this with relative units e_a, the argument stops at

$$e_a * - - a = a * e_a.$$

If we want groups to have this (G5) "double-inverse" property, then we must assume it independently. Doing so fits with methodology in nonclassical logic, say where we drop full classical negation, and then need more negation axioms to fill in the finer-grained work. When it comes to calculating with real numbers, the motivation that $a - a$ does not always vanish seems separate from the question of whether $- - a$ is a, in just the same way that (paraconsistent) negation eschews ex falso quodlibet while still validating double negation introduction and elimination.

Similar comments apply to the property

(G6) $(-a * -b) = -(a * b)$.

The standard proof in this case (assuming commutativity) uses associativity and inverses:

$$(a * b) * (-b * -a) = ((a * b) * -b) * -a$$
$$= (a * (b * -b)) * -a$$
$$= (a * e_b) * -a$$
$$= (a * -a) * e_b$$
$$= e_a * e_b.$$

By transitivity, $(a * b) * (-b * -a) = e_a * e_b$ and hence

$$(-b * -a) * e_{a*b} = -(a * b) * e_a * e_b.$$

The classical proof then assumes that $e_a = e_b = e$ and gets (G6). For us, it will need to be assumed separately.[14]

The upshot is that, in (G5) and (G6), some instances of (absolute) cancellation will be part of the explicit characterization of a group, rather than consequences thereof (cf. Axiom 1 for the real numbers in the next chapter). Just because one goes subclassical does not make one a Luddite. The real cost (or is it a gain?) from adding these seemingly innocuous extra assumptions, aside from some diminished presentational elegance, is that fewer structures satisfy the additional group axioms (G5) and (G6).

7.3.2 Rings, Fields, and Vector Spaces

For the rest of the chapter, we go back to more concrete notation, using $+$ and $\overset{\bullet}{\times}$ for operations and 0, 1 for units. Still, although these symbols are suggestive, they are not here meant to be the real thing. We already know from set theory that $0 = 1$ is absurd, with $0 = \varnothing$ and $1 = \langle \varnothing, \varnothing \rangle$. But just algebraically, with 0 and 1 as symbols, the absurdity of $0 = 1$ isn't forced on us by the structure of a ring or field. Even if we add the assumption that $0 \neq 1$, they could still be identical, paraconsistently speaking. This introduces some intriguing flexibility.[15]

7.3.2.1 Rings

A *ring* is a structure $\langle R, +, \times \rangle$ such that

(R1) $\langle R, + \rangle$ is a (commutative) group, with absolute unit 0, and inverses $-a$ and relative units $\mathbf{0}_a$ (one for each $a \in R$).

---

[14] Leaving the equation as it is, noise and all, has the virtue of honesty, but it seems unduly messy. It would be nice, for instance, at least to consolidate the notation, so, for example $e_a * e_b = e_{a*b}$, but doing this would exactly assume the property in question at the last step of $(a * -a) * (b * -b) = (a * b) * (-a * -b) = (a * b) * -(a * b)$.

[15] Compare this section with [Birkhoff and Maclane, 1999, p. 85].

(R2) $\langle R, \times \rangle$ is a *monoid*:

  (i) Associativity of $\times$ holds, $a \times (b \times c) = (a \times b) \times c$;
  (ii) there is a $1 \in R$ such that $a \times 1 = a$.

(R3) Distribution holds, $a \times (b + c) = (a \times b) + (a \times c)$.

Calculations are fine-grained, as we've already started to see. For example, let R be a ring and ask what, for any $a \in R$, is the product of a with the absolute zero? One might guess that the answer is 0, but facts about cancellation make it not quite so.

Proposition 43. *In a ring,*

$$(a \times 0) + \mathbf{0}_{a \times a} = \mathbf{0}_{a \times a}.$$

Proof From the fact that $\langle R, + \rangle$ is a group,

$$a + 0 = a.$$

Then multiplying by both sides (valid just by substitution of identicals),

$$a \times (a + 0) = a \times a.$$

Distributing with (R3),

$$(a \times a) + (a \times 0) = a \times a,$$

at which point one might wish to subtract $a \times a$ from both sides, which results in, not $a \times 0 = 0$, but rather the more richly nuanced result. □

This will be confirmed again with a slightly different argument in Section 8.1.

7.3.2.2 Fields

A field is a ring with a few more elements: a neutral element for multiplication (a *unity*), and multiplicative inverses. That is, fields are structures with two operations, $+$ and $\times$, with unit 0 for addition, but with the further property that elements also have an inverse for $\times$ *as long as they are not* 0. So classical fields have multiplicative cancellation. This immediately leads to the Dunn–Mortensen problem from Section 7.1, since the same "spread" argument will go through mutatis mutandis: say just for $x, y > 0$, and x^{-1} an inverse,

$$x < x \Rightarrow x \times y < x \times y$$
$$\Rightarrow x^{-1} \times (x \times y) < x^{-1} \times (x \times y)$$
$$\Rightarrow (x^{-1} \times x) \times y < (x^{-1} \times x) \times y$$
$$\Rightarrow 1 \times y < 1 \times y$$
$$\Rightarrow y < y.$$

That's bad. The same sort of solution suggests itself, then: for each a there is an "intrinsic" unity $\mathbf{1}_a$ that cancels only in the special case that $\mathbf{1}_a = 1$, where 1 is the element such that $x \times 1 = x$.

A *field* is a structure $\langle F, +, \times \rangle$ such that

(F1) $\langle F,+ \rangle$ is a (commutative) group, with absolute unit 0 and for each $a \in F$ relative units 0_a.
(F2) $\langle F \backslash \{0\}, \times \rangle$ is a (commutative) group, with absolute unit 1, and relative 1_a for each $a \in F$, such that $1_a = a \times a^{-1}$.
(F3) $a \times (b + c) = (a \times b) + (a \times c)$.

Context should make clear when "unit" and variations thereon refer to 0 or 1.

7.3.2.3 Vector Spaces

A vector is any element of a *vector space* V, which is a (commutative) group over a field F such that all $\mathbf{a}, \mathbf{b} \in V$ and $s, t \in F$ behave according to the following laws:

(V1) $\langle V,+ \rangle$ is a group.
(V2) $s \times \mathbf{a} \in V$.
(V3) $s \times (\mathbf{a} + \mathbf{b}) = (s \times \mathbf{a}) + (s \times \mathbf{b})$.
(V4) $\mathbf{a} \times (s + t) = (\mathbf{a} \times s) + (\mathbf{a} \times t)$.
(V5) $\mathbf{a} \times (s \times t) = (\mathbf{a} \times s) \times t$.
(V6) $1 \times \mathbf{a} = \mathbf{a}$.

Since a vector space is a group, it has a (unique) zero vector $\mathbf{0}$, where $\mathbf{a} + \mathbf{0} = \mathbf{a}$ for all $\mathbf{a} \in V$, just as with groups. Every $\mathbf{a} \in V$ has an (additive) inverse, $-\mathbf{a} \in V$, and a relative $\mathbf{0_a} \in V$ with $\mathbf{a} + -\mathbf{a} = \mathbf{0_a}$. And as F is a field, we have the following. For a vector space V, for absolute $\mathbf{0} \in V$ and $0 \in F$,

$$\mathbf{a} \times 0 = \mathbf{0} \text{ and } s \times \mathbf{0} = \mathbf{0}$$

for all $\mathbf{a} \in V$ and $s \in F$. And

$$(-1) \times \mathbf{a} = -\mathbf{a}$$

in the case of absolute $1 \in F$. For relative units, these results will be messier, as we will see in the next chapter.

As an example of a vector space, consider the set V^n of all n-tuples $\langle x_0, \ldots, x_n \rangle$ with the x_i elements of some field X and $n \in \mathbb{N}$ fixed. Then for all such n-tuples $\mathbf{a}, \mathbf{b}$, by the law of ordered pairs, we have $\mathbf{a} = \mathbf{b}$ iff $x_i = y_i$ for each $x_i \in \mathbf{a}$ and $y_i \in \mathbf{b}$. Define

$$\mathbf{a} + \mathbf{b} := \langle x_0 + y_0, \ldots, x_n + y_n \rangle$$
$$x \times \mathbf{a} := \langle x \times x_0, \ldots, x \times x_n \rangle$$
$$\mathbf{e} := \langle e, \ldots, e \rangle$$
$$-\mathbf{a} := \langle -x_0, \ldots, -x_n \rangle.$$

Let's check commutativity of addition:

$$\mathbf{a} + \mathbf{b} = \mathbf{b} + \mathbf{a}$$

is true because X is a field, so $x+y = y+x$ for each $x, y \in X$. Thus $\langle x_0+y_0, \ldots, x_n+y_n \rangle = \langle y_0+x_0, \ldots, y_n+x_n \rangle$ by substitution. Contraction is not needed either because $=$ contracts. Checking inverses, $\mathbf{a} - \mathbf{a} = \langle x_0 - x_0, \ldots, x_n - x_n \rangle = \langle e_{x_0}, \ldots, e_{x_n} \rangle = \mathbf{e_a}$. Thus V^n is a vector space.

The pictures from Section 7.2 are subsumed into formalism. The transition from intuitive visual representation to dry algebra has its costs, but is worthwhile. In the next two chapters, the formalism will provide new insights, new windows, pictures that can only be drawn in invisible platonic sand.

7.3.3 One-Element Structures

The Dunn–Mortensen triviality result threatens implosion, to reduce the field of, e.g., real numbers to a single point. This is by now a familiar threat; I've rejected the notion that $0 = 1$ and in this chapter and the next do a lot of work to avoid it, via relative zeros. But to end this chapter, it is worth briefly considering the alternative. Is it incoherent to work in some very small algebraic structures, say those with only one element?

A ring may be defined (even in classical mathematics) on a singleton $\{\bullet\}$. This is called the trivial ring, or *zero ring*. All the requirements on a ring may be satisfied by fixing the equations:

$$\bullet + \bullet = \bullet$$

$$\bullet \times \bullet = \bullet$$

Both operations commute just because of the reflexivity of identity. So then associativity of both operations is by substitution, e.g., for $+$,

$$\bullet + (\bullet + \bullet) = \bullet + \bullet = (\bullet + \bullet) + \bullet$$

The unit element(s) coincide, $0 = \bullet = 1$, and $\bullet = -\bullet$ is its own inverse.[16]

Fields are not groups or rings, though. Classically, the smallest finite field is the *galois field of two elements*, F_2. The operations can be defined by the following tables:

+	0	1		×	0	1
0	0	1		0	0	0
1	1	0		1	0	1

Thus in F_2 it is the case that $1 + 1 = 0$ and $1 = -1$.

The real numbers, following Dedekind, are a field along with an ordering relation that is total, and complete, meaning it "has enough points." Classically, there is a structure

---

[16] Classically, there is only one trivial ring, up to isomorphism. It would be worth thinking about zero rings built off inconsistent or substructural objects, e.g., some $a \neq a$, and whether they can be meaningfully distinguished. Could a group with one element have a different shape than another group with one element? This could have applications in algebraic topology, and perhaps real estate fraud – both beyond the scope of this book. For some ideas about the "not-so-fundamental group," see [Deloup, 2005].

answering to this description, and essentially only one: any two real-closed ordered fields are isomorphic. The two element field F_2 cannot be a model of the reals, $\mathbb{R}$, classically [Spivak, 2006, p. 573]. But the classical argument that there is no totally ordered two-point field relies on some assumptions that are not appropriate here. We can see this just by looking at the usual reason that there is no classical field with only one element: because the additive identity must be different from the multiplicative identity. This is mandated either by simply asserting that $0 \neq 1$ as a field axiom, or, what comes to the same thing, saying that all the members of the field that are not 0 form a group, as we've done with (F2) in Section 7.3.2.2, with $F \backslash \{0\} = \{x \in F : x \neq 0\}$. But of course, again taking these purely as symbols, there is nothing to rule out that also $0 \neq 0$, or $0 = 1$. So, for anyone who has ever wondered, however fleetingly, if it shouldn't be possible to *divide by zero* after all, we ask: could there be a *one-element field*?

A one-element field is not completely implausible in context, since by comparison there are very small finite models of other bits of paraconsistent mathematics.[17] For all we've said in this chapter, indeed such a structure can exist, by taking some $\bullet \neq \bullet$ as the single element of the field $F_\bullet$. We've already seen that this (without any inconsistency) supports the zero ring. So $\langle \{\bullet\}, + \rangle$ is a commutative group. But also since $\bullet \in \{\bullet\} \cap \overline{\{\bullet\}}$ (assuming $\bullet \neq \bullet$), we have $\langle \{\bullet\} \backslash \{\bullet\}, \times \rangle$ forming a commutative group over $\varnothing_\bullet$. Since in this field the additive and multiplicative inverses coincide, it is possible to "divide by 0" but harmlessly: the result is $\bullet$. To the very brave, there may be some alluring possibilities to working in a one-element field – especially if that field turned out to be $\mathbb{R}$! Perhaps some open problems in functional analysis could be solved? It would save the trouble of using relative unit elements $\mathbf{0}_x$ to avoid the Dunn–Mortensen result. At the least, it would be a fertile source of counterexamples.

However, to end this discussion on a conservative note, there are hard limits on what can be seriously done with such structures. The field of reals will need to be bigger than a point, once a strict ordering relation $<$ is imposed on it – as we are going to do next.

7.4 A Short Conclusion to a Short Chapter

The introduction of relative complements into the group structure ostensibly solves the Dunn–Mortensen triviality problem, and underwrites some appealing (once you get used to it) intuitions about inconsistent vectors. But this is only so much algebra in the void, so to speak. Let us now take some of this abstract machinery and turn it *on*, to see what it does.

---

[17] E.g. set theory in LP [Weir, 2004], and arithmetic in R [Meyer and Mortensen, 1984]; and recall theorem 22, suggesting a way to 'project' large structures into points. For some more mainstream suggestions, see [Tits, 1957]. Thanks to George Lazou here.

8

Real Analysis

Some things appear to be *continuous*, like the sky.[1] The case for the sorites and the boundary paradoxes of Chapter 1 being genuine sites for contradiction comes down to continuity – that nature makes no jumps, *natura non facit saltus*. The target of this chapter is to give some rudiments of (paraconsistent) real analysis (cf. Section 7.1), a dissection of the linear continuum as that which has no gaps. To keep things manageable, a great many topics (convergence, limits, differentiation, integration, etc.) are not considered; I keep the focus on the topology of the real line, establishing the general principle at stake: that if a change occurs, it must occur somewhere. This is confirmed at the intermediate value theorem (Theorem 52). Along the way, we recapture the (continuous) sorites paradox (Theorem 46), flirt with the existence of infinitesimal quantities, and prove a theorem on "splitting" geometric points, in anticipation of the next chapter.[2]

Leibniz offers us an ominous invitation:

Only geometry can provide a thread for the labyrinth of the composition of the continuum ... and no one will arrive at a truly solid metaphysics who has not passed through that labyrinth.[3]

8.1 Into the Labyrinth: Real Numbers

The following axioms describe a set $\mathbb{R}$ called the *real numbers* that is a totally densely ordered closed field. As with arithmetic in Chapter 6, we defer foundationalist issues, such as construction of the reals from set theory, and take these as postulates. The operations $+, \times$ and relation $<$ are axiomatized in the following discussion, extending their behaviors from $\mathbb{N}$ in light of the discussion in the last chapter. Note that all these axioms are classically

---

[1] Example from Bell in [Bell, 2014].

[2] Thanks to Maarten McKubre-Jordens for previous collaboration on the topic of this chapter – in particular, very insightful suggestions on the formulation of the least upper bound axiom (8.1.4) and the proof of the nested intervals Theorem 48, going back to [McKubre-Jordens and Weber, 2012].

[3] [De Usu Geometriae, 1676] [Leibniz, 2002, p. xxiii]. He also warned of the other labyrinth "where our reason very often goes astray," namely "the origin of Evil."

true, when the logical connectives are read as classical and $\mathbf{0}_x$ is interpreted as 0. Therefore, this chapter is sound with respect to classical real analysis as a model.[4]

8.1.1 Axiom 0

The natural numbers are all real numbers, $\mathbb{N} \subseteq \mathbb{R}$.

8.1.2 Axiom 1: $\mathbb{R}$ Is a Field

Denote addition by the symbol $+$ and multiplication by $\times$. $\mathbb{R}$ is closed under $+$ and $\times$: for all $x, y, z \in \mathbb{R}$, then $x + y \in \mathbb{R}$ and $x \times y \in \mathbb{R}$. Addition and multiplication are associative, commutative, and distribute:

- $x + y = y + x$
- $x \times y = y \times x$
- $x + (y + z) = (x + y) + z$
- $x \times (y \times z) = (x \times y) \times z$
- $x \times (y + z) = (x \times y) + (x \times z)$

There exists in $\mathbb{R}$ elements 0 and 1, and for each x, elements $-x$, $\mathbf{0}_x$, x^{-1} and $\mathbf{1}_x$ (see Section 7.3) such that

- $x + 0 = x$
- $x \times 1 = x$
- $x + (-x) = \mathbf{0}_x$
- $(x = 0 \Rightarrow \bot) \Rightarrow x \times x^{-1} = \mathbf{1}_x$

Some usual notational conventions are: $x - y := x + -y$, and $xy := x \times y$, and $x/y := x \times y^{-1}$.

Following on from the discussion about cancellation in Section 7.3.1, further assumptions are added that would conventionally follow as consequences of the axioms already given, but here do not without additive cancellation:

- $0 = \mathbf{0}_0$
- $x = - - x$
- $x \times -y = -(x \times y) = -x \times y$

---

[4] From an undergraduate textbook: "It may seem surprising that something so fundamental as consensus on the treatment of the real numbers is to some extent a matter of faith and/or hope – hope that the discovery of some hidden inconsistency does not bring the whole structure tumbling down. Given enough assumptions about sets and rules for manipulating them, generations of mathematicians have been able to persuade themselves that it is possible to construct a complete ordered field. ... There have always been dissenters of genius to the general consensus" [Berberain, 1977, p. 12].

8.1.2.1 Properties of Axiom 1

The reals are a field. Let's tease out some of the basic consequences, splitting attention between classical truisms and the more baroque generalizations that obtain using relative zeros and ones.

The final axiom just stated means we can regard additive inverses as products with the unital inverse:

$$-x = x \times -1.$$

Proof: $-x = -x \times 1 = -(x \times 1) = x \times -1$.

In particular, $-(-1) = 1 = -1 \times -1$. Therefore, what we assumed as group axiom (G6) in the previous chapter,

$$-x + -y = -(x + y)$$

is derivable from distribution. Proof: $-x + -y = (x \times -1) + (y \times -1) = -1 \times (x + y)$.

The zero of x may be represented:

$$\mathbf{0}_x = (x + -x)$$
$$= (x \times 1) + (x \times -1)$$
$$= x \times (1 + -1)$$
$$= x \times \mathbf{0}_1.$$

If $1 - 1 = 0$, then this all collapses back into classical field theory.

If $a = a + b$, then $\mathbf{0}_a = b + \mathbf{0}_a$. If $a + b = 0$, then $a + \mathbf{0}_b = -b$, and (repeating, in the absence of contraction) if $a + b = 0$, then $b + \mathbf{0}_a = -a$. More simply, since $0 + a = a$, by transitivity of identity,

$$0 + a = 0 \implies a = 0. \qquad (*)$$

Then these follow immediately:

$$0 = -0 \qquad (8.1)$$
$$\mathbf{0}_x = -\mathbf{0}_x \qquad (8.2)$$
$$-\mathbf{0}_x = \mathbf{0}_{-x} \qquad (8.3)$$
$$\mathbf{0}_x + \mathbf{0}_x = \mathbf{0}_{\mathbf{0}_x} \qquad (8.4)$$
$$\mathbf{0}_x + \mathbf{0}_y = \mathbf{0}_{x+y} \qquad (8.5)$$
$$x \times \mathbf{0}_y = \mathbf{0}_{x \times y} = y \times \mathbf{0}_x \qquad (8.6)$$

For the first, we've assumed $0 + -0 = \mathbf{0}_0 = 0$ with Axiom 1, so apply $(*)$. The others are an exercise in instantiating the axioms, especially making use of the property $-x + -y = -(x+y)$. (For example, for (8.2), you find $-\mathbf{0}_x = -(x - x) = -x + --x = x - x = \mathbf{0}_x$.) The last, (8.6), is true because $x\mathbf{0}_y = xy\mathbf{0}_1 = yx\mathbf{0}_1 = y\mathbf{0}_x$ and allows simplifying a little bit. This also means that $2 \times \mathbf{0}_x = \mathbf{0}_{\mathbf{0}_x}$. Further "laws of zeros" will be developed as they are needed.

As foreshadowed in Section 7.3.1, a few basic calculations from the axioms give some sense of what they mean. Compare this section with [Spivak, 2006, p. 7].

Multiplicative cancellation holds (up to unities) when x is not 0:

$$(x = 0 \Rightarrow \perp) \ \& \ xy = xz \Rightarrow x^{-1}(xy) = x^{-1}(xz)$$
$$\Rightarrow (x^{-1}x)y = (x^{-1}x)z$$
$$\Rightarrow \mathbf{1}_x y = \mathbf{1}_x z.$$

In particular, if $x = 0 \Rightarrow \perp$, then $xy = 0 \Rightarrow \mathbf{1}_x y = \mathbf{1}_x 0$.

What about the property $x \times 0 = 0$? Since $x0 = x(0 + 0) = x0 + x0$, it follows that

$$x0 + \mathbf{0}_{x0} = \mathbf{0}_{x0}, \tag{8.7}$$

which means that multiplication by 0 does not destroy all information as it does classically. More generally,

$$(x \times \mathbf{0}_y) + \mathbf{0}_{xy} = \mathbf{0}\mathbf{0}_{xy}. \tag{8.8}$$

The conditional

$$x - y = y - x \ \Rightarrow \ x + \mathbf{0}_{y/2} = y + \mathbf{0}_{x/2} \tag{8.9}$$

holds; prove it by reasoning as so (assuming $2 = 1 + 1 > 0$):

$$x - y = y - x \Rightarrow (x - y) + y = (y - x) + y = y + (y - x)$$
$$\Rightarrow x + \mathbf{0}_y = y + y - x$$
$$\Rightarrow x + x + \mathbf{0}_y = (y + y - x) + x = y + y + \mathbf{0}_x$$
$$\Rightarrow 1x + 1x + \mathbf{0}_y = 1y + 1y + \mathbf{0}_x$$
$$\Rightarrow x(1 + 1) + \mathbf{0}_y = y(1 + 1) + \mathbf{0}_x$$
$$\Rightarrow 2^{-1}(x(2) + \mathbf{0}_y) = 2^{-1}(y(2) + \mathbf{0}_x)$$
$$\Rightarrow x + 2^{-1}\mathbf{0}_y = y + 2^{-1}\mathbf{0}_x$$
$$\Rightarrow x + \mathbf{0}_{y/2} = y + \mathbf{0}_{x/2}.$$

The last lines use the fact that $\frac{\mathbf{0}_x}{2} = \mathbf{0}_{\frac{x}{2}}$, by calculation: $2^{-1}\mathbf{0}_x = 2^{-1}(x - x) = 2^{-1}x - 2^{-1}x = \mathbf{0}_{2^{-1}x}$.

For the product of two negative numbers, the result is surprisingly clean:

$$(-x \times -y) = (x \times -1) \times (y \times -1)$$
$$= (x \times y) \times (-1 \times -1)$$
$$= x \times y \times 1$$
$$= x \times y.$$

A negative times a negative is a positive – and for once, with no noise.

8.1.3 Axiom 2: $\mathbb{R}$ *Has a Dense Linear Order*

Binary $<$ is such that for each x, y, z in $\mathbb{R}$:

- $0 < 0 \Rightarrow \bot$
- $x < y \Rightarrow \neg(y < x)$
- $x < y \Rightarrow (y < x \Rightarrow y = x)$
- $x < y \Rightarrow (y < z \Rightarrow x < z)$
- $x < y \lor y < x \lor x = y$ (trichotomy)
- $x < y \Rightarrow \exists z(z \in \mathbb{R} \,\&\, x < z < y)$ (density)

The first proposition carries on the "consistency at zero" idea from arithmetic (Section 6.1.3), and rules out $\mathbb{R}$ being a one-point field (Section 7.3.3). The third proposition is a (classically vacuous) form of antisymmetry. The last, density, is classically a consequence of the completeness of the reals (Axiom 3 in Section 8.1.4); it (or the statement that between any two reals are an *infinite* number of other reals) was one of Dedekind's three basic conditions on the real line. For all x, y, z in $\mathbb{R}$, it is assumed that addition and multiplication respect the following order:

- $x < y \Rightarrow x + z < y + z$
- $x < y \,\&\, z > 0 \Rightarrow xz < yz$

8.1.3.1 Properties of Order

From Axiom 2, $x \not< x$, by asymmetry and reductio.

Observe the equivalence:

$$(x = 0 \Rightarrow \bot) \Leftrightarrow (x < 0 \lor 0 < x). \tag{8.10}$$

Left to right: use trichotomy; two of the cases already give what we want. And if $x = 0$, then ex hypothesi $\bot$, so it is in particular what we want. *Right to left:* Suppose $x = 0$. Then by substitution on hypothesis, $0 < 0 \lor 0 < 0$, which implies $\bot$. Therefore, the restriction on multiplicative inverses (that x not be identical to 0, on pain of absurdity) may be restated as

$$x < 0 \lor 0 < x \Rightarrow x \times x^{-1} = \mathbf{1}_x.$$

From the ordering relation $<$ on $\mathbb{R}$, the usual notation can be used:

$$x \leqslant y := x < y \lor x = y.$$

Then $\leqslant$ will have the expected properties: reflexivity,

$$x \leqslant x \tag{8.11}$$

because $x = x$; antisymmetry

$$x \leqslant y \,\&\, y \leqslant x \Rightarrow x = y \tag{8.12}$$

because

$$(x < y \vee x = y) \ \& \ (y < x \vee y = x) \Rightarrow (x < y \ \& \ y < x) \vee x = y$$
$$\Rightarrow x = y \vee x = y$$
$$\Rightarrow x = y$$

(using Axiom 2 and argument by cases); and linearity

$$x \leqslant y \vee y \leqslant x \tag{8.13}$$

follows from trichotomy by $\vee$-introduction: since $x < y \vee y < x \vee x = y$, then $(x < y \vee y < x \vee x = y) \vee x = y$, and therefore by associating, $(x < y \vee x = y) \vee (y < x \vee x = y)$. And it follows that for all x, y, z in $\mathbb{R}$:

$$x \leqslant y \Rightarrow x + z \leqslant y + z \tag{8.14}$$
$$x \leqslant y \ \& \ z \geqslant 0 \Rightarrow xz \leqslant yz \tag{8.15}$$

mirroring the final clauses of Axiom 2 for $\mathbb{R}$.

It is useful to record that the following hold:

$$0 \leqslant a \Rightarrow b \leqslant b + a$$
$$0 \leqslant a \Rightarrow b - a \leqslant b + \mathbf{0}_a$$
$$0 \leqslant a \Rightarrow -a \leqslant \mathbf{0}_a.$$

Then similarly $a \leqslant 0$ implies $\mathbf{0}_a \leqslant -a$, and so forth, and also all of these for $<$.

8.1.3.2 Containing Inconsistency

Let's check that the Dunn–Mortensen triviality problem introduced in Section 7.1 (extending to the spread $x < x \Rightarrow y < y$) is avoided. In classical real analysis (but not necessarily here), one would have

$$x < y \Rightarrow \exists z (z \in \mathbb{R} \ \& \ z > 0 \ \& \ x + z = y). \tag{8.16}$$

An instance of this would be if $x < x$, then $x + z = x$ for some $z > 0$. Then using conventional subtraction, cancelling x from both sides yields $z = 0$, and since $z > 0$, then $0 < 0$. This is bad, as forecast. But we've now seen many times over how intrinsic zeros ensure that the difference between two unequal numbers remains localized. For how would one even arrive at (8.16) in the first place? One would argue that if $x < y$, then $x - x < y - x$, so $\mathbf{0}_x < y - x$. And $x + (y - x) = \mathbf{0}_x + y$. But that means we have only

$$x < y \Rightarrow \exists z \in \mathbb{R}(z > \mathbf{0}_x \ \& \ x + z = y + \mathbf{0}_x). \tag{8.17}$$

which seems harmless. And even if we *could* get as far as (8.16), then subtracting x from both sides, we would still only derive $(\mathbf{0}_x + z = \mathbf{0}_x) \ \& \ (z > 0)$. In terms of the original Dunn–Mortensen problem, we've made it so that even when $x \neq y$, we don't unreflectively divide by $x - y$.

Unless $\mathbf{0}_x = 0$, intrinsic zeros are irreducible: they don't cancel. This means each of them is, in some sense, the "origin for their own number line." If η is some inconsistent number, $(\eta < \eta)$ & $(\eta > 0)$, then $\eta - \eta < \eta - \eta$, so $\mathbf{0}_\eta < \mathbf{0}_\eta$. From there, for example, $\mathbf{0}_\eta + 6 < \mathbf{0}_\eta + 6$ by adding 6 to both sides, but this does not make 6 inconsistent. We should regard any inconsistency about quantities such as $6 + \mathbf{0}_x$ as being about $\mathbf{0}_x$, not about 6, in the same way that p & (the liar sentence is true) is not about p; or how the inconsistent Routleyization $\mathcal{Z}(X)$ of any set X does not make X itself inconsistent (Section 5.2.2.1). Inconsistent expressions track back to an inconsistent starting point. We should also reiterate that talk about inconsistency here is all hypothetical. The axioms for $\mathbb{R}$ being given are all classically true, and nothing said so far suggests they are inconsistent.

There is more to say about the $\mathbf{0}_x$s interaction with order, deferred until infinitesimals are explicitly floated in Section 8.3.1.

8.1.3.3 Order and Signed Numbers

One property that seems wise to *avoid* is the "contrapositive" rule $x \leqslant y \Rightarrow -y \leqslant -x$. The closest we can prove to this is

$$x \leqslant y \Rightarrow -y + \mathbf{0}_x \leqslant -x + \mathbf{0}_y$$

by adding inverses to both sides of the antecedent inequality. And that seems good; this "contraposition" property would be a poor fit with the assumption that $\mathbf{0}_x = -\mathbf{0}_x$ and linearity, by the following line of argument:

1   From linearity, either $0 \leqslant \mathbf{0}_x$ or $\mathbf{0}_x \leqslant 0$.
2   If $\mathbf{0}_x \leqslant 0$, then $-0 \leqslant -\mathbf{0}_x$, by "contraposition."
3   Then $0 \leqslant \mathbf{0}_x$, since $\mathbf{0}_x = -\mathbf{0}_x$ and $0 = \mathbf{0}_0$.
4   So $0 \leqslant \mathbf{0}_x$, from 1 by cases.
5   But then, $-\mathbf{0}_x \leqslant -0$ by "contraposition" again.
6   And $\mathbf{0}_x \leqslant 0$, as in 4.
7   So $0 = \mathbf{0}_x$ for all x, by antisymmetry on 4, 6.

Collapsing all zeros together (how gauche!) would be to go in a different direction than the solution to the Dunn–Mortensen spread problem I've proposed.[5] There are other options, but we forgo assuming "contraposition" in general, retaining it only in the conditional sense

$$\mathbf{0}_x = 0 = \mathbf{0}_y \Rightarrow (x < y \Rightarrow -y < -x)$$

as with most of the preceding expressions. (But remembering again Belnap and Dunn's warning that one should not confuse this added condition as some sort of classical recapture – "*adding* premises cannot possibly *reduce* threat" [Anderson et al., 1992, p. 503].)

---

[5] This derivation could indicate that either linearity or antisymmetry are inappropriate. The total ordering of all reals is not accepted in constructive real analysis, for instance. And maybe it is even possible to try to accept *all* of this: if there are quantities $\eta < \eta$, then it seems reasonable (?) to think that $\mathbf{0}_\eta$ is *different* from $-\mathbf{0}_\eta$. We might say that $\mathbf{0}_x$ has both a negative and a positive aspect, *signed zeros*. There are uses for such things in computer science, in particular numerical representation in "floating arithmetic" [Kahan, 1987].

As a consequence, we lose the proof that changing sign toggles over zero: if a is positive, $-a$ might also be positive! To prove otherwise would ordinarily use "contraposition" on the special case $0 < a \Rightarrow -a < -0$. Without that, we argue using trichotomy. If $0 < a$, then $-a \leqslant \mathbf{0}_a$, and there are three possibilities: $\mathbf{0}_a < 0$ or $\mathbf{0}_a = 0$ or $0 < \mathbf{0}_a$. In the first two cases, by transitivity $-a < 0$, but if $a - a$ is positive, then it is possible that $0 < -a < \mathbf{0}_a$. And that's as much as we can say. This ultimately undermines the triangle inequality (Section 8.3.1.2), meaning that the space being described here is a *semimetric* space [Wilson, 1931]. The vector pictures of Chapter 7 might have already clued us in: the shortest distance between two points may not be a straight line, because the straight line might be shorter than itself, and so not be the shortest.[6]

8.1.4 Axiom 3: $\mathbb{R}$ Is Closed

8.1.4.1 Least Upper Bound Principle

Let X be any set. If $\exists x (x \in X \cap \mathbb{R})$, and some $z \in \mathbb{R}$ is an *upper bound*,

$$\forall x (x \in X \cap \mathbb{R} \Rightarrow x \leqslant z).$$

Then $X \cap \mathbb{R}$ has a *least* upper bound $\xi \in \mathbb{R}$:

$$\forall x (x \in X \cap \mathbb{R} \Rightarrow x \leqslant \xi)$$
$$\& \quad \forall x (\xi < x \Rightarrow x \notin X)$$
$$\& \quad \forall y (y < \xi \Rightarrow \exists x (x \in X \cap X \cap \mathbb{R} \,\&\, y < x)).$$

In words, ξ is an upper bound for the reals in X; anything above ξ is not in X; and anything below ξ is not an upper bound. Any such ξ is a *supremum* for $X \cap \mathbb{R}$, and we write

$$\xi \in \sup(X \cap \mathbb{R}).$$

Axiom 3 asserts the conditions on which sup X is not empty,[7] and all of its members ξ are least, in the sense that anything less than ξ is dominated by some real from $X \cap X$.

The doubling up of $X \cap X$ in Axiom 3 is not a typo. That's noncontraction making itself known. For the least upper bound principle to work as intended, it needs to guarantee that any $y < \xi$ is dominated by some real from the "inner core" of X, some x that is a member of X *and* a member of X. The proofs of Proposition 44(ii) and Theorems 45 and 48 all require this. I suggest that this is a *discovery* about the nature of set of reals and the substructure of their relationship to upper bounds.

---

[6] Classically, any metric space (supporting the triangle inequality) is a Hausdorff space, where points are sufficiently "separated" [Willard, 1970, §13.6]. As we'll explore in later sections, this space is not, if there can be $x \neq y$ but every epsilon radius around x contains y – they are infinitesimally close.

[7] Terminology as in Chapter 5: "non-empty" and its variants mean "occupied," $\exists x (x \in X)$. The less informative notion of "not identical to the empty set" $X \neq \varnothing$ will be expressed explicitly if at all.

8.1.4.2 Greatest Lower Bound Principle

Every nonempty set X bounded from below has a greatest lower bound or *infimum*, $\zeta \in$ inf $X \cap \mathbb{R}$, with

$$\forall x (x \in X \cap \mathbb{R} \Rightarrow x \geqslant \zeta)$$
$$\&\quad \forall x (x < \zeta \Rightarrow x \notin X)$$
$$\&\quad \forall y (y > \zeta \Rightarrow \exists x (x \in X \cap X \cap \mathbb{R} \;\&\; x < y)).$$

A greatest lower bound of X is a lower bound of X; anything lower than it is not in X and anything higher than it is not a lower bound. The existence of greatest lower bounds is assumed independently, as they do not appear to follow as simple duals from the conditions for suprema.

Unfurling the meaning of this axiom is the main task of the next section.

8.2 Dedekind Cuts

$\mathbb{R}$ is intended to be the linear continuum. It expresses a geometric line as a set of *points*. Before getting back into the weeds of theorems and proofs, let us cast our minds back to some of the paradoxes in Chapter 1 and reflect.

8.2.1 Points and Lines

A point is the smallest unit of reality. Points are so small as to have no extension. They take up no space at all. Points are *in* space without *filling* any space. From Zeno onward, many have urged that this idea makes only tenuous sense, especially if we are thinking of space as entirely composed of points (or punctiform). Analyze down a line, via infinitely many bisections, and you get (we are told) to points at the limit; but to synthesize a line back up from points, intuitively they would need to have to have some "shape." How can many nothings add up to something? How do we start with a something and end up analyzing it down to a nothing?

The idea that continuous objects can always be divided, further and further, making divisibility constitutive of continuity, goes back to Aristotle.[8] The limits of an infinite number of divisions are points. And so, built up from points and divisible down to them, a *continuum* has been thought to be many (incompatible) things. Two core features are salient:[9]

---

[8] [Hellman and Shapiro, 2018, ch. 1], cf. [Priest, 2002a, ch. 2, 2006b, ch. 11]. Some have suggested that everything is indefinitely divisible, if not physically then at least conceptually; even an indivisible mereological atom has a left side and a right side: "The left hand side of an object, even [if] it could not exist physically on its own, is still a perfectly good object conceptually" [Priest, 1994b, p. 6].

[9] The scholarship question of who held these views (Bolzano? Dedekind? Brouwer? etc.), as well as the conceptual question of whether they are, in the final analysis, independently coherent or mutually incompatible, or even jointly desirable, we leave aside. These are fairly traditional intuitions about the geometric line.

- NO GAPS – if you pass *through* a continuum, you must *touch* it, at a point.
- UNITY – you can't separate a continuum exclusively and exhaustively into two closed parts.

Both these conditions are captured in the modern Dedekind–Cantor answer to "how can many nothings make a something": the received answer is is that there are so *many* points – uncountably infinitely many – that all together they are a cohesive something.

But that doesn't answer the question.[10] If a point takes up *no* space, then intuition cries out that many points together – even uncountably many – can't suffice to form a spatially extended line. The basic units of reality must be so small as to be smaller than any other part of reality, but large enough to take up some portion of reality. They must be so small as to not have parts, but big enough to be parts. This relates back to the reasons for the inconsistency of the early calculus outlined in Section 7.1. This is a dark passageway in the labyrinth Leibniz warns about.

Dedekind teaches us to think of problems as definitions (see the Introduction). His proposed 1872 definition [Dedekind, 1901], of points as cuts, is a solution (to NO GAPS), and a new problem (UNITY, which conflicts with SYMMETRY from Section 1.3.1.1). Let us recall, then, that a Dedekind cut (*schnitt*) is as follows:

(1) An exhaustive division of the line into two parts, $\langle X, Y \rangle$.
(2) Every point in X is strictly to the left of every point in Y.
(3) There is exactly one point, at the cut, that is either the greatest member of X, or the least member of Y.

This requires an arbitrary and capricious assignment of the "cut" k to the left or the right side of the line, because if the cut belonged to both the left *and* the right, then $k < k$ on the assumption that every point in X is strictly to the left of every point in Y. Dedekind's line cannot be cut perfectly in half – consistently.

There is much to like about thinking of the line as composed of cuts; any potential "gap" is simply rechristened as a point. There are also problems, as we've seen. One new way of putting the problem is that Dedekind's points are *isolated*: as Weyl puts it [Weyl, 1919], no matter how many points there are or how densely they are packed, points as individuals are just as separated from each other as the natural numbers are. Points are discrete, like pebbles, or grains of sand. And yet, the intuition is that a continuum has unity, cohesion: it is "all in one piece." That is why it "separates" badly, asymmetrically. How can unextended objects that never touch each other form a smooth total?

Questions such as these motivate the intuitionistic conception of the continuum, due to Brouwer in the 1920s, wherein the reals simply cannot be decomposed into disjoint parts.[11] As Shapiro puts it, following van Dalen, on this conception if you try to cut the line with a knife, there is always a strand of "syrup" that sticks to the knife. To "take the bull by the

---

[10] As pointed out by [Black, 1951].
[11] [van Dalen, 1997] and [van Atten, 2007].

horns and try to turn this idea into a definition" [Bell, 2014, p. 8], constructive analysis says that $\mathbb{R}$ is "cohesive" iff, when $\mathbb{R} = A \cup B$ and $A \cap B = \varnothing$, then either $\mathbb{R} = A$ or $\mathbb{R} = B$. Equivalently, for any partition of the continuum into two parts A and B, either $A = \varnothing$ or $B = \varnothing$. In the intuitionistic conception, you can't exhaustively separate proper nonempty parts of the continuum; doing so would in fact imply the law of excluded middle.

How does the intuitionistic conception inform a paraconsistent approach? Again there is a lot to like, but the Brouwerian approach can't meet the goals set back in the sorites/vagueness discussion Section 3.3.3.2. In constructive analysis, even in its less revisionary forms such as Bishop's, the intermediate value theorem *fails*.[12] A change occurs without occurring specifically *somewhere*. (Intuitionism seems to *embrace* what Williamson calls a fallacy (Section 1.2.3), that without an algorithm for how *I* might see something, I can't commit to it being there.) A divisible, punctiform-yet-cohesive continuum *is* puzzling, but that is the problem to be explained, not explained away or left indeterminate.[13] Points would seem to be one of the most basic and ubiquitous concepts in all of mathematics (and physics beyond), even if they are more something we get used to than understand. Occurring somewhere *means* occurring at a point. As I've been trying to do from the start, I want to push through the puzzlement to try to learn more about this, to work with the world as made up of points in space, not to make up a less puzzling story about someplace else.

No GAPS calls for a punctiform object, whereas UNITY (along with some SYMMETRY intuitions) calls for a sticky or "pointless" one. The intuitionistic conception does not deal in open or closed sets, which is the subject of the next chapter; but already a way to take the dual approach to Brouwer et al. suggests itself. Together, No GAPS and UNITY call for "pointless points." The idea is that if the punctiform continuum *is* completely separated into two closed nonempty disjoint parts, that are still somehow stuck together, then *some point is "ripped" in two.* (Note that everyone can agree with this conditional statement, if the antecedent is impossible.) This appears later as Theorem 45; working through this idea will be the culmination of the chapter.

8.2.2 Schnitts

In this section, we prove several closely related results about Dedekind cuts, to view the situation from several angles. Classically, describing the line in terms of Dedekind cuts is equivalent to the least upper bound axiom; we will see how our version, Axiom 3, develops.[14] We will be dividing, if not the entire universe, then a bit of it.

---

[12] See [Bishop and Bridges, 1985].

[13] For example, by throwing out the notions of divisibility, or point, say via the older Aristotelian conception of continuity, or an excursion into nonpunctiform (or pointless) geometry, as in [Arntzenius, 2012, ch. 4]. Such theories include "gunk": the line is made up of *finite* parts, each of which has finite proper parts, each of which has parts . . . such that zero-dimensional points are never reached. As [Hellman and Shapiro, 2018] show, less turns on the existence (or not) of points than it might appear, since suitably nested equivalence classes of spatial regions lead to (classically) equivalent results. See [Clarke, 1985].

[14] Compare the following material with the conventional versions, e.g., in [Buskes and van Rooij, 1997, pp. 29–31]. The standard source is [Rudin, 1953].

Some things to notice right away are how the conditions for separating a line play out (cf. Section 5.1.5). *On the one hand,* if $A \cup B = \mathbb{R}$ and $a < b$ for all $a \in A$ and all $b \in B$, this already implies that A and B must be different sets, in the strong sense that $A = B \Rightarrow \bot$. (Proof: For any $z \in A \cap B$, it will be that $z < z$; if this were true for all of $\mathbb{R}$, then it would be that $0 < 0$, which is impossible.) So 0 is in one and only one of A or B. *On the other hand,* the possibility of nonempty intersection, some $z < z$, between A and B in general is not ruled out (in the sense of $\bot$). *And on the other other hand,* from the set theory, if B is the complement of A in $\mathbb{R}$, then $\mathbb{R} \equiv A \cup B$; but if $\mathbb{R} \equiv A \cup B$ and a real r is not in A, then unless B is just *defined* as $\mathbb{R}\backslash A$, do not infer (without disjunctive syllogism) that $r \in B$.

On suprema (and infima), two initial remarks are needed.

First, least upper bounds are not necessarily *unique*, in the sense that

$$k_0 \in \sup X \ \& \ k_1 \in \sup X$$

can hold *without*

$$k_0 = k_1$$

as the first theorem (Proposition 44) illustrates. This is in keeping with the motivations from vagueness and the nature of well-orderings already developed in previous chapters. However, if one adds the condition that every least upper bound of A is a member of A, then uniqueness follows:

$$\sup A \sqsubseteq A \Rightarrow (k_0, k_1 \in \sup A \Rightarrow k_0 = k_1)$$

by antisymmetry: every $x \in A$ is below every $k_i \in \sup A$. But without that extra restriction, if any part of $\sup A$ is not in A, then bets are off.

Second, since $\sup A$ is in general a *set*, and a nonempty one bounded from above at that, there is a question of what to make of iterations, $\sup(\sup A)$, and so forth. Again, if $\sup A \sqsubseteq A$, then $\sup \sup A \equiv \sup A$ (the least upper bounds of the least upper bounds). Otherwise, suppose $k \in \sup A$ and $k' \in \sup \sup A$. Then $k \leq k'$; if $k = k'$, we may stop, but if $k < k'$, then $\exists z(z \in \sup A \ \& \ k < z)$, so there is a possibility of an open sequence of suprema – and sets of suprema, $\sup \sup \sup \cdots \sup A$ – fading into the limit of A.

Now, theorems.

Proposition 44. *Let X be a nonempty subset of $\mathbb{R}$ bounded from above, and let $\zeta \in \sup X$.*

(i) A supremum is a least upper bound: if $\xi \in \mathbb{R}$ is an upper bound for X, then $\zeta \leq \xi$, or else $\xi < \xi$.

(ii) If X and Y are nonempty sets of reals with $Y = \mathbb{R}\backslash X$, and every member of X is strictly less than every member of Y, then for all $z \in \mathbb{R}$, if $z \leq \zeta$, then $z \in X$.

Proof (i) Suppose there is another upper bound of X, some $\xi \geq x$ for all $x \in X$. By trichotomy, either $\xi = \zeta$, or $\zeta < \xi$, or $\xi < \zeta$. In the first two cases, stop. In the last case, since ζ is a least upper bound in the sense of Axiom 3, then $\exists z(z \in X \ \& \ \xi < z < \zeta)$.

But for this z, we also have that $z \leqslant \xi$ on hypothesis. (That's the first time we've appealed to the hypothesis.) By transitivity, $z \leqslant \xi < z$, which by transitivity again means $\xi < \xi$, as stated.

(ii) Suppose $x < y$ for all $x \in X$ and all $y \in \mathbb{R} \backslash X$, and that $z \leqslant \zeta \in \sup X$. If $z \in X$, we are finished. Otherwise, $z \in \mathbb{R}$ and $z \notin X$, i.e., $z \in \mathbb{R} \backslash X$, whereby $x < z$ for all $x \in X$, since every member of X is below every member of $\mathbb{R} \backslash X$. Recalling that $\zeta \in \sup X$, by the least upper bound property, if $z \leqslant \zeta$ (which we assumed but haven't used yet), then by Axiom 3 there is some $x_0 \in X$ such that

$$z < x_0.$$

Then

$$x_0 < z$$

since $x < z$ for all $x \in X$; so then $x_0 = z$ by antisymmetry. Noting for a second time (recalling the phrasing with $X \cap X$ of the least upper bound Axiom 3) that $x_0 \in X$, we use substitution of identicals to conclude that $z \in X$.   □

Officially, for any $A, B \sqsubseteq \mathbb{R}$, the pair $\langle A, B \rangle$ is a *Dedekind cut* iff

- $A \cup B \equiv \mathbb{R}$
- $\exists x (x \in A)$ and $\exists x (x \in B)$
- and $\forall x \forall y (x \in A \;\&\; y \in B \Rightarrow x < y)$

Since $\mathbb{R} \equiv A \cup \mathbb{R} \backslash A$, this may replace $A \cup B = \mathbb{R}$ and still yield a Dedekind cut.

Theorem 45. *The following is true about Dedekind cuts.*

(i) *For any Dedekind cut $\langle A, B \rangle$, there is some $c \in \mathbb{R}$ such that for every $a \in A$ and $b \in B$*

$$a \leqslant c \leqslant b.$$

(ii) *For any Dedekind cut $\langle A, B \rangle$, for any $\xi \in \sup A$ and $\zeta \in \inf B$, for all $c \in \mathbb{R}$*

$$\xi < c < \zeta \;\Rightarrow\; c = \xi \vee c = \zeta.$$

(iii) *For any Dedekind cut $\langle A, B \rangle$, if $\sup A \sqsubseteq A$ and $\inf B \sqsubseteq B$, then there is some c such that*

$$c < c.$$

In particular, $\xi < \xi$ or $\zeta < \zeta$.

Proof (i) On assumption, $\xi \in \sup A$ exists. Every $a \in A$ is bounded above by ξ, $a \leqslant \xi$. Let $b_0 \in B$. If $b_0 < \xi$, then $b_0 \in A$ by Proposition 44(ii). Then since b_0 is in both A and B, for every $a \in A$ and $b \in B$,

$$a < b_0 < b$$

with $b_0 < b_0$. On the other hand, if $\xi \leqslant b_0$, then $a \leqslant \xi \leqslant b_0$. Either way, there is a witness for c.

(ii) This is a conditional about a hypothetical, that expresses how continuous the line so imagined really is. As proof, assume the antecedent. For any $x \in \mathbb{R}$, either $x \in A \vee x \in B$. If $x \in A$, then $x \leqslant \xi$; if also $x > \xi$, then $x = \xi$ by antisymmetry. If $x \in B$, then $x > \zeta$; if also $x < \zeta$, then $x = \zeta$ by antisymmetry. So the supremum and infimum of a cut are close enough together that, if they are not identical, then there is nothing between them but themselves.

(iij) Let $\xi \in \sup A$, $\zeta \in \inf B$. As $\xi \in A$ and $\zeta \in B$, we have $\xi < \zeta$. Then by density of order,

$$\xi < c < \zeta$$

for some c. But $c \in A$ or $c \in B$. In the first case, $c \leqslant \xi < c$, and in the second case, $c < \zeta \leqslant c$. Either way, $c < c$. $\square$

Classically, the strictness of $<$ is enough to enforce the nonoverlap of two sides of a Dedekind cut. Here, the proofs keep showing that the conditions have been recast: in the event that a strict ordering of the line into two halves has a nonempty overlap, then what is in the overlap will be "ripped" or *split*, $a < a$. That is, if $\langle A, B \rangle$ is a Dedekind cut, then any part of B falling below the least upper bound of A will be self-separated.

In the event that $\langle A, B \rangle$ is a Dedekind cut in which $a \leqslant b$ for every $a \in A$ and $b \in B$ (but not necessarily *strictly*), much of the preceding will still go through. One ends up with the case that $\xi \leqslant \zeta$, and then either $\xi = \zeta$ or $\xi < \zeta$.

Continuous Sorites Regained

This much in hand, we reach the milestone of recovering the *continuous sorites* (Section 3.3.3.2). Recall the intuition: for φ to be vague, something must be φ, something is not, and it is not the case that something is both "near" a φ but not itself a φ, as in Chapter 1. The preferred phrasing here is extensional, with material $\supset$ (and does not involve talk of limits, Cauchy sequences, or the like):[15]

Tolerance, Continuous Version: For $A \subseteq \mathbb{R}$, $\neg(\forall x(x \in A \supset \varphi(x))$ & $\neg\varphi(\sup A))$.

That is, it is not the case that a vague property holds everywhere in A up to a least upper bound but not at the least upper bound. Since $\sup A$ is not unique, $\varphi(\sup A)$ is taken to mean $\varphi(\xi)$ for all $\xi \in \sup A$. Now we consider a tolerant property mapped exhaustively over the reals.

Theorem 46 (Chase). *Let $A = \{x \in \mathbb{R} : \varphi(x)\}$ and $B = \{x \in \mathbb{R} : \neg\varphi(x)\}$ both be nonempty, with $a < b$ for every $a \in A$ and $b \in B$. Suppose that φ satisfies continuous-tolerance, so*

---

[15] In the context of analysis, this is related to the *Leibniz Continuity Condition* [Mortensen, 1995, ch. 6].

$\exists x(x \in A \ \& \ \neg\varphi(x)) \lor \varphi(\xi)$, *for* $\xi \in \sup A$ *and*
$\exists x(x \in B \ \& \ \varphi(x)) \lor \neg\varphi(\zeta)$, *for* $\zeta \in \inf B$.

Then $\varphi(z) \ \& \ \neg\varphi(z)$ *for some* $z \in \mathbb{R}$.

Proof Consider the disjuncts of the preceding supposition. If there is some $x \in A$ that is $\neg\varphi(x)$, then, since all members of A are φ, that x is the contradictory element we seek, and similarly for some $x \in B$ that is φ. So take the other disjuncts, and suppose $\varphi(\xi)$ for $\xi \in \sup A$ and $\neg\varphi(\zeta)$ for $\zeta \in \inf B$. Use trichotomy:

- If $\xi = \zeta$, then they are both φ and not φ.
- If $\zeta < \xi$, then $\varphi(\zeta)$ as in Proposition 44(ii).
- If $\xi < \zeta$, then by density of order there is some $z \in \mathbb{R}$ between them, $\xi < z < \zeta$. Either $\varphi(z)$ or $\neg\varphi(z)$. If $\varphi(z)$, then $z \in A$, and then, since $z > \xi$, also $\neg\varphi(z)$, by stipulation on least upper bounds. Similarly if $\neg\varphi(z)$.

The only sticking point is that in the third case, to apply density would require several repetitions of the assumption. Since density and trichotomy are both axioms of $\mathbb{R}$, though, it may be repeated as needed. A contradiction follows no matter what.   □

This last theorem, note, is classically true; it would just indicate that, classically, the supposition is impossible. The message is topological. If $\mathbb{R}$ is divided into two sets via a vague property, then, since some point is very close to both sets, some point is in both sets. The contradiction is due to the further assumption that the sets are exclusive, in the sense of one being the (relative) complement of the other. More of this to come.

8.2.3 Intervals

A set I is an *interval* if and only if

- $I \subseteq \mathbb{R}$.
- $\exists a, b(a \in I \ \& \ b \in I)$.
- $\forall a, b(a \in I \ \& \ b \in I \Rightarrow \forall x(a \leqslant x \leqslant b \Rightarrow x \in I))$.

On this definition, a point $\{a\} \subseteq \mathbb{R}$ qualifies as an interval. So we say that a *proper* interval satisfies the additional constraint that

- $\exists a, b(a \in I \ \& \ b \in I \ \& \ a < b)$.

A proper interval I contains some distinct a, b and also every x between a and b. By trichotomy, X is an interval iff X is either a proper interval or a point. We will see later, though, that there may be points that count as proper intervals: for any real number a, if $a < a$, then $\{a\}$ is a proper interval.

Define the notation for *closed* and *open* intervals, respectively:

$$[x, y] := \{z : x \leqslant z \ \& \ z \leqslant y\}.$$
$$(x, y) := \{z : x < z \ \& \ z < y\}.$$

Then the mixed cases $(x, y]$ and $[y, x)$ are as expected.

8.2.3.1 Adherent Points

Intuitively, a point is very nearby a set if any space around that point, however small, overlaps (has a nonempty intersection) with the set. The idea can be captured with some careful definitions as follows.

For $X \subseteq \mathbb{R}$ and $x \in \mathbb{R}$, x is *adherent* to X iff either $x \in \sup X \cup \inf X$; or else for every $\varepsilon \in \mathbb{R}$, if $\varepsilon > 0$ and (repeating) $\varepsilon > 0$, then

$$\exists y (y \in \mathbb{R} \quad \text{and} \quad y \in [x - \varepsilon, x + \mathbf{0}_\varepsilon] \cup [x, x + \varepsilon] \quad \text{and} \quad y \in X).$$

That is, an adherent point x of X, if not a sup or inf of X, is such that no matter how small an interval one considers around x, that interval has a nonempty overlap with X. Here are a few observations about adherent points.

First, definitionally, if X has a least upper bound ξ, then ξ is adherent to X. Second, every member of X is adherent to X. This is because $x - \varepsilon \leqslant x \leqslant x + \mathbf{0}_\varepsilon$ for any $\varepsilon > 0$, and if $\varepsilon > 0$, then $x < x + \varepsilon$. So $x \in X$ and $x \in [x - \varepsilon, x + \mathbf{0}_\varepsilon] \cup [x, x + \varepsilon]$, thereby satisfying the definition. And as a third preliminary, we can show that if x is adherent to $X \cap A$, then x is adherent to A. That's because if x is adherent to $X \cap A$, then take any small interval around x, with nonempty overlap of X and A; then by conjunction elimination, it has nonempty overlap of just A.

These notions can be used to show that the reals are connected (though that has yet to be defined), as follows:

Theorem 47. *Let A be a nonempty subset of $\mathbb{R}$ and $B = \mathbb{R} \backslash A$ nonempty, too. There is some $\xi \in \mathbb{R}$ that is adherent to both A and B.*

Proof For $a \in A$ and $b \in B$, without loss of generality suppose $a < b$, so $X := \{x : x \in A \ \& \ x \leqslant b\}$ is nonempty and bounded from above by b. Thus its supremum ξ exists and is adherent to A. For $\varepsilon > 0$, consider $\xi + \varepsilon$. Since $\xi < \xi + \varepsilon$, then $\xi + \varepsilon \notin X$, by definition in Axiom 3; since it is in $\mathbb{R}$, it is in B. Then ξ is adherent to both A and B. □

The same argument works with $\mathbb{R}$ replaced by an arbitrary closed interval $[a, b]$ of $\mathbb{R}$.

8.2.3.2 Nests

A *nest of intervals* is a set of nonempty closed intervals, $\{[a_n, b_n] : n \in \mathbb{N}\}$, such that

$$\ldots \subseteq [a_2, b_2] \subseteq [a_1, b_1] \subseteq [a_0, b_0]$$

so that $[a_{n+1}, b_{n+1}] \subseteq [a_n, b_n]$ for all n. The following, which is classically equivalent to the least upper bound principle (as is almost everything we've proved in this section), is sometimes attributed to Cantor.

Theorem 48 (Nested Intervals Theorem). *The intersection of a nest of intervals in $\mathbb{R}$ is nonempty.*

Proof Suppose ... $\subseteq [a_1, b_1] \subseteq [a_0 \subseteq b_0]$ is a nest of intervals. The goal is to show there exists $x \in \mathbb{R}$ such that $x \in [a_n, b_n]$ for all n. Since the set of left endpoints

$$(a_n) := \{a_0, a_1, \ldots\}$$

is bounded above by any member of the set of right endpoints

$$(b_n) := \{b_0, b_1, \ldots\},$$

then $\xi \in \sup(a_n)$ exists. Since (b_n) is bounded below by any member of (a_n), $\zeta \in \inf(b_n)$ exists. By trichotomy,

$$\xi = \zeta \quad \vee \quad \xi < \zeta \quad \vee \quad \zeta < \xi.$$

We argue cases.

Since $a_n \leqslant \xi$ and $\zeta \leqslant b_n$ for all n, if $\xi = \zeta$ or $\xi < \zeta$, then either

$$a_n \leqslant \xi = \zeta \leqslant b_n$$

or

$$a_n \leqslant \xi < \zeta \leqslant b_n$$

for each n. Either way, $\xi, \zeta \in [a_n, b_n]$ for each n.

In the third case, we've seen the idea of the proof before. If $\zeta < \xi$, then because $\xi \in \sup(a_n)$, there is some $a \in (a_n)$ such that $\zeta < a$. But then because $\zeta \in \inf(b_n)$, this implies some $b \in (b_n)$ is such that $b < a$. (One could argue, mutatis mutandis, for some $b < \xi$.) But $b \geqslant a$ just by construction of (b_n) and (a_n). So $a = b$ by antisymmetry, with

$$a_n < a = b < b_n$$

for all n. Again, there is a member in every interval in the nest: for some c,

$$c \in \bigcap_{n \in \mathbb{N}} [a_n, b_n]$$

as required.   □

8.3 Continuity; *or,* "Amongst the Ghosts of Departed Quantities"

Having arranged the real line so that it is (or, with some poking, seems to be) continuous, and affirmed, within a paraconsistent framework, that the conditions for a continuous sorites do indeed give rise to an inconsistency, we might fairly say that our official task for the chapter is done. But this only scratches the surface of the continuum and the nature of points that compose it, and does not yet prove the intermediate value theorem. In this section, we will explore some tentative ideas for how to formulate infinitesimals, which lays the groundwork for studying topology proper in the next chapter, and gets a deeper grip on continuity. The target is an improvement on Theorem 46, proving the intermediate value theorem.

8.3.1 Infinitesimals

A few times now, the possibility of some quantity a such that $a < a$ has arisen. This suggests some "distance" between a and itself, much the way relative self-complements and relative zeros suggest some remnant of a without itself. But it seems plausible – to comically understate the matter – that the distance between a thing and itself is less than any nonzero quantity. Putting these last two sentences together, seasoned with the philosophical hunch from Section 8.2.1 that points need to be somehow both extended and not, we now consider infinitesimals. (See also Section 7.1.)

8.3.1.1 The Idea

The basic conception of an *infinitesimal* that we will toy with is the simplest, most "naive" one: that they are infinitely small real numbers. So they are (1) infinitely small, and (2) real numbers. Let's focus on (2) first.

Cantor proposed that transfinite numbers be treated as far as possible like finite numbers [Hallett, 1984, p. 7] (noting that Cantor himself was opposed to infinitesimals). Applying the same principle here, and to go further than possible, we would like to say that infinitesimals are just numbers like any other. The motivation is by analogy to dialetheias: some propositions are contradictory, but they are not a different type of thing than any other proposition. So infinitesimals are quantities that are, yes, smaller than any quantity, but unlike other approaches they would not be sequestered into their own class of "bad" objects. They are just more real numbers.[16] Since they are the inconsistent difference between something and itself, then in some sense – the sense in which no contradictions are true – they do not even exist (and nothing that follows shows they do, only what it would be like if they did). Are they a glimpse of Leibniz's fulgrations, "continual flashes of silent lightning"?

8.3.1.2 Fluxions

To have quantities that are infinitely small means quantities smaller than any finite nonzero real number. To express this, we need a little standard notation: for any quantities a, b,

$$|a - b| = \begin{cases} a - b & \text{if } a > b \text{ or } b = a \\ b - a & \text{if } b > a \end{cases}.$$

The notation relies on trichotomy, and is unambiguous, even given the possibility of inconsistency, since if $b < a$ and $a < b$, then $b = a$, so $b - a = a - b$ (and not, of course). Note that if $a < b$, then $\mathbf{0}_a < b - a$, so $\mathbf{0}_a < |a - b|$.

---

[16] Most famously, Robinson's nonstandard analysis uses the expressive weakness of the language of first-order logic and its model theory to simulate a class of hyperreals [Robinson, 1966; Goldblatt, 1998] and a transfer theorem to import results back to the reals. In the more natural approach of *smooth infinitesimal analysis* [Bell, 2008], infinitesimals are intermediary between points and lines, a "smudged" point, an instant with some extension less than any finite extension. Some of these "linelets" cannot be reduced to exact points. In this context (Bell, 2008, p. 22), the closed interval $[0, 0]$ will have a proper subset that does not reduce to $\{0\}$; Bell's infinitesimals are neither 0 nor not 0. See also the *surreal numbers*, based on generalized Dedekind cuts [Conway, 1976].

Our weaker setup creates space for possibilities.[17] In classical real analysis, if the distance between points is infinitely small – approaches 0 – then those points are identical; that is, if $|x - y| < \varepsilon$ for all $\varepsilon > 0$, then $x = y$. Without assuming consistency, we have a more general fact: by trichotomy,

Proposition 49. *If $|x - y| < \varepsilon$ for every $\varepsilon \in \mathbb{R}$ such that $\varepsilon > \mathbf{0}_x$ and $\varepsilon > \mathbf{0}_y$, then $x = y$ or $|x - y| < |x - y|$.*

Proof Suppose $|x - y|$ is less than *every* quantity greater than $\mathbf{0}_x$ and $\mathbf{0}_y$. Either $x = y$, or $x < y$, or $y < x$. In the latter two cases, either $|x - y| > \mathbf{0}_x$ or $|x - y| > \mathbf{0}_y$. Then on supposition, $|x - y| < |x - y|$.   □

The cases where x is distinct from y but $|x - y| < \varepsilon$ are standardly ruled out by consistency requirements, but not here. Is this where to find infinitesimals?

Let us use the symbol $\mathbf{0}$ as a variable ranging over all zeros; so $\varepsilon > \mathbf{0}$ is short for $\forall x(\varepsilon > \mathbf{0}_x)$. A proposed (paraconsistent) definition: a real number $\eta \in \mathbb{R}$ is *infinitesimal* if it is nonzero and yet smaller than any nonzero number:

$$\eta \in \mathsf{Inf} \leftrightarrow \eta \in \mathbb{R} \,\&\, \eta > \mathbf{0} \,\&\, \forall \varepsilon(\varepsilon \in \mathbb{R} \,\&\, \varepsilon > \mathbf{0} \Rightarrow \eta < \varepsilon).$$

If follows immediately that if η is infinitesimal, then

$$\eta < \eta.$$

It then follows that there are no infinitesimals, since if $\eta \in \mathbb{R}$ and $\eta > \mathbf{0}$ and $\eta \not< \eta$, then by counterexample $\exists \varepsilon \neg(\varepsilon \in \mathbb{R} \,\&\, \varepsilon > \mathbf{0} \Rightarrow \eta < \varepsilon)$. Since $x \not< x$ for all x, it follows that $x \notin \mathsf{Inf}$ for all x. It then follows (in contradistinction to either the nonstandard or smooth approaches) that in particular 0 is *not* infinitesimal, since $\mathbf{0}_0 = 0$ and $0 < 0$ is forbidden by Axiom 2 (Section 8.1.3). It is not yet *ruled out*, though, whether Inf has any members or not.

If Inf does have members, then there is, in a sense, at most *one*:

$$\eta_0, \eta_1 \in \mathsf{Inf} \Rightarrow \eta_0 = \eta_1. \tag{8.18}$$

Proof If $\eta_0, \eta_1 \in \mathsf{Inf}$, then $\eta_0 > \mathbf{0}$ and $\eta_1 > \mathbf{0}$. But also $\eta_0 > \mathbf{0} \Rightarrow \eta_1 < \eta_0$, and $\eta_1 > \mathbf{0} \Rightarrow \eta_0 < \eta_1$. Then

$$\eta_0 < \eta_1 \,\&\, \eta_1 < \eta_0.$$

By antisymmetry, $\eta_0 = \eta_1$. This has some affinity to the notion of a *shadow* in Robinson's non-standard analysis [Goldblatt, 1998, p. 53].   □

Thus we may refer to all infinitesimals (if there are any) with the name η. What should we make of this? When we *count* the number of infinitesimals, the answer is "(at most)

---

[17] At this point, one would standardly check the triangle inequality, $|x - y| + |y - z| \geq |x - z|$, but as mentioned earlier, this does not appear to work if the space is only semimetric. Usually Axiom 3 would imply Archimedes' axiom, that no point is infinitesimally smaller than another: if $x < y$, then there is some $n \in \mathbb{N}$ such that $y < xn$. However, the usual argument breaks down in the present setting. See theorem 5 in [McKubre-Jordens and Weber, 2012, p. 910].

one"; distinctly, following Varzi ([Varzi, 2014], as in Chapter 5), when we *countenance* the set of all infinitesimals, and note that

$$\ldots < \eta < \eta < \eta < \eta < \eta < \eta < \ldots,$$

then there are *many* of them.

What, if anything, is/are these infinitesimals? Leibniz sets the mood when he says that each simple substance or monad, although it has no parts, is a "perpetual living mirror of the universe" [Leibniz, 1714, §56]. Let's squint closely.

8.3.1.3 Imploded Intervals and Monads

Infinitesimals have internal structure; each one is a continuum "viewed on end." When $\eta \in \mathsf{Inf}$, all the points strictly between η and itself

$$]\eta[\quad := \quad \{x : \eta < x < \eta\}$$

is, if nonempty, an "imploded" interval (checking the definition back at Section 8.2.3). The interval $]\eta[$ is in fact a *nest* of intervals, with a nonempty intersection:

$$\ldots \sqsubset [\eta, \eta] \sqsubset [\eta, \eta] \sqsubset [\eta, \eta]$$

by repeated application of $\varphi \Rightarrow \varphi$. If η exists, the inclusion relation here is *proper*. Then Theorem 48 applies and the intersection of this nest is not empty. In the same way, Dedekind-cut theorems also apply: assuming η exists, there are nonempty sets A, B, such that $A \cup B =]\eta[$, where every member of A is strictly less than every member of B. Then Theorem 45 is triggered and we have a point whose internal structure mirrors the line.[18] This interval has no endpoints, and so is "open" inside itself.

Along similar lines, the *monad* of x is the set of all reals closer to x than any positive real number,[19]

$$\mu(x) := \{y \in \mathbb{R} : \forall \varepsilon(\varepsilon > \mathbf{0} \Rightarrow |x - y| < \varepsilon)\}.$$

Loosely speaking, the monad of x picks out members of the internal structure of x. The monad of x is classically identical to $\{x\}$. Here, since sup is a relation, the monad can be nondegenerate. If $y \in \mu(x)$, then $y = x$ or $|y - x| = \eta$. An imploded interval is properly included in its monad but likely not vice versa, since the monad always contains at least x and may be a completely classical object.

8.3.1.4 Closeness

Now we apply this microdigression to thinking about the theory of continuity itself. Write

$$x \approx y := \forall \varepsilon(\varepsilon \in \mathbb{R} \,\&\, \varepsilon > \mathbf{0} \Rightarrow |x - y| < \varepsilon)$$

---

[18] "Every portion of matter can be thought of as a garden full of plants or a pond full of fish. But every branch of the plant, every part of the animal, ... is another such garden or pool" Leibniz [1714, §67]. But I should probably say explicitly that, unlike [Brown and Priest, 2004], none of this is intended as a historically serious or even revisionist reconstruction of Leibniz or Newton, and that if per impossible either of them saw this material, I suspect they would dislike it very much.

[19] The name "monad" is from Robinson, channeling Leibniz [Robinson, 1966, p. 57]; later schools of non-standard analysis called this equivalence class a halo [Goldblatt, 1998, p. 52].

to mean that x is (infinitely) *close* to y. This is a way of saying with a two-place relation what we've already said with the one-place relation μ, that y is in the monad of x, which we can now rewrite $\mu(x) = \{y \in \mathbb{R} : x \approx y\}$.

The monad of x is a similarity class, because the closeness relation $\approx$ is reflexive ($x \approx x$ because $x = x$) and symmetric ($x \approx y \Rightarrow y \approx x$ because $|x - y| = |y - x|$ and $x = y \Rightarrow y = x$). Transitivity would require the triangle inequality. We have some considerable evidence that transitivity for $\approx$ will not be forthcoming. Sorites-type considerations of a long chain such as

$$x_0 \approx x_1 \approx \cdots \approx x_n$$

show that there is no transitivity of closeness.

If x and y are infinitely close, then either $x = y$ or $|x - y| < |x - y|$. In the latter case, assuming that $y < x$, then we may further infer at least one of the following:

$$x + \mathbf{0}_y < x + \mathbf{0}_y \quad \text{or} \quad \mathbf{0}_x - y < \mathbf{0}_x - y \quad \text{or} \quad \mathbf{0}_{x+y} < \mathbf{0}_{x+y}.$$

If $\mathbf{0}_u = \eta$ for all u, this disjunction is true just because $\eta < \eta$ (but I am agnostic as to whether relative zeros are infinitely small or not). If $x \approx y$, then $[x, y]$ is either infinitesimal (an imploded interval) or a point.

Closeness brings us now to continuity.

8.3.2 Continuous Mappings

Insofar as our target is the intermediate value theorem – a continuous mapping from one part of $\mathbb{R}$ to the other must cross a line – then we need to say what a continuous mapping is. We begin as ever with the constraints balanced against the demands.

A completely orthodox definition of continuous functions (e.g., using ε/δ) won't do, because (a) we are using relations rather than functions, and want to take seriously the idea of mapping to more than one place; and that's because (b) part of our purpose here is to develop mathematics that makes sense of the vagueness story in Chapter 1. If some value can continuously transition from being high up – represented as *above* a line – to *not* being high up – represented as *below* a line – then some single point is both high up and not, both above and below the line, continuously. The definition of a continuous mapping requires, in Meyer's phrase, "an expansion of the pragmatic imagination" [Meyer, 1975, p. 5].

Setting aside uniqueness and consistency, then, what is important is that continuous mappings preserve closeness. For a mapping f, this means at least that if x is infinitesimally close to y, then some $u \in f(x)$ is infinitesimally close to some $v \in f(y)$. But this condition is a bit too loose. Each $f(x)$ should be connected to the images of points nearby x, not "teleporting." Merely requiring that *some* values of $f(x)$ and $f(y)$ be close would allow the following counterexample:

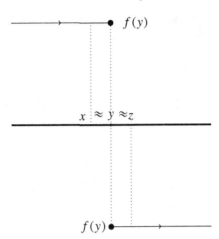

Here, y is very close to both x and z, and f is taking y to one location very close to where it takes x, and another location very close to where it takes z. But f "blips" at y, from above the axis to below. It is clearly (!) not continuous (to say nothing of failing the "vertical line test").

A middle way is to consider the following (recalling the meaning of "interval" from Section 8.2.3, and that an interval is either a proper interval or a point).

Definition 3. *A mapping f is* continuous *iff for any X, Y such that $f : X \longrightarrow Y$; if $A \sqsubseteq X$ is an interval, then $f''(A) \sqsubseteq Y$ is an interval.*

(The image $f''(A) = \{z : \exists u (u \in A \ \& \ z \in f(u))\}$ is as in Section 5.2.2.3.) Definition 3 is a basic property of continuous functions f from orthodox mathematics: *the continuous image of an interval is an interval.* We expand the definition as a proposition.[20]

Proposition 50. *If $f : X \longrightarrow Y$ is continuous, then*

$$x \approx y \Rightarrow (u \in f(x) \ \& \ v \in f(y) \ \Rightarrow \ [u, v] \sqsubseteq f''[x, y]),$$

i.e., if $u < u_0 < v$, then $u_0 \in f''[x, y]$.
A fortiori, if $u \in f(x) \ni v$, then $[u, v] \sqsubseteq f(x)$.

Proof The first claim is a restatement of the definition, noting that (vacuously) if $x \approx y$, then $[x, y]$ is an interval, and that $f''[x, y] = \{z : \exists u (z \in f(u) \ \& \ u \in [x, y])\}$. For the second claim, take $x \approx x$, and observe that $f(x) = f''\{x\} = f''[x, x]$. □

---

[20] To work with continuous relations further than we will here, one might demand that if relations f are constrained to be functions, then the definition of continuity delivers an ordinary continuous function. See [Grimeisen, 1972], where a relation is defined as continuous just if it includes a continuous function.

This illustrates how our previous counterexample is dealt with by this definition: if f takes y to two different (apart) places, then it also takes y to *all points in between*:

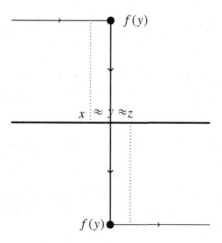

No teleportation! If $x \approx y$, then $u, v \in f[x, y] \Rightarrow \forall u_0(u < u_0 < v \Rightarrow \exists z(z \in [x, y]$ & $u_0 \in f(z)))$. This means that infinitesimally close points continuously map to infinitesimally close points (among others): abusing notation (violently),

$$x \approx y \Rightarrow f(x) \approx f(y), \tag{8.19}$$

where $f(x) \approx f(y)$ is slang for $\exists u \exists v(u, v \in f''[x, y]$ & $u \approx v)$.

This also reveals that our definition of continuity is (designed to be) more flexible than the classical notion, and so it harbors a surprising possibility: that a continuous mapping may take values *outside its own range*. If y maps continuously to 0 and y maps continuously to 1, then y also maps to all the points between 0 and 1, even when those points are not in the intended range of f. Now, we know (Theorem 13) that there are fixed points for any mapping – and sometimes points coming from outside the range of the mapping. Soon we will see this happen (Theorem 53). But, if this seems (too) impossible, note that the impossible only happens when we ask f to do something impossible. That is, just classically speaking, if a continuous function is applied in a situation where it cannot be applied, then the function will "misbehave" dramatically. Our continuous maps may take values outside their own range *if* they are asked to do something "discontinuous." In fact, here we are exerting considerably *more* control than in the ex falso, anything-goes case. And because f is a mapping as per Section 5.2.2.3, $u \in f(x) \Rightarrow (u \neq v \Rightarrow v \notin f(x))$. This "functional" property means that any true account of the movement of f will not have x going to strongly separated places (such as 0 and 1) at the same time (even if it also does).

The definition of continuity gives rise to a nice property:

Proposition 51.   *The composition of continuous maps is continuous.*

Proof Let f, g be continuous maps, and $h = g \circ f$. Map f takes intervals to intervals; map g takes intervals to intervals, so h does, too. ☐

8.3.3 The Intermediate Value Theorem

It is time to vindicate the intuition: if a change occurs, it must occur somewhere. Day does not become night without passing through twilight. How do we prove this – Bolzano's intermediate value theorem? As with Dedekind, the main work of this chapter has been a reverse-engineering project to make the following immediate.

Theorem 52 (Intermediate Value Theorem (Bolzano 1817)). *Let $f : [a, b] \longrightarrow \mathbb{R}$ be a continuous map. Let $x \in f(a)$ and $y \in f(b)$, and suppose $x < 0 < y$. Then*

$$0 \in f(c)$$

for some $c \in [a, b]$.

Proof Since $[a, b]$ is an interval, $f''[a, b]$ is an interval, by continuity of f (Definition 3). With $x \in f(a)$ and $y \in f(b)$, if $0 \in (x, y)$, then $0 \in [x, y]$, and then $0 \in f''[a, b]$, by Proposition 50. Then by definition of images, there is some $c \in [a, b]$ with $0 \in f(c)$ as required. ☐

Classically, the functional version of the Intermediate Value Theorem (IVT) has as a corollary a fixed point theorem (as pictured way back in Section 0.3): If $f : [0, 1] \longrightarrow [0, 1]$ is continuous, then there is some $z \in [0, 1]$ such that z is a fixed point for f. For we would take $g : [0, 1] \longrightarrow \mathbb{R}$ to be a function with $g(x) = x - f(x)$, so that if $g(z) = 0$, then $f(z) = z$. This argument relies on much more than we have at our disposal, but it is not the only argument.

Let's think about a contrapositive: if there is *no* fixed point for f, then there is a continuous mapping from the interval onto its boundary. This is, classically, for reductio, since there is no continuous mapping of an interval onto its endpoints. Why not? Because any such mapping will rip the interval, and so not be continuous. But – what if it is still also continuous? What if instead of a reductio, we finish thinking through the case and see if the truth will still emerge? Here we ask of a notion of continuity to do more than just keep up with the intuitive arguments that motivate the paraconsistent approach to begin with; alternative reasons for classical theorems can be found if the paraconsistent mathematician can say a bit *more* than others, e.g., about what happens in "impossible" cases. Our last result of the chapter considers a continuous transformation of the interval onto its boundary – a *retraction*, to be officially defined in the next chapter. The beauty here is that such an action can still count as continuous; it will just be inconsistently so. And in this continuous, discontinuous ripping, the dueling NO GAPS and UNITY properties discussed in Section 8.2.1 are vindicated, prefiguring the paraconsistent proof of Brouwer's fixed point theorem.

When a map f takes some p to two distant points, $a \in f(p)$ & $b \in f(p)$, with a apart from b, it will be fun to say that the map *splits* point p. The intuition is that, if f is continuous on a space (like $[0, 1]$) but pulling in two directions (like onto $\{0, 1\}$), then some point will be split. (In the wider context of this study, think about a mapping of all propositions onto the endpoints $\{\text{true}, \text{false}\}$, and how some propositions are pulled in two directions.) The following is, again, classically true, just by ex falso quodlibet.

Theorem 53. *Let $f : [0, 1] \longrightarrow \{0, 1\}$ be a continuous mapping of the interval* onto *its boundary.*

(i) There is some nonempty $X \sqsubset [0, 1]$ such that $[0, 1] \sqsubseteq f''X$.
(ii) If there is some some "split" $p \in [0, 1]$ such that

$$0 \in f(p) \ni 1,$$

then $p \in f(p)$.

Proof For (i), if f is "onto," then every point in the interval maps to either 0 or 1, and there are points that map to 0, and others that map to 1. So there are $p_0, p_1 \in [0, 1]$ with $0 \in f(p_0)$ and $1 \in f(p_1)$. We assume without loss that $p_0 \leqslant p_1$. Then by Proposition 50, for all $x \in [0, 1]$ there is some $y \in [p_0, p_1]$ such that $x \in f(y)$.

For (ii), we are in effect considering the case from (i) where $p_0 = p_1$. Again by Proposition 50, for any $x, y \in f(p)$, $[x, y] \sqsubseteq f(p)$. If $0, 1 \in f(p)$, then because $p \in [0, 1]$,

$$p \in f(p)$$

as required.   □

This shows that for some nonempty subset X of the interval, on assumption, $X \sqsubset f''X$.

Given a continuous surjection of an interval onto its boundary, this theorem does not guarantee the existence of a split point. Thinking of the interval as two pieces, $A = \{x \in [0, 1] : 0 \in f(x)\}$ and $B = \{x \in [0, 1] : 1 \in f(x)\}$, the points where A breaks from B are not necessarily close enough together (by Theorem 45) to coincide and split. What we have in full generality is a "syrupy" strand that connects the ripped parts together. Unless we already know that $x = y$, then if $0 \in f(x)$ and $1 \in f(y)$, no matter how close together x and y may be; even if closer than any $\varepsilon > 0$, there still may be some infinitesimal space between them, some strand of viscous goo. But what (ii) of the theorem shows is that, when that strand is between a point and itself, a point ripped in half, that is where a fixed point can be found. Paradoxes arise in full when there is something lurking in the inky spaces between.

8.4 Out of the Labyrinth: The Topology of a Point

Transitioning from metric space in this chapter to topology in the next, let us take a moment to reflect on the nature of *points*.

Some people think there can be extended simples – or at least, some people argue that there isn't enough reason (yet) to think there *can't* be extended simples [McDaniel, 2007]. To take the most extreme example, monists hold that the entire universe is one big extended simple, a whole with no parts.[21] What about the other way? Having now seen many hypothetical examples, I suggest *unextended complexes*, extensionless things with proper parts. If there were such things, they would be a counterexample to a (usually unstated) assumption, that if something has (proper) parts, it is extended. But the mathematics we are looking at suggests again and again that unextended points may have internal structure. Points have parts. So what is a point?

Euclid says in Book 1, definition 1 of the *Elements* that "a point is that which has no parts" [Euclid, 1956]. It would be classically equivalent, if a bit odd, to say it this way:

A point is that which has only itself as a part

or a bit more formally:

point$_1$: α is a point iff (if x is a part of α then $x = \alpha$).

More dramatically,

A point is that which all its proper *parts are identical to itself*

or

point$_2$: α is a point iff (if x is a proper part of α then $x = \alpha$).

Classically, the latter implies that a point has no proper parts, since otherwise it would be a proper part of itself; and going back, by ex falso, since it has no proper parts, "all" of them are just identical to the point. Even more dramatically,

A point is that which all of its proper parts have the point itself as a proper part

or:

point$_3$: α is a point iff (if x is a proper part of α then α is a proper part of x).

And so forth. Each suggests a nontrivially different possible internal structure of a point, invisible in the antecedents of classically equivalent conditionals. This sort of mathematical slip of the tongue is where there are opportunities to investigate a new structure: the inner space of points.

On, then, to point-set topology.

---

[21] Proponents of extended simples need to get around the "geometric correspondence principle": that any subregion of a spatially extended object corresponds to some part of that object [Simons, 2004, p. 371]. Very small physical objects such as Planck squares may plausibly violate this principle [Braddon-Mitchell and Miller, 2006].

9

Topology

In the previous chapter, we studied the shape of the linear continuum, focusing on what happens when one tries to cut or tear – divide – it in two. In this penultimate chapter, we develop a qualitative generalization of these ideas, about *connectedness*, but now without any metrics. Topology is the abstract geometry of closeness that manages to capture properties of nearness without any appeal to distance. We will focus on closed sets, boundaries, and eventually, continuous transformations and their fixed points, bringing ideas from analysis back to set theory.[1]

9.1 Closure Spaces

9.1.1 Closure Operators

A topological *closure* operator

$$c : \mathscr{P}X \longrightarrow \mathscr{P}X$$

on subsets $x, y \in \mathscr{P}(X) = \{z : z \subseteq X\}$ is assumed to satisfy the following properties.

(C1) $x \subseteq c(x)$
(C2) $c(c(x)) \subseteq c(x)$
(C3) $c(x \cup y) \subseteq c(x) \cup c(y)$
(C4) $x \sqsubseteq y \Rightarrow c(x) \subseteq c(y)$
(C5) $c(\varnothing_X) = \varnothing_X$

(recalling that if $x \subseteq X$ then $x \sqsubseteq X$). The closure axioms[2] are formulated to be strong, in terms of logic. For example, from (C4), since $x \subseteq y$ implies $x \sqsubseteq y$, we also have $x \subseteq y \Rightarrow c(x) \subseteq c(y)$.

---

[1] For this chapter, I have consulted [Armstrong, 1979; Hocking and Young, 1961; Jänich, 1984; Mendelson, 1975; Munkres, 2000; Steen and Jr, 1978], among others. For some reconnaissance about inconsistent topology, see Mortensen [1995], ch. 10, [2010, chs.1, 2]; and on paraconsistent mereotopology, [Weber and Cotnoir, 2015]. A topic that is *not* addressed is the notion of topological equivalence, via *homeomorphisms*, which rely on the (functional) notion of bijection; see the Excursus of Chapter 5. A weaker notion, *homotopic maps*, might work better in future for a sort of topological equivalence; cf. Section 7.3.3.

[2] In this chapter, I will follow standard practice of referring to constraints on topological spaces as "axioms," but as with group theory "axioms" and the like, they are more like definitions than assumptions.

A set $x \sqsubseteq X$ is *closed* in X iff

$$x \equiv c(x),$$

that is, if x is a fixed point for the closure operator. It follows that $c(x)$ is itself closed in X, since $c(x)$ is a subset of X by construction and

$$c(x) = c(c(x)) \tag{9.1}$$

by (C1), (C2), and (C4). The idea is to think of a point z as "nearby" x iff $z \in c(x)$; the closure of a set is all the points nearby it. A closed set contains all its nearby points.

9.1.1.1 Why These Axioms?

There are many ways to get topology up and running; classically, they are all equivalent. Nonclassically, different approaches will yield different results.[3]

Hausdorff developed point-set topology in the latter parts of his famous 1914 book [Hausdorff, 1957], where he presented the idea of a topology in terms of axioms for "neighborhoods," open sets of nearby points. It has become more common since then to define a topology on X very abstractly, as a collection of subsets of X that contains all unions and finite intersections of open sets, and among which are the whole space X and the empty set. Elements of a topology are called "open" sets, but this is only a heuristic to begin with.[4] A third alternative is to try to capture the "closed" subsets of X, as all intersections, finite unions, and X and $\varnothing$. On any of these three approaches, one is then tasked with proving that the elements of any topology, so defined, do indeed accord with the intuitive ideas about nearness, e.g., that the union of closed sets is closed. See Section 9.1.3. My experience has been that approaching topology from any of these directions, paraconsistently, is very hard going. Too much relies on artifacts of classical quantifier logic/set theory, as opposed to the meaning of the axioms, to get from a "topology" to the target of a *topology*.

Already in the 1920s, though, Kuratowski saw that one could go about this the other way: start by defining the properties we want closed sets to have, via an abstract closure operator, and then show that any such operator generates a topology.[5] We follow the latter approach. When topology is reconstructed paraconsistently in this way, two methodological themes emerge. First, duality between open and closed sets breaks down, except in special cases. And second, furthermore, closed sets seem better to work with than open sets.[6]

---

[3] See Jänich [1984], pp. 5–7, for comparisons.

[4] "The 'reasonableness' of [this definition of topological space] ... can be justified only by reading the forty-two sections which follow it" [Willard, 1970, p. 23].

[5] See, e.g., Kelley [1955, p. 43]; cf. [Kuratowski, 1962]. Thanks to Cotnoir for suggesting this.

[6] Cf. closed set logic as paraconsistent logic [Mortensen, 2010]. This is comparable to the situation in constructive, intuitionisitic topology, where the fundamental notion of "apartness" turns out to be better to work with than closeness (cf. Bridges and Vita, 2011). We will have some shared experience with Bishop, when he says: "Very little is left of general topology after that vehicle of classical mathematics has been taken apart and reassembled constructively. With some regret, plus a large measure of relief, we see this flamboyant engine collapse to a constructive size" [Bishop, 1967, p. 63].

9.1.1.2 Notes on the Closure Axioms

The Kuratowski properties, as in orthodox mathematics, do not determine a unique closure operator on X. Being closed is a relative notion; a set may be closed under one closure operator but not under another. Similarly, a set may be closed in X but not in Y. When needed, we can index the closure operator: $c_X(x) \equiv x$ would mean that x is closed *in X*. As it happens, though, much of the time we can ignore context and just talk about closed sets simpliciter.

If sets are equivalent, they have the same closure:

$$x \equiv y \implies c(x) = c(y), \tag{9.2}$$

even though $\equiv$ does not substitute (is not a congruence) in general. That's because, by (C4) and factor, $x \sqsubseteq y \,\&\, y \sqsubseteq x \implies (c(x) \subseteq c(y)) \,\&\, (c(y) \subseteq c(x))$. If $\equiv$ did substitute, then we would also immediately have that, if sets are equivalent, and one is closed, then so is the other; this is true, but we will wait until Section 9.1.2 to prove it without substitution or contraction.

For any closure operator on X, the whole space X is closed in itself,

$$X \equiv c(X),$$

by (C1) and the fact that $c : \mathscr{P}X \longrightarrow \mathscr{P}X$, i.e., $c(x) \sqsubseteq X$ for any $x \sqsubseteq X$. From (C5), on the closure of empty sets, notice the special case when c is over the entire universe; since $\emptyset_V = V - V = \emptyset$, then

$$c(\emptyset) = \emptyset.$$

And then $c(\emptyset) \sqsubseteq x$ for *all* x. But in general (C5) is stated in terms of relative zeros (Definition 1), as should now be familiar from previous chapters.

Classically, C4 (monotonicity) is redundant, as it is derivable from C1–C3. Here, we *can* use C1–C3 to prove something very close: $x \sqsubseteq y \Rightarrow c(x) \sqsubseteq c(y)$. To do this, we show that $x \cup y \equiv y \implies c(x) \sqsubseteq c(y)$, by the following argument; cf. Section 5.1.6.

1. $x \cup y \equiv y$ (assume)
2. $c(x \cup y) = c(y)$ (1, by (9.2))
3. $c(x \cup y) \equiv c(x) \cup c(y)$ (by C3 and Proposition 54(o) below)
4. $c(x) \cup c(y) \equiv c(y)$ (2, 3, transitivity)
5. $c(x) \sqsubseteq c(y)$ (4, because $(p \vee q \Rightarrow r) \Rightarrow (p \Rightarrow r)$)

And then since $x \sqsubseteq y \Rightarrow x \cup y \equiv y$, we have (nearly) C4. But we assume the slightly stronger (C4) version of monotonicity directly.

A space that satisfies (C1), (C2), and (C4) is called a *closure space*. The axioms for a closure space may be stated more succinctly, in the axiom

(C0) $x \subseteq c(y) \Leftrightarrow c(x) \subseteq c(y)$.

We prove a substructural version of the folklore result (following [Martin and Pollard, 1996, pp. 7–8]) that axioms (C1), (C2), and (C4) imply (C0), and *two copies* of (C0) imply (C1), (C2), and (C4).

First suppose (C1), (C2), and (C4). There are two directions of the biconditional of (C0) to show. Suppose $x \subseteq c(y)$. Then applying (C4), $c(x) \subseteq c(c(y))$. By (C2), $c(c(y)) \subseteq c(y)$. Then by transitivity of the subset relation, $c(x) \subseteq c(y)$ showing

$$x \subseteq c(y) \Rightarrow c(x) \subseteq c(y).$$

Now for the other direction of the biconditional, suppose $c(x) \subseteq c(y)$. By (C1), $x \subseteq c(x)$, so transitivity again shows

$$c(x) \subseteq c(y) \Rightarrow x \subseteq c(y).$$

That's half the proof, with each of (C1), (C2), and (C4) used just once to get (C0). Now suppose (C0), and observe the following:

- $x \subseteq c(x)$ because $c(x) \subseteq c(x)$, using (C0) right to left. That's (C1).
- $c(c(x)) \subseteq c(x)$ because $c(x) \subseteq c(x)$, using (C0) left to right. That's (C2).
- $x \subseteq y \Rightarrow c(x) \subseteq c(y)$ because if $x \subseteq y$, then from (C1) (which we've now proved), it follows that $x \subseteq c(y)$, and now apply (C0) left to right. That's $c(x) \subseteq c(c(y))$, and then by (C2), which is also now proved, we have (C4).

The whole proof is done after some substructural accountancy. (C1) and (C2) follow from one assumption of (C0), using each of its directions separately. Then we used (C0) left to right along with (C1) (which itself needed (C0) right to left) to get (C4). So that was another assumption of (C0).

9.1.1.3 A Separation Axiom

Spaces may be imbued with additional structure, via *separation axioms* or *T axioms* (for *Trennungsaxiom*, introduced by Alexandroff and Hopf in 1935). For example, a space is *Hausdorff* or $T2$ if, roughly, for any two distinct points, there is some non-overlapping space around each of the points. Now, in Chapter 8 we saw that nothing forced the distance metric on our $\mathbb{R}$ to satisfy the triangle inequality; from a topological viewpoint, it is well known that the triangle inequality would imply that $\mathbb{R}$ is Hausdorff. That adds up to very diminished expectations that the topology corresponding to our $\mathbb{R}$ of Chapter 8 would satisfy the $T2$ axiom, or anything stronger.

A weaker $T1$ axiom defines a *Fréchet space*,[7] wherein for any two points x, y, point x has some space around it not including y, and y has some space around it not including x (though the spaces could still overlap on other points). A classical equivalent of the $T1$ axiom is the stipulation that all singletons are closed: $\{x\} \equiv c(\{x\})$. That amounts to saying

---

[7] Not to be confused with a Fréchet-V space, a closure space satisfying also (C5), identified as a good background for topological sorites [Dzhafarov and Dzhafarov, 2010].

that anything "near" a point x is identical to x. Even this is a bit strong in general, given the monads of Section 8.3.1 (though see Section 9.3.3).

In terms of rehabilitating topology with paraconsistent logic, however, it is a minimum requirement that closure spaces be closed under intersections – that is, *the intersection of closed sets is closed*. This requires a very weak separation axiom – really more of a *non*separation axiom – which is that *the closure of an intersection contracts*:

(C6) $c(x \cap y) \sqsubseteq c(x \cap y) \cap c(x \cap y)$.

Roughly, anything nearby an intersection is *very* nearby it. Later, we will also mention a less modest version of this "separation" axiom, that anything nearby an intersection is *infinitely* nearby too.

9.1.2 Closed Sets

Call $c(x)$ the *closure* of x. The closure of x is closed ((9.1) in Section 9.1.1). Let's check that the closure of x is the *smallest closed set that includes x*:

$$\forall y(y \equiv c(y) \Rightarrow (x \sqsubseteq y \Rightarrow c(x) \sqsubseteq y)). \tag{9.3}$$

Proof: If $x \sqsubseteq y$, then $c(x) \subseteq c(y)$. If $c(y) \sqsubseteq y$, which is true when y is closed, then transitivity gives the result. More, if x is closed, then if $u \in x$, then by (9.3), $\{u\} \sqsubseteq x \Rightarrow c(\{u\}) \sqsubseteq x$, by (5.3) of Section 5.1.2.

On this basis, we can prove as promised earlier, that

$$x \equiv y \Rightarrow (x \equiv c(x) \Rightarrow y \equiv c(y)). \tag{9.4}$$

Proof: Suppose $x \equiv c(x)$. Then by (9.3), if $y \sqsubseteq x$, then $c(y) \sqsubseteq x$. But now if also $x \sqsubseteq y$, then $c(y) \sqsubseteq y$. Since $y \subseteq c(y)$ by (C1), we have shown that $(x \equiv c(x) \,\&\, x \sqsubseteq y \,\&\, y \sqsubseteq x) \Rightarrow y \equiv c(y)$, as required.

For points, we note the following properties. If x is closed, then

$$c\{x\} \sqsubseteq c\{x\} \cap c\{x\}.$$

In a slogan, anything nearby a closed point is very, *very* nearby that point. This is because identity contracts; and for the same reason, since $\{x\} \sqsubseteq \{x\} \cap \{x\}$, then (C4) entails $c(\{x\}) \sqsubseteq c(\{x\} \cap \{x\})$. A few more elementary consequences of the closure axioms are now collected.

Proposition 54. *The following are true.*

(o) $c(x) \subseteq c(x \cup y)$.
(i) $c(x) \cup c(y) \sqsubseteq c(x \cup y)$.
(ii) $c(x \cup y) \equiv c(x) \Rightarrow y \subseteq c(x)$.
(iii) $y \subseteq c(x) \Rightarrow c(x \cup y) = c(x)$.
(iv) $c(x \cup \emptyset_x) = c(x)$.

Proof (o) It is the case that $x \subseteq x \cup y$, so $c(x) \subseteq c(x \cup y)$ by monotonicity (C4).

(i) Use (o) twice, for $c(x) \subseteq c(x \cup y)$ and $c(y) \subseteq c(x \cup y)$. Then use argument by cases for $c(x) \cup c(y) \sqsubseteq c(x \cup y)$.

(ii) From (o), $c(y) \subseteq c(x \cup y)$. Then $y \subseteq c(x \cup y)$ by (C1), so if $c(x \cup y) \equiv c(x)$, then $y \subseteq c(x)$.

(iii) Assume $y \subseteq c(x)$. Then $y \cup c(x) \equiv c(x)$ (from (5.7) in Section 5.1.4). Then

$$c(y \cup c(x)) \sqsubseteq c(x)$$

by (C4) and then (C2). Since $x \cup y \subseteq c(x) \cup y$ using (C1) and $\vee$-introduction, (C4) gives

$$c(x \cup y) \subseteq c(c(x) \cup y).$$

Then $c(x \cup y) \subseteq c(x)$. Using (o), we have $c(x \cup y) = c(x)$.

(iv) follows from (iii), since $\varnothing_x \subseteq x \subseteq c(x)$. $\qquad\square$

Note that (iv) is also true of *the* empty set, $c(x \cup \varnothing) = c(x)$, but not for arbitrary $\varnothing_y$ with y different from x (since $\varnothing_y \subseteq x$ isn't true in general).

Proposition 55. *The (binary) union of closed sets is closed.*

Proof By (C1), $x \cup y \sqsubseteq c(x \cup y)$. By (C3), $c(x \cup y) \sqsubseteq c(x) \cup c(y)$. If $x \equiv c(x)$ and $y \equiv c(y)$, then $c(x) \cup c(y) \sqsubseteq x \cup y$. Altogether, when x and y are closed,

$$x \cup y \equiv c(x \cup y)$$

as required. $\qquad\square$

The preceding argument will work for any finite union $x \cup \cdots \cup y$. This means the next theorem holds for finitely many unions of closed sets.

Proposition 56. *A set Y is closed, $Y \equiv c(Y)$, iff all members of Y are contained in finitely many closed subsets of Y, i.e.,*

$$y \in Y \Rightarrow \exists U_0 \ldots \exists U_m ((U_0 \equiv c(U_0) \ \& \ U_0 \sqsubseteq Y)$$
$$\& \ldots \& (U_m \equiv c(U_m) \ \& \ U_m \sqsubseteq Y) \ \& \ y \in U_0 \cup \ldots \cup U_m)$$

for some $m \in \mathbb{N}$.

Proof Left to right is immediate by the fact that $Y \subseteq Y$ and is closed. (Here, $m = 0$.)

Right to left, consider m closed subsets $U_0, \ldots, U_m$ of Y. Assume $y \in Y \Rightarrow y \in U_0 \vee \cdots \vee y \in U_m$. Then

$$Y \equiv U_0 \cup \cdots \cup U_m. \tag{9.5}$$

But all the U_i are closed, so by Proposition 55, their union is closed too:

$$c(U_0 \cup \cdots \cup U_m) \equiv U_0 \cup \cdots \cup U_m$$

Then Y is closed by (9.4) and (9.5). $\qquad\square$

Proposition 57.   *The (binary) intersection of closed sets is closed.*

Proof   Suppose that $x \equiv c(x)$ and $y \equiv c(y)$. Since $x \cap y \subseteq x$ and $x \cap y \subseteq y$, then $c(x \cap y) \subseteq c(x)$ and $c(x \cap y) \subseteq c(y)$. From this, since x and y are closed, $c(x \cap y) \sqsubseteq x$ and $c(x \cap y) \sqsubseteq y$, so

$$c(x \cap y) \cap c(x \cap y) \sqsubseteq x \cap y$$

by factor. Then by (C6),

$$c(x \cap y) \sqsubseteq x \cap y,$$

so by (C1) the intersection is closed.   □

From Proposition 57, it follows that

$$c(x \cap y) \sqsubseteq c(x) \cap c(y). \tag{9.6}$$

For $c(x) \cap c(y)$ is the intersection of closed sets, hence closed. Since $x \cap y \sqsubseteq c(x) \cap c(y)$ by two uses of (C1) and factor, then by (9.3), the result holds.

From classical topology, one would expect that, while only finite unions of closed sets are closed, any *arbitrary* intersection of closed sets is closed. Amplifying Proposition 57 to the general case is achieved by amplifying (C6) to the general case: the closure of an intersection is included in any number of intersections of it with itself (recalling the picture back in Section 5.1.6.2):

(C6$^\infty$): $c(x \cap y) \sqsubseteq \bigcap_{\alpha \in On} c(x \cap y)_\alpha,$

which looks slightly ridiculous and is classically trivially true.

9.1.3 Inducing Topology: Interiors and Open Sets

Let X be a set and c a closure operator on subsets of X satisfying (C1)–(C6). Let $\mathcal{T} \sqsubseteq \mathscr{P}(X)$ be the subsets of X that are closed for c:

$$\mathcal{T} = \{x : x \sqsubseteq X \ \& \ x \equiv c(x)\}.$$

The collection $\mathcal{T}$ is called a *topology* on X. The work up to this point is enough for the following almost-classical result.

Theorem 58.   *A topology on X satisfies the following conditions:*

- $\varnothing_X \in \mathcal{T}.$
- $X \in \mathcal{T}.$
- *If $x \in \mathcal{T}$ and $y \in \mathcal{T}$, then $x \cup y \in \mathcal{T}$,*
- *If $x \in \mathcal{T}$ and $y \in \mathcal{T}$, then $x \cap y \in \mathcal{T}$,*

Proof $\varnothing_X$ is closed (C5) and X is closed, as in Section 9.1.1.2. Proposition 55 proves that $\mathcal{T}$ is closed under finite unions, and Proposition 57 proves it is closed under finite intersections.   □

For every topological closure operator on X, then, there is a corresponding topology, in the sense just given. Again, a stronger theorem holds for closure operators that obey [C6$^\infty$]; then a topology is closed under any intersection of closed sets, i.e., if $Y \in \mathcal{T}$, then $\bigcap Y \in \mathcal{T}$.

It is more common in topology to focus on *open* sets. Definitionally, it usually doesn't matter: open sets are just the dual of closed sets, and, in classical settings, all the work in the previous section could be repeated, or simply had for free, for open sets (e.g., the finite intersection of open sets are open). But matters are more subtle here. Say that an *interior operator* is a mapping $i : \mathcal{P}(X) \longrightarrow \mathcal{P}(X)$ such that

(I1) $i(x) \subseteq x$.
(I2) $i(x) \subseteq i(i(x))$.
(I3) $i(x) \cap i(y) \subseteq i(x \cap y)$.
(I4) $x \subseteq y \Rightarrow i(x) \subseteq i(y)$.
(I5) $i(X \cup \overline{X}) \equiv X \cup \overline{X}$.

If defined as $i(x) := \overline{c(\overline{x})}$, then these properties follow from the closure axioms. The first four are just by contrapositions: if $x \subseteq y$, then $\overline{y} \subseteq \overline{x}$. (That's one of the nicest things about intensional subsets.) For example, $\overline{x} \subseteq c(\overline{x})$ is an instance of (C0); then $\overline{c(\overline{x})} \subseteq \overline{\overline{x}} = x$, proving (I1). For (I4), $\overline{y} \subseteq \overline{x} \Rightarrow \overline{c(\overline{y})} \subseteq \overline{c(\overline{x})}$. For (I5), just note that $\overline{\varnothing_X} = X \cup \overline{X}$. So then we can say x is *open* if it is equivalent to its interior, $x \equiv i(x)$, and prove the interior of x is the largest open set included in x,

$$\forall y(y = i(y) \ \& \ y \sqsubseteq x \Rightarrow y \sqsubseteq i(x)),$$

and so forth, for the other interior properties.

Duality is not as simple as all this seems, though. First, for genuine duality one would expect to say that x is open iff $\overline{x}$ is closed, and $\overline{x}$ is open iff x is closed. But being closed (resp., open) has been defined extensionally, where x is closed when $x \equiv c(x)$. Since $\equiv$ does not substitute, some otherwise natural reasoning breaks down (described later). This might be remedied if we say that x is *intrinsically closed* when $x = c(x)$, and *intrinsically open* when $x = \overline{c(\overline{x})}$, and try to work with these. But this would only paper over the deeper issue with duality. For second and more basically, a more familiar form of (I5), the identity $i(X) = X$, is not recoverable from (C5), suggesting the loss of duality is really to do with complementation. When reasoning about topological spaces, one tends to pretend that the space X in which a topology exists is "the universe," as a kind of restricted quantification. Unless the topological space in question really is the entirety of the cosmos, though the appropriate complement of $x \sqsubseteq X$ is the complement *in* X, or $X \backslash x$, not the unrestricted complement $\overline{x}$. We are working with absolute generality (Section 4.3.2.3), so the unrestricted complement of some closed sets in, e.g., $\mathbb{R}$ will include open sets in $\mathbb{R}$, but also many things not related to $\mathbb{R}$ at all.

In light of this, the i operator defined $\overline{c(\overline{x})}$ is much too general, as would be its corresponding notion of "open." Our official definition of open sets, to capture what is meant in mathematical practice, will need to be as the *relative* complement of a closed set: for $A \sqsubseteq X$,

$$A \text{ is } open \text{ iff } X \backslash A \text{ is closed.}$$

This notion of being open is better, in terms of capturing "working" usage, but duality with closed sets is broken. If $X \backslash A$ is open, then $X \backslash (X \backslash A)$ is closed, but that is not to say that A is closed, because $X \backslash (X \backslash A)$ does not reduce to A (Section 5.1.5). And if A is closed, we don't know that $X \backslash A$ is open. And then, even if $i(A) := X \backslash c(X \backslash A)$, being open in that sense does not coincide with the property of A being identical to its interior, $A \equiv i(A)$, and no longer gives (I1)–(I5) for free. You get what you pay for.

For all these reasons, then, we tend to stick with closures and give arguments directly about the closure operator, rather than taking contrapositive detours through considerations about interiors and open sets.

Excursus: Consequence as Closure

In this brief digression, we will look at how well closure space connects to logical space in a paraconsistent substructural setting.

A *theory* is a set of sentences closed under logical consequence: anything that follows from the theory is also in the theory. To what extent is this notion of closure the same as the topological closure operator studied in Section 9.1.1? In 1930, Tarski investigated how logic could be viewed via a closure operator [Tarski, 1956b, pp. 60–109]; cf. [Martin and Pollard, 1996], and in particular, closure spaces satisfying (C1), (C2), and (C4). (If we read "$x \in c(X)$" as "x is a logical consequence of X," thinking of x as a sentence and X a set of sentences, then no one thinks (C3) or (C5) should be satisfied.) Without contraction, whether consequence is "tarskian" becomes more subtle.[8] Let's lay out two sets of conditions that reasonably capture consequence and closure, and show that they link up.

Let $c : \mathscr{P}(Z) \longrightarrow \mathscr{P}(Z)$ be an operator on Z satisfying three closure axioms:[9]

Increasing: $X \sqsubseteq c(X)$.
Idempotent: If $c(X) = \{y_0, \ldots, y_n\}$, then $c(y_0) \cup \ldots \cup c(y_n) \sqsubseteq c(X)$.
Monotone: $X \sqsubseteq Y \Rightarrow c(X) \sqsubseteq c(Y)$.

Let $\vdash$ be a subset of $\mathscr{P}(Z) \times Z$, called a *consequence relation*, such that for all $x, y \in Z$, and all $X, Y \subseteq Z$,

Reflexive: If $x \in X$, then $X \vdash x$,
Transitive: If $X \vdash x$ and $Y \cup \{x\} \vdash y$, then $X, Y \vdash y$.
Monotone: If $X \vdash x$ and $X \sqsubseteq Y$, then $Y \vdash x$.

---

[8] See [Ripley, 2015b; Cintula and Paoli, 2016].
[9] The idempotence condition borrows from [Cintula and Paoli, 2016]. We could have obtained the same shaped result by taking as our axiom for idempotence that $x \in c(X) \Rightarrow c(Y \cup \{x\}) \sqsubseteq c(X \cup Y)$. See Galatos et al. [2007, p. 63].

The consequence relation of our paraconsistent master theory subDLQ satisfies these requirements.

Now define

$$c_\vdash(X) := \{x \in Z : X \vdash x\}$$

and

$$\vdash_c := \{\langle X, x \rangle : x \in c(X)\}$$

with the following nice result:

$c_\vdash$ is a closure operator induced by consequence, and $\vdash_c$ is a consequence relation induced by closure.

To show that $c_\vdash$ is a closure operator:

- For increasingness, let $x \in Z$ and suppose $x \in X$. Then $X \vdash x$. Then $x \in Z$ & $X \vdash x$, so $x \in c_\vdash(X)$. Therefore, $X \sqsubseteq c_\vdash(X)$. Note that we need to make the explicit stipulation that $x \in Z$ in the definition of closure; even though $X \subseteq Z$, we can only use the assumption that $x \in X$ once.
- For idempotence, let $z \in c_\vdash(y_0) \cup \ldots \cup c_\vdash(y_n)$. Then $(y_0 \vdash z) \vee \ldots \vee (y_n \vdash z)$. But if $(X \vdash y_0)$ & $\ldots$ & $(X \vdash y_n)$ $(X \vdash z) \vee \ldots \vee (X \vdash z)$ by transitivity, so $X \vdash z$. That means $z \in c(X)$.
- For monotonicity, let $x \in c_\vdash(X)$. Then $X \vdash x$. Suppose $X \sqsubseteq Y$. Then then $Y \vdash x$. Then $x \in c_\vdash(Y)$.

To show that $\vdash_c$ is a consequence relation:

- For reflexivity, as $X \subseteq c(X)$, if $x \in X$, then $X \vdash_c x$.
- For transitivity, suppose $X \vdash_c y$ and $y \vdash_c z$. Then as $y \in c(X)$, and therefore $c(y) \sqsubseteq c(X)$, we have $x \in c(X)$, whereby $X \vdash_c z$.
- For monotonicity, suppose $X \vdash_c x$. As $X \subseteq X \cup Y$, then $c(X) \subseteq c(X \cup Y)$. Then $x \in c(X \cup Y)$, so $X \cup Y \vdash_c x$.

This completes the proof.

This means that we can use the language of closure operators to talk about sets of propositions ordered by consequence, as step toward topological semantics for paraconsistent logic.[10]

9.2 Boundaries and Connected Space

A basic platform for topology in place, we commence abstract investigation of key concepts behind the spatial paradoxes of Chapter 1.

---

[10] Cf. Mortensen [2010, 1§2].

9.2.1 Boundaries

A boundary of an object is, intuitively, the closest thing to that object which at the same time is *also* the closest thing to the complement of the object. As understood in classical topology, boundaries have the following properties:

- A set shares a boundary with its complement.
- Every boundary is the boundary of itself.
- The boundary of $\emptyset$ is empty.
- A set is both open and closed iff its boundary is empty.

In this section and the next, we check how the notion works out in a paraconsistent setting, via the following definition. For $A \sqsubseteq X$, the *boundary* of A

$$\partial(A) := c(A) \cap c(X \backslash A)$$

is the intersection of A's closure and the closure of its complement in X. Since closures are closed and a boundary is the intersection of two closures, it is closed (Proposition 57). Immediately by conjunction elimination,

$$\partial(A) \sqsubseteq c(A) \qquad \text{and} \qquad \partial(A) \sqsubseteq c(X \backslash A). \tag{9.7}$$

Other properties are as follows.

Proposition 59. *Let $A \sqsubseteq X$. The following conditions hold:*

(i) The boundary of $\emptyset_X$ is empty,

$$\emptyset_X \cap \emptyset_X \sqsubseteq \partial(\emptyset_X) \sqsubseteq \emptyset_X.$$

(ii) The boundary of a set is included in itself,

$$\partial(\partial(A)) \sqsubseteq \partial(A).$$

(iii) Every set shares a boundary with its complement,

$$\partial(A \cap A) \sqsubseteq \partial(X \backslash A) \sqsubseteq \partial(A \cup \emptyset_X).$$

Proof For (i), there are two directions to prove. To show that the boundary of $\emptyset_X$ is included in $\emptyset_X$, just notice

$$\partial(\emptyset_X) = c(\emptyset_X) \cap c(X \backslash \emptyset_X)$$
$$= \emptyset_X \cap c(X \backslash \emptyset_X)$$
$$\sqsubseteq \emptyset_X$$

by using (C5). To show that $\partial(\emptyset_X)$ includes two copies of itself, we have $x \notin X \Rightarrow (x \notin X \vee x \in X)$, so $x \in X \ \& \ x \notin X \Rightarrow x \in X \ \& \ (x \notin X \vee x \in X)$, so by factor

$$\emptyset_X \sqsubseteq c(X \backslash \emptyset_X).$$

By (C1), $\emptyset_X \sqsubseteq c(\emptyset_X)$. So by factor, we have (i).

For (ii), compute the definitions:

$$\partial(\partial(A)) = c(\partial(A)) \cap c(X \backslash \partial(A))$$
$$= c(c(A) \cap c(X \backslash A)) \cap c(X \backslash \partial(A))$$
$$\sqsubseteq c(c(A) \cap c(X \backslash A))$$
$$\equiv c(A) \cap c(X \backslash A)$$
$$= \partial(A).$$

The second-to-last step is justified because boundaries are closed. Then $\partial(\partial(A)) \sqsubseteq \partial(A)$.

For (iii), there are two directions. The first shows that the boundary of the "core" of A is included in the boundary of the complement of A. So

$$\partial(A \cap A) = c(A \cap A) \cap c(X \backslash (A \cap A))$$
$$\sqsubseteq c((X \cap A) \cup \varnothing_X) \cap c(X \backslash A)$$
$$= \partial(X \backslash A).$$

In line 2, $c(A \cap A)$ is replaced by $c((X \cap A) \cup \varnothing_X)$ using the facts that $A \sqsubseteq X$, so $A \cap A \sqsubseteq X \cap A \sqsubseteq (X \cap A) \cup \varnothing_X$, and then applying monotonicity. Also in line 2, $c(X \backslash (A \cap A))$ is replaced by $c(X \backslash A)$ using the fact that $X \backslash (A \cap A) = X \cap \overline{(A \cap A)} = X \cap (\overline{A} \cup \overline{A}) \equiv X \backslash A$, and applying (9.2). Then line 3 is arrived at from Lemma 8, $X \backslash (X \backslash A) = (X \cap A) \cup \varnothing_X$, and the definition of ∂.

For the other direction of (iii), again from Lemma 8,

$$\partial(X \backslash A) = c(X \backslash A) \cap c((A \cap X) \cup \varnothing_X).$$

Now, $(A \cap X) \cup \varnothing_X \sqsubseteq A \cup \varnothing_X$, so by monotonicity, $c((A \cap X) \cup \varnothing_X) \sqsubseteq c(A \cup \varnothing_X)$. By the same token, $X \cap \overline{A} \sqsubseteq X \cap \overline{(A \cup \varnothing_X)}$, because $\overline{A} \sqsubseteq \overline{A} \cap (X \cup \overline{X}) = \overline{A \cup \varnothing_X}$; so by (C1) again $c(X \cap \overline{A}) \sqsubseteq c(X \cap \overline{(A \cup \varnothing_X)})$. And so since

$$\partial(A \cup \varnothing_X) = c(A \cup \varnothing_X) \cap c(X \cap \overline{(A \cup \varnothing_X)}),$$

we have the result. □

The boundary of the complement of a set A is *between* the "inner" and "outer" boundaries of A. The situation in this proposition is a finer-grained view of the orthodox understanding of how a set shares a boundary with its complement. What we have is a roundabout expression of the noncontractive idea that $A \cap A \subseteq A$ but not vice versa; a corollary is

$$\partial(A \cap A) \sqsubseteq \partial(A \cup \varnothing_X)$$

but not likely vice versa.[11] This is the sort of thing we are looking for in reconstructing mathematics in this framework: some insight into the paradoxes, from the fine details of the substructure of space.

---

[11] In [Weber and Cotnoir, 2015], we proved $\partial(A \cup \varnothing_X) = \partial(X \backslash A)$, which has a similar "asymmetry." (They share a boundary modulo inconsistency.) But that result rested on using a stronger logic, where, e.g., $X \cap A = A$ when $A \sqsubseteq X$.

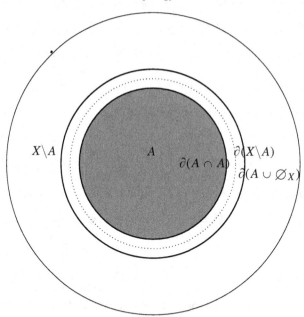

Proposition 59(i) is a limiting case of (iii). Indeed, from (i), in the special case where $\emptyset_X = \emptyset$, then we have the much "cleaner" $\emptyset \subseteq \partial(\emptyset)$, too, so $\partial(\emptyset) = \emptyset$. In fact, this gives us a special case of (iii), that *the universe shares a boundary (perfectly) with the empty set*:

$$\partial(\mathcal{V}) = c(\mathcal{V}) \cap c(\mathcal{V}\backslash\mathcal{V})$$
$$= c(\mathcal{V}) \cap c(\emptyset)$$
$$= c(\mathcal{V}) \cap \emptyset$$
$$= \emptyset$$
$$= \partial(\emptyset)$$

since $\mathcal{V} \cap \emptyset = \emptyset$.

What these theorems show – in particular, the observation that the universe shares a boundary with the empty set – is that the idea that an object shares its boundary with its complement is a *global* intuition: it is only vindicated by a totally unrestricted complement.[12] In the case of restricted complements, we have from Proposition 59(iii)

$$\partial((X\backslash A) \cap (X\backslash A)) \sqsubseteq \partial(X\backslash(X\backslash A)) \sqsubseteq \partial((X\backslash A) \cup \emptyset_X).$$

---

[12] So with respect to Section 9.1.3, we might say that the *metaphysical boundary* of A, $\mathrm{bdy}(A) := c(A) \cap c(\overline{A})$, is what is both closest to A and to *everything else*. Then $\mathrm{bdy}(\overline{A}) = c(\overline{A}) \cap c(\overline{\overline{A}})$, so $\mathrm{bdy}(A) = \mathrm{bdy}(\overline{A})$, as classically predicted. The context-sensitive boundary ∂ is more suited for proving theorems in topology, but the absolute boundary bdy is on hand too to impress your friends.

Since $X \backslash (X \backslash A)$ does not reduce to A, describing the boundaries between a set and its complement become complicated quickly. In general, an object *does* share a boundary with its complement (in a space), but in a substructured way.

Boundaries give us an alternate characterization of closure:

$$c(A) \equiv A \cup \partial(A). \tag{9.8}$$

For if $x \in A \cup \partial(A)$, then $x \in A \vee x \in \partial(A)$; then by cases, $x \in c(A)$ by (C1), or $x \in c(A)$ by the definition of boundary and conjunction elimination. In the other direction, take $x \in X$ and suppose $x \in c(A)$. Either $x \in A$ or not. If $x \notin A$, then $x \in X \backslash A$, so $x \in c(X \backslash A)$ by (C1), and then on assumption (which we have not used yet) $x \in c(A) \cap c(X \backslash A) = \partial(A)$, as needed.

The next theorem is important for paradox recapture, recalling the discussion of the Great Red Spot in Chapter 1. Say (officially now) that A is *open* iff $X \backslash A$ is closed.

Theorem 60. *The following are true.*

(i) A is closed iff $\partial(A) \sqsubseteq A$.
(ii) If A is open, then $\partial(A \cap A) \sqsubseteq X \backslash A$.
(iii) If A is both open and closed, then $\partial(A) \cap \partial(A \cap A) \sqsubseteq A \cap (X \backslash A)$.

Proof (i) Left to right, suppose $A \equiv c(A)$. If $x \in \partial(A)$, then $x \in c(A)$ & $x \in c(X \backslash A)$. Then $x \in c(A)$ and, as we've assumed, $c(A) \sqsubseteq A$, then $x \in A$. (Note that the stronger result, that $\partial(A) \subseteq A$, follows if A is intrinsically closed, $A = c(A)$.) For (i) right to left, if $\partial(A) \sqsubseteq A$, then $\partial(A) \cup A \equiv A$. But from (9.8), transitivity of $\equiv$ is enough to show that A is closed.

(ii) If A is open, then $X \backslash A$ is closed. Then apply (i), for $\partial(X \backslash A) \sqsubseteq X \backslash A$, and use Proposition 59(iii) and transitivity.

And (iii) is (i) and (ii) together. □

Call $\partial(A) \cap \partial(A \cap A)$ the "inner" boundary of A. Then Theorem 60(iii) shows that if an object A is both open and closed, anything in its inner boundary is inconsistent, both in A and not in A. Classically, that means that nothing is in the boundary of a clopen object. But it just means that anything in its inner boundary is inconsistent.

We now have more insight into the nature of boundaries, as substructurally characterized. But we haven't got to the heart of the matter. For problems with boundaries are downstream effects of problems with *points*. Take as a special case of the SYMMETRY puzzle from Section 9.2.1 the problem of slicing an orange in half (paradoxes at breakfast!): the cut will leave two faces that shared a boundary when the orange was whole, but on the classical story, the boundary needs to belong only to one face or the other. With (paraconsistent) topology, we can now provide a way (at Theorem 61) for the boundary to go with both halves, and still the whole original orange was connected. Nevertheless, if a change occurs, it must occur somewhere, and so the original puzzlement is shifted down to the level of points; look how Casati and Varzi relocate the problem [Casati and Varzi, 1999, pp. 87–88, emphasis in the original text]:

Topologically, the cutting of an object is no bloodstained process – there is no question of which severed halves keeps the boundary, leaving the other open and bleeding (as it were). Rather, topologically, the explanation is simply that the outer surface of the sphere is progressively deformed until the sphere separates into two halves. Of course, there is something deeply problematic about the magic moment of separation. However, this is true of every topological change .... *But nothing depends on the boundary issue.*

The two faces of the orange can keep sharing their boundary right until the last moment, when they are connected at a single hinge, a point – and then there is a cut. *That* is the site of our puzzlement. One may defer the global problem until it is a local one, but even a point may be regarded as a very small boundary. The smallness of the point does not remove its mystery, as Max Black observed. The question then is: what happens at the point?

We turn to how boundaries affect the ambient space in which they exist.

9.2.2 Connectedness

Connectedness is a property of a whole space – of being *all* together in one piece. It is a global property. Being separated, on the other hand, is local: one break is all it takes for a space not to be connected. Classically these are dual – a whole space X is connected iff it cannot be separated – and so it is relatively easy to prove global results about connectedness through local results about separation. This is much the way, classically, statements of the form $\forall x (\varphi(x)\ implies\ \psi(x))$ are established by proving that the statement $not\,(\exists x (\varphi(x)\ and\ not\ \psi(x)))$. In the paraconsistent case, the latter is not sufficient for the former. (Intensionality is good for sameness, and extensionality is good for difference.) We will tend to need to work with the two notions of connection and separation independently.

Sets A, B are *separated* iff they do not touch – neither overlapping (having nonempty intersection) nor even abutting – so that for all x,

$$x \notin c(A) \cap B \text{ and } x \notin A \cap c(B). \tag{9.9}$$

Since $x \notin c(A) \rightarrow x \notin A$ by (C1), if A and B are separated, then[13] by (9.9),

$$\forall x (x \notin A \vee x \notin B). \tag{9.10}$$

Define a pair A, B as a *separation* of X iff

(1) $\exists x (x \in A)$ and $\exists x (x \in B)$, and
(2) $X \equiv A \cup B$ and
(3) $\forall x (x \notin A \cap B)$ and
(4) A and B are both closed.

---

[13] This is not to say that if two sets are totally separated, they share no members. If $x \in A$, then it would take disjunctive syllogism to infer that $x \notin B$.

Then X is *separated* iff there is a separation of it.[14] Separated spaces are always nonempty because they can be split into nonempty parts.

So a space is *not* separated iff for any pair A, B, either one of the pair is empty, or they fail to exhaust the space, or they overlap, or one of them is not closed. Evidently, there are many ways to be not separated. If we said simply that a space is *connected* iff it is not separated, then it would be easy to be connected, but only because such nonseparated spaces are not connected in any conditional ($\rightarrow$ or $\Rightarrow$) sense. There are many nonequivalent arrangements that could define a stronger notion of connectedness; take any one of the four conditions for a separation, negate it, and make it the consequent of a conditional.[15] For any of these conditionals, a separation of a space X is a counterexample, so the existence of a separation implies that X is not connected – but not vice versa: assuming that X is not connected is not equivalent to assuming it is separated. Connection and separation are compatible, in some sense. Because my concern is how the world may be "glutty," overlap is the preeminent notion, so we will focus on the following as the definition of connected space:

Definition 4. X *is* connected *iff for any A, B, if*

$\exists x (x \in A)$ *and* $\exists x (x \in B)$
and $X \equiv A \cup B$
and A and B are both closed

then $\exists x (x \in A \cap B)$.

An immediate lemma: if X is connected and $X \equiv Y$, then Y is connected, too. This is not by (invalid) substitution on $\equiv$, but rather the fact that $\equiv$ is transitive. And another lemma: for any nonempty A, B such that $A \cup B \equiv X$, if either

$$\exists x (x \in A \cap c(B)) \text{ or } \exists x (x \in c(A) \cap B),$$

then X is connected. For if A and B are both closed, then this amounts to the conditions for Definition 4.

As intended, the definition of connectedness delivers the following. If a connected space X is partitioned into $A = \{x \in X : \varphi(x)\}$ and $B = \{x \in X : \neg\varphi(x)\}$ where both A, B are nonempty and closed, then $A \cup B \equiv X$, and therefore there is some $c \in A \cap B$, and $\varphi(c) \,\&\, \neg\varphi(c)$.

All spaces with one point $\{\bullet\}$ are connected. Let A, B be nonempty, closed subsets of the one-point space. So $z \in A \rightarrow z = \bullet$ and so too for B. Since we've assumed A, B are

---

[14] Since it invokes a closure operator, separation (and connectedness) is relative to a topology. In most of the discussion here, though, it doesn't matter.
[15] For example, recalling from earlier that A is closed iff $\partial(A) \sqsubseteq A$,

> **Connected 1:** For any exhaustive division of X into nonempty A and B, with $\partial(A) \sqsubseteq A$ and $\partial(B) \sqsubseteq B$, then $\exists x (x \in A \cap B)$.
> **Connected 2:** For any exclusive and exhaustive division of X into nonempty A and B, then for one of A or B, there is something in their boundary they do not contain, $x \in \partial(A) \,\&\, x \notin A$ or $x \in \partial(B) \,\&\, x \notin B$ for some x.
> **Connected 3:** For any exclusive and exhaustive division of X into closed A and B, then one of A, B is empty.
> **Connected 4:** Any exclusive, nonempty closed sets A, B, always fail to exhaust X.

occupied, their intersection is too, and by weakening the space is connected q.e.d. On the other hand, some spaces with one point could be separated, like if $\bullet \neq \bullet$. This depends on whether singletons are closed. If so, then $A = \{\bullet\}$ and $B = \{\bullet\}$ are a separation. Cf. Section 9.3.3.

Less pathologically, $\mathbb{R}$ is connected, by Theorems 45 and 47 in the last chapter, if we treat the supremum of a set as the boundary. By the same token, the interval $[0, 1]$ (and indeed, any interval) is connected.

9.2.3 Topological Sorites Regained

Our task from Section 9.2.1 was to explain how a space X may be connected, and nevertheless be divided exhaustively into nonempty proper subsets that both include their boundaries. The next theorem does this, and at the same time achieves the "line drawing" form of the TOPOLOGICAL SORITES.

Theorem 61.   *Some connected spaces X have a nonempty proper subspace A that is closed, and nonempty $X \backslash A$ closed too, with the boundary of A inconsistent.*

Proof   Let X be a space. By (C5), $\varnothing_X$ is closed, and X is closed. But $X \backslash \varnothing_X$ is closed too, because $X \equiv (X \backslash \varnothing_X) = X \cap (X \cup \overline{X})$. If $\varnothing_X$ is not empty, then $X \backslash \varnothing_X$ is not empty either. But $\varnothing \subset \varnothing_X \subset X$. So for the sake of the theorem, $\varnothing_X$ is a good candidate for the subspace we are looking for.

The argument just given goes for any nonempty relative zero. Let X be connected. Consider the space of its intersection with the Routley set, $\mathcal{Z}(X) = X \cap \mathcal{Z}$. The zero of this space $\varnothing_{\mathcal{Z}(X)}$ is not empty, since every member of X is both in $\mathcal{Z}(X)$ and also not (Section 5.2.2.1). This is a non-empty subset of X, with $X \equiv \mathcal{Z}(X)$. Then take any nonempty, closed subsets $A \cup B \equiv \mathcal{Z}(X) \equiv X$. Because X is connected, there is some point of X in the intersection of A and B; but this is also a point of $\mathcal{Z}(X)$. Therefore, $\mathcal{Z}(X)$ is connected too.

This does not prove that all connected spaces are also disconnected. It proves that from any connected space, a witness for the theorem (in the form of the Routleyization of that space) may be constructed. This is more than enough for the result. $\square$

That's how things sit in space. How do they move? A topological characterization of continuous maps is next.

9.3 Continuity

Leibniz says that nature makes no jumps. What happens when nature also makes jumps? We come to a final sequence of theorems, culminating in a fixed point result. As before, following standard practice, notation will be abused: when f is a map from X to Y, and $A \subseteq X$, then we will often write $f(A)$ instead of $f''(A)$ for $\{y \in Y : \exists x(x \in X \, \& \, y \in f(x))\}$, the image of A under f.

9.3.1 Continuous Mappings

As with connectedness, there are several options for defining continuous mappings topologically. Here we will take the tack of simply adopting our definition from metric space, which is weaker than a classical definition.

Definition 5. *A mapping $f : X \longrightarrow Y$ is continuous if and only if: if $A \sqsubseteq X$ is connected, then $f(A)$ is connected.*

Continuous images of connected spaces are connected: a connected space is "cohesive" in the sense that if it is continuously ripped apart, there is some viscous connective strand that remains. The following are two immediate examples: (i) the *inclusion* function is continuous; (ii) any *constant* mapping from a set into $\{c\}$, to some constant, is continuous. An immediate consequence of the definition:

Proposition 62. *The composition of continuous maps is continuous.*

Proof Let $f : X \longrightarrow Y$, $g : Y \longrightarrow Z$ be continuous maps. By continuity, if $X_0 \sqsubseteq X$ is connected, then so is $f(X_0) \sqsubseteq Y$; then so is $g(f(X_0)) \sqsubseteq Z$. The composition $g \circ f$ is continuous. $\qquad\square$

Continuity relates to conventional definitions of continuity as follows.

Proposition 63. *Let $f : X \longrightarrow Y$ be continuous, and suppose that f preserves boundaries:*

$$f(\partial(A)) \sqsubseteq f(A) \cup \partial(f(A)).$$

Then for any $A \sqsubseteq X$, $f(c(A)) \sqsubseteq c(f(A))$.

Proof We use the fact that $c(A) \equiv A \cup \partial(A)$. Then by (5.34), $f(c(A)) \equiv f(A \cup \partial(A))$. Let $x \in A \cup \partial(A)$. Let $y \in f(x)$. If $x \in A$, then $y \in f(A)$; if $x \in \partial(A)$, then $y \in f(\partial(A))$. Therefore,

$$f(A \cup \partial(A)) \sqsubseteq f(A) \cup f(\partial(A)).$$

But ex hypothesi f preserves limits, so $f(c(A)) \sqsubseteq f(A) \cup f(A) \cup \partial(f(A)) \equiv c(f(A))$ and the result follows. $\qquad\square$

Now, in search of some fixed points, let's look at a kind of continuous mapping that transforms a space into one of its subspaces in a way that preserves structure.

9.3.2 Retractions

Given a space X and a (nonempty) subspace $Y \sqsubseteq X$, a *retraction*

$$r : X \longrightarrow Y$$

is a continuous mapping of X into Y such that $\forall y(y \in Y \Rightarrow y \in r(y))$. Points in X are moved into Y; points in Y, called the *retract* of X, are not moved, so a retraction maps each $y \in Y$ *at least* to itself.[16] And while in general we have been working with relations rather than functions, in the case of retractions we will take on the stronger (and, one would hope, uncontroversial) assumption that the output of a retraction is unique: they map *only* to themselves: assume for $r : X \longrightarrow Y$ that there is some $z \in Y$ such that

$$r''\{x\} = \{y : y \in r(x)\} = r(x) = \{z\}$$

for every $x \in X$.

Classically, a continuous function r is a retraction iff $r \circ r = r$. Here, we have the following.

Proposition 64. *If $r : X \longrightarrow Y$ is a retraction into Y, then*

$$r(x) \cap r(x) \sqsubseteq r(r(x)) \sqsubseteq r(x).$$

That is, $r(x) \equiv r(r(x))$ up to contraction.

Proof For all $x \in X$, if $z \in r(x)$, then z is a member of Y. If z is a member of Y, then $z \in r(z)$. Thus

$$z \in r(x) \,\&\, z \in r(x) \Rightarrow z \in r(x) \,\&\, z \in r(z).$$

Then by existential generalization,

$$z \in r(x) \,\&\, z \in r(x) \Rightarrow \exists u(\langle x, u\rangle \in r \,\&\, \langle u, z\rangle \in r).$$

But checking definitions,

$$r(r(x)) = (r \circ r)(x) = \{z : \exists u(\langle x, u\rangle \in r \,\&\, \langle u, z\rangle \in r)\}.$$

So we've showed $r(x) \cap r(x) \sqsubseteq r(r(x))$.

For the other direction, $r(r(x)) \sqsubseteq r(x)$, this is obtained by assuming r is "functional." Letting $z \in r(r(x))$, then $\langle x, u\rangle \in r$ and $\langle u, z\rangle \in r$ for some u; as $u \in Y$, we have $\langle u, u\rangle \in r$; as the output of r is unique, and $\langle u, z\rangle \in r$, then $u = z$. Therefore, $\langle x, z\rangle \in r$, i.e., $z \in r(x)$. $\qquad\square$

A space X has the *fixed point property*[17] iff every continuous map f from X to X has a fixed point, some $z \in X$ such that $z \in f(z)$. A fundamental property of retractions is that retractions preserve the fixed point property.

Theorem 65. *If X has the fixed point property, any retract of X has the fixed point property too.*

Proof Suppose every continuous map from X to X has a fixed point. Let $r : X \longrightarrow Y$ be a retraction from X onto $Y \sqsubseteq X$, and

$$i : Y \longrightarrow X$$

---

[16] Being fancy about it, a retraction is a continuous transformation (a homotopy) of the identity map on X, $\mathbb{1}_X$, to the identity map on Y, $\mathbb{1}_Y$. See [Armstrong, 1979].

[17] From Section 0.3. Cf. Schröder [2003, 1§4 and ch. 4].

the inclusion map from Y to X, $i = \{\langle y, x \rangle : y \in Y \,\&\, x \in X \,\&\, x = y\}$. Consider any continuous

$$g : Y \longrightarrow Y.$$

Since r, i, and g are all continuous, their composition

$$h = i \circ g \circ r$$
$$= \{\langle x, y \rangle : \exists u(\langle x, u \rangle \in r \,\&\, \exists v(\langle u, v \rangle \in g \,\&\, \langle v, y \rangle \in i))\}$$

is continuous (Proposition 62), with $h : X \longrightarrow X$ depicted:

$$
\begin{array}{ccc}
X & \xrightarrow{\ \ r\ \ } & Y \\
\downarrow{\scriptstyle h} & & \downarrow{\scriptstyle g} \\
X & \xleftarrow{\ \ i\ \ } & Y
\end{array}
$$

Since X has the fixed point property, there is some $p \in h(p)$, that is, $\langle p, p \rangle \in h$. Then

$$\exists u(\langle p, u \rangle \in r \,\&\, \exists v(\langle u, v \rangle \in g \,\&\, \langle v, p \rangle \in i)).$$

By the definition of i, if $\langle v, p \rangle \in i$, then $v = p$, so

$$\exists u(\langle p, u \rangle \in r \,\&\, \langle u, p \rangle \in g \,\&\, \langle p, p \rangle \in i).$$

Since $\langle p, p \rangle \in i$, we know that $p \in Y$, and since r is a retraction, $\langle p, p \rangle \in r$. Since retractions are functional, if $\langle p, u \rangle \in r$, then $u = p$. Therefore,

$$\langle p, p \rangle \in r \,\&\, \langle p, p \rangle \in g \,\&\, \langle p, p \rangle \in i.$$

This produces

$$p \in g(p),$$

which is a fixed point for g as required. Then Y has the fixed point property. $\qquad\square$

The proof expresses the following idea. We presume every map from a space X to itself has some property $\mathcal{F}$. The existence of a retraction of X onto a subspace Y means that we can regard any map from Y to Y

$$Y \xrightarrow{\ g\ } Y$$

as a composite map from X to X

$$X \xrightarrow{\ r\ } Y \xrightarrow{\ g\ } Y \xrightarrow{\ i\ } X.$$

Then this map has property $\mathcal{F}$. But since the r and i portions of the map are making no difference, then property $\mathcal{F}$ must show up in the g portion.

Upon reflection, the proof of this theorem calls us back to the question: why are there paradoxes? For we know that the universe $\mathcal{V}$ has a fixed point property (Theorem 13). Now, if $\mathcal{V}$ retracts onto any subspace of itself, would *that* be why all maps have fixed points? Classically, there cannot be such a mapping (even overlooking the classical nonexistence of the universe), because one subset is $Y = \{0\} \cup \{1\}$, which cannot be continuously covered. But we illustrated just what such a mapping would look like in Theorem 53. So in paraconsistent mathematics – is there a retraction from the universe onto *any* nonempty set?

There is a map that is *classically equivalent* to a such a retraction. Say that r is a *soft retraction* from X onto Y iff r is continuous and $y \notin r(y) \Rightarrow y \notin Y$. This is just the contrapositive of the standard definition; cf. Theorem 22.

Theorem 66.   *Every set has a subset that is a soft retract of the universe* $\mathcal{V}$.

Proof Every Y has beneath its surface $\mathcal{Z}(Y) = \{y : y \in Y \ \& \ y \notin \mathcal{Z}(Y)\}$. This is a subset of Y, and every member of Y is both in it and not. Then the universe $\mathcal{V}$ softly retracts into $\mathcal{Z}(Y)$, because $y \notin \mathcal{Z}(Y)$ for all y, so by weakening, $y \notin r(y) \Rightarrow y \notin \mathcal{Z}(Y)$, for any continuous r. That shows a soft retraction. $\square$

For any Y and Z and f, the map

$$Y \xrightarrow{\ f\ } Z$$

can, in a weak sense, be regarded as a map over the universe

$$\mathcal{V} \xrightarrow{\ r\ } \mathcal{Z}(Y) \xrightarrow{\ i\ } Y \xrightarrow{\ f\ } Z \xrightarrow{\ i\ } \mathcal{V},$$

which suggests how any map might inherit some of the universe's mapping properties. It's not quite the answer to why there are paradoxes, since contrapositive "soft" retractions (even functional ones) don't preserve fixed points; the proof of Theorem 65 does not go through for soft retracts, paraconsistently. Nevertheless, from a classical perspective soft retracts just are retracts, so this theorem whispers why every mapping in the naive universe has a fixed point.

Finally we arrive at the following.

9.3.3 The Disc

"*. . . which has but one side . . .*" (*Borges*)
A wheel is spinning. All its points go round and round. Or do they? What about the point at the very center of the wheel – does it spin too? Points can't spin! If the center point of a spinning wheel is removed, then it is obvious that all the points move. But the world is missing no points.

Let $\mathbb{D}$ be a (nonempty) connected, closed, proper subspace of $\mathbb{R}^2 = \mathbb{R} \times \mathbb{R}$. Call $\mathbb{D}$ the *disc*. We will consider only the cases of topologies of the disc in which *all singletons are closed*. This is (classically) equivalent to the T_1 separation axiom (Section 9.1.1.3), making $\mathbb{D}$ a Fréchet space. The standard topology on $\mathbb{R}^2$, comprising all subsets that are unions of closed sets, is (classically) such a topology.

The property we wish to discern from the disc is that any continuous transformation (in an attenuated sense) from $\mathbb{D}$ to itself has a fixed point. This concerns connectedness. However, while the theorem carries Brouwer's name, there is no constructive proof for it. The arguments are indirect, starting from the supposition that there are no fixed points, and leading to a contradiction. We make the indirect argument geometrically, without any detours through algebraic topology;[18] we focus mainly, then, on discussing *separation* and how it handles inconsistency, since that is what comes up in the indirect reductio proof.

Let's get paraconsistent.

Lemma 67. *If X is connected and separated, then X has an inconsistent part.*

Proof If X is separated (and therefore *not* connected), consider a separation of it, $U, V \sqsubseteq X$ such that U, V are both nonempty, closed, exclusive, and $U \cup V \equiv X$. By connectedness, there is some $p \in U \cap V$. By separation, $p \notin U \cap V$, so $p \notin U$ or $p \notin V$. But $p \in U \cup V$, so either

$$p \in U \not\ni p \qquad or \qquad p \in V \not\ni p,$$

i.e., p is both in and not in U, or both in and not in V, for some U, V, as required. □

Inconsistency is thus a *local* property: if X is both connected and disconnected, it is inconsistent *somewhere*.

In light of this locality lemma, alongside the last theorems of the previous chapter, it seems reasonable to conjecture that a direct expression of the *no retraction* theorem is true: *If* $r : \mathbb{D} \longrightarrow \partial(\mathbb{D})$ *is a continuous mapping of the disc to its boundary circle, then* $\mathbb{D}$ *has an inconsistent part, and so r is not a retraction.* The idea would be that some set of points are "ripped" in order to stretch the interior of the disc onto the boundary, violating part of the definition of being a retraction. But there are a lot of points on the disc, and proving this "geometrically" looks to be more involved than we are ready for. (The argument would likely involve at least path connectedness, as well as the property that the union of overlapping connected spaces is connected, neither of which we have addressed.) I leave it open for now and move directly into the final sequence of proofs, which reasons deep into an impossible case.

A space is *totally separated* iff for every $x, y \in X$, when $x \neq y$ there is a separation U, V such that $x \in U$ and $y \in V$. Similarly, a space is *totally disconnected* iff every connected subset of the space is a singleton. That much is classical.[19] This much is not:

Definition 6. *A space X is* absolutely separated *iff for every* $x \in X$ *there is a separation of X into closed, disjoint, nonempty subsets* $U \cup V \equiv X$, *such that both U and V contain x.*

---

[18] The classical arguments – especially for n-dimensions, rather than just 2 as we are targeting – are conducted through algebraic concepts, e.g., that the supposition of the theorem would imply, impossibly, that the fundamental group of the circle is trivial [Hatcher, 2001, p. 32]; cf. Section 7.3.3. That will have to wait for another day.

[19] See [Steen and Jr, 1978, p. 31; Willard, 1970, p. 210].

In an absolutely separated space X, even the singletons are separated (which in echo of classical definitions could be called *absolute disconnection*). For let $\{a\}$ be a singleton included within X; then $a \in X$; then there are closed, disjoint, nonempty subsets $U \cup V \equiv X$, with $U \ni a \in V$; and then the two sets $U \cap \{a\}$ and $V \cap \{a\}$ separate $\{a\}$.

As usual, explosive $\varnothing$ satisfies the conditions for being absolutely separated. More interestingly, a classical equivalent of the (nonempty) empty set, $\mathcal{O} = \{x : x \neq x\}$, with some natural closure operator on it, would be an absolutely separated space. An absolutely separated space is a separation of itself:

(1) $\exists x (x \in \mathcal{O})$ and $\exists x (x \in \mathcal{O})$, and
(2) $\mathcal{O} \equiv \mathcal{O} \cup \mathcal{O}$ and
(3) $\forall x (x \notin \mathcal{O} \cap \mathcal{O})$ and
(4) $\mathcal{O}$ and $\mathcal{O}$ are closed

as per the definition of a separation. And observe that for each $a \in \mathcal{O}$ that

$$\mathcal{O} \backslash \{a\} \equiv \mathcal{O} \cap \mathcal{O}$$

since $a \in \mathcal{O} \leftrightarrow a \neq a$, and $a \in \mathcal{O} \leftrightarrow a \in \mathcal{O}$, and factor.

In the next lemma, we establish a sufficient condition for being absolutely separated.

Lemma 68. *If every member of a space X is non-self-identical, then X is absolutely separated.*

Proof Let $x \neq x$ for all $x \in X$. Let $a \in X$. Then (much as in the previous example of $\mathcal{O}$), take

$$U = \{x \in X : x = a\}$$
$$V = \{x \in X : x \neq a\}.$$

This is a separation of X, each part containing a, as follows:

- First, $X \equiv U \cup V$, because from right to left, both U and V are subsets of X, and from left to right, since

$$(x = a \vee x \neq a) \Rightarrow (x \in X \Rightarrow (x = a \vee x \neq a) \,\&\, x \in X)$$

  is just an instance of the logical axiom for &-intro, then by the law of excluded middle and modus ponens, we have

$$x \in X \Rightarrow (x = a \vee x \neq a) \,\&\, x \in X.$$

  Then by distribution, $x \in X \Rightarrow (x \in X \,\&\, x = a) \vee (x \in X \,\&\, x \neq a)$, showing $X \subseteq U \cup V$.
- Second, U and V are both nonempty (since a is a member of both) and exclusive (since nothing is both identical to a and not identical to a).
- Third, U and V are both closed because X is closed, and $U \cup V$ is equivalent to X.

Thus X is absolutely separated.   □

Now, suppose $X \sqsubseteq \mathbb{D}$ is a closed, nonempty subset of the disc that is *both connected and everywhere-non-self-identical* (and hence absolutely separated). It is all in one piece, and yet even its points are broken into pieces. What can we say about its structure? The answer looks to be surprisingly (?) classical: *a connected, totally separated space collapses to a point*. It is a classical theorem that the only totally separated connected space is a point, but the reason is that if a totally separated space has more than one point, it can be separated, contradicting connectedness. We will look for better reasons.

This "implosion" property is nearly the case already just for absolutely separated spaces alone, irrespective of connectedness, by the following exhaustion-style argument. First, it seems reasonable to think that if X is absolutely separated, then every subset of X is absolutely separated too. (For let $Y \sqsubseteq X$. If $y \in Y$, then $y \in X$, so there is a separation U, V of X with $U \ni y \in V$. But then, $U \cap Y$ and $V \cap Y$ are a separation of Y.) This in hand, let a be a member of an absolutely separated space X; then there are $U_0 \ni a \in V_0$ that separate X; but since $a \in U_0$ and every subset of X is absolutely separated, too, then there are $U_1 \ni a \in U_2$ that separate U_0, and mutatis mutandis $V_1 \ni a \in V_2$ that separate V_0; but then ... Each "smaller" space containing a is divided in half, and since each half again contains a, they are divided in half, and divided in half, until intuitively we reach some "dust" at the smallest limit – presumably, singletons – with a a member of every singleton, making every member of X identical to a. But, however suggestive, there are many gaps in this argument (that probably cannot [and should not be] be filled).

The idea for the argument(s) we will use goes in the other direction. Rather than analyze down to points, we start from thinking about X as a union of its singletons, and then use connectedness: non-self-identity will let us rearrange the points of X in such a way that, by connectedness, all the points become one.

Let $X \sqsubseteq \mathbb{D}$ be nonempty and closed, with all its members non-self-identical. Let $a \in X$, and like all members of X, $a \neq a$. Some reconnaissance to get oriented. First, if every member of X is non-self-identical, then

$$X \equiv X\backslash\{a\}. \tag{9.11}$$

(Right to left is conjunction elimination; left to right is from $x \neq a \Rightarrow (x \in X \Rightarrow x \in X \cap \overline{\{a\}})$ and the fact that $\forall x (x \neq a)$ by Lemma 7.) Recall that, even without substitution, if $X \equiv Y$ and X is closed, then so is Y, by (9.4). Thus since X is closed, $X\backslash\{a\}$ is closed. Second, by a similar token, for each $x \in X$ we consider singletons $\{x\} \sqsubseteq X$, and the union of those singletons:

$$X \equiv \bigcup_{x \in X}\{x\}. \tag{9.12}$$

Then $\bigcup_{x \in X}\{x\}$ is closed since X is closed. Third, then, from (9.11) and $a \in X$,

$$X \equiv (X\backslash\{a\}) \cup \{a\} \tag{9.13}$$

is true. Both $X\backslash\{a\}$ and $\{a\}$ are nonempty – the former at least because $a \in X\backslash\{a\}$ and the latter because $a = a$ like everything else in X. And they are both closed, since we suppose singletons are closed in the topology of the disc, and $X\backslash\{a\} \equiv X$, with X closed on assumption.

With that much clear, here is a *Gedankenexperiment*.

Since X may be regarded as the union of its singletons (9.12), then for some arbitrary but fixed $a \in X$, in (9.13), by connectedness the two parts of $(X\backslash\{a\}) \cup \{a\}$ overlap and there is some

$$b_0 \in X\backslash\{a\} \cap \{a\}$$

with $a = b_0$. Now repeat this for

$$(X\backslash\{a,b_0\}) \cup \{a\} \equiv X,$$

noting that $\{a,b_0\} = \{a\} \cup \{b_0\}$ is the finite union of closed sets hence closed, and $X\backslash\{a,b_0\} \equiv X$ again; by connectedness, there is some

$$b_1 \in (X\backslash\{a,b_0\}) \cap \{a\},$$

with $a = b_0 = b_1$. And so forth: at each step, some member b_k of X is identified with a, then b_k is removed for the next step, where some other member b_{k+1} of X is identified with a ... Since at each stage $X\backslash\{a,b_0,b_1,\ldots\} \equiv X$, then we are assured of connectivity and closure, and hence that the overlap of this set with $\{a\}$ is nonempty. Iterate this process until every member b_i of X has been successively removed after being identified with a, with $a = b_i$.

Another way less finitary way to put this is via each way of associating the union that singles out one point;

$$X \equiv \{a\} \cup \left(\left(\bigcup_{x \in X} \{x\} \right) \backslash \{a\} \right). \tag{9.14}$$

We have that $(\bigcup_{x \in X} \{x\})\backslash\{a\}$ is closed, because it too is equivalent to X. A fortiori, since we assume that singletons in the topology on $\mathbb{D}$ are closed, then the right-hand side of (9.14) is a *separation* of X, because it is two nonempty, exclusive, exhaustive, closed parts of X. From (9.14), by distributing, we get

$$X \equiv \{a\} \cup \left(\bigcup_{x \in X} \left(\{x\} \cap \overline{\{a\}} \right) \right), \tag{9.15}$$

that is, all the points that are either a, or else some point in X that is *not* a. By the associativity of unions in general,[20] (9.15) holds for every $a \in X$. By connectivity, there must be some element in the intersection of those two parts. The totality of all these elements amounts to a single self-identity. For any $a,b \in X$, given all the ways of associating the union of singletons together and the fact that each must overlap, it follows that $a = b$.

---

[20] If $(A_i)_{i \in I}$ and $(I_\alpha)_{\alpha \in \kappa}$ are indexed sets of sets, with $\alpha, \kappa \in On$, and $I = \bigcup_{\alpha \in \kappa} I_\alpha$, then $\bigcup_{i \in I} A_i \equiv \bigcup_{\alpha \in \kappa} (\bigcup_{i \in I_\alpha} A_i)$.

From these ideas, I propose that every member of a closed, connected everywhere-non-self-identical set is identical. That is, the following (classically valid) statement is true:

Lemma 69. *If nonempty closed $X \sqsubseteq \mathbb{D}$ is connected, and $x \neq x$ for all $x \in X$, then X is a singleton.*

This brings us to the final theorem: a continuous shift (in a restricted sense) of an entire bounded space into itself must leave at least one point unmoved, even if every point is also moved. To fill it in, we need to specify what the restricted sense of continuity under discussion will be.

With loss of generality, define an *ultracontinuous mapping* $f : X \longrightarrow Y$ to be the following:

(1) a continuous mapping, such that
(2) if $U \sqsubseteq X$ is closed, then $f(U) \sqsubseteq Y$ is closed, *and* $Y \backslash f(U)$ is closed;
(3) and for each $x \in X$ there is a z such that $f''\{x\} = \{y \in Y : y \in f(x)\} = \{z\}$, i.e., f is "functional," taking a point from X to (a range of points that has) just one member from Y.

So all ultracontinuous maps are continuous maps, which are univocal relations. As with the results about retractions in Section 9.3.2 (and Brouwer's fixed point theorem is, classically, no more than a restatement of a theorem about [the nonexistence of one type of] retractions), the result is proved only about continuous mappings $f : \mathbb{D} \longrightarrow \mathbb{D}$ with a *unique* output.

Clause (2) in the definition of ultracontinuity says that these mappings f shift regions so that both they and their complements under f retain their boundaries. This isn't necessarily the case for continuous maps in general. It is the case for maps that respect SYMMETRY (Section 1.3.1.1), and more broadly, what I called Premise 0 way back in the Preface – the Spinoza–Leibniz principle of sufficient reason. Leibniz says that Archimedes

takes it for granted that if there is a balance in which everything is alike on both sides, and if equal weights are hung on the two ends of that balance, the whole will be at rest. That is because *no reason can be given why one side should weigh down rather than the other.*[21]

For a continuous image of a closed set, when there is no reason why one side should get the boundary rather than the other, then we want a name for this class of mappings. Ultracontinuous mappings are exactly those continuous shifts for which SYMMETRY holds.

Here is what we will now show. *Even if* an ultracontinuous shift of a connected bounded space moves every point, and in doing so absolutely shatters the space down to its atoms, even shatters the points themselves, then *still* some point remains anchored in place.

Theorem 70. *Let $f : \mathbb{D} \longrightarrow \mathbb{D}$ be ultracontinuous. Then there is some x such that $x \in f(x)$.*

---

[21] In the Leibniz–Clarke correspondence, emphasis added; see Della Rocca [2008, ch. 8].

Proof Argue by cases. Either some $x \in f(x)$, or else every $x \notin f(x)$. In the second case, we are to suppose the set $\{x : x \notin f(x)\}$ is the whole disc. So for every $a \in \mathbb{D}$, the set $\mathbb{D} \backslash f(a)$ is not empty, because at least $a \in \mathbb{D} \backslash f(a)$. By definition of a (total) mapping, $f(a)$ is not empty, either. Then $f(a)$ and $\mathbb{D} \backslash f(a)$ are a nonempty, exhaustive, and exclusive partition of the disc into two sets. Both the image of $\{a\}$ and $\mathbb{D} \backslash f(a)$ are closed, by definition of ultracontinuity.

Since the disc is connected, the two sets $f(a)$ and $\mathbb{D} \backslash f(a)$ overlap: there is a point z on the disc such that

$$z \in f(a) \not\ni z.$$

Then $f(a) \neq f(a)$. Since ultracontinuous maps are functional, i.e., $f(a) = \{z\}$ for some $z \in \mathbb{D}$, then $z \in \{z\}$ and $z \notin \{z\}$, and therefore $z \neq z$. Since a was arbitrary, ex hypothesi this is so for *every* point on the disc under the image of f,

$$x \in f(\mathbb{D}) \Rightarrow x \neq x.$$

The disc is connected, but its image under f is *absolutely separated* by Lemma 68.

But the continuous image of a connected set is connected. So $f(\mathbb{D})$ is connected. And because $\mathbb{D}$ is closed, by ultra-continuity its image under f is closed, too. By lemma 69, then, the everywhere-nonselfidentical continuous image of a connected disc is a point, some fixed $\{\mathfrak{p}\} \equiv f''\mathbb{D}$, for which

$$x \in \mathbb{D} \Rightarrow \mathfrak{p} \in f(x)$$

Since $\mathfrak{p} \in \mathbb{D}$,

$$\mathfrak{p} \in f(\mathfrak{p})$$

which is a fixed point as required.   □

And yet, as Galileo never said, this point does not move—*e pur non si muove*.

Part IV
Why Are There Paradoxes?

The Targus is more beautiful than the river that flows through my village,
But the Targus is not more beautiful than the river that flows through my village
Because the Targus is not the river that flows through my village.
– Pessoa XX

10

Ordinary Paradox

Last night the first stars began to appear when there was still light in the sky. For a few moments, it was both night and not night. There was a first moment when it was night; that moment was also not the first moment; there were others; and not. We were standing in the doorway while we watched the sky. We were indoors and not indoors. There were paradoxes everywhere. And it was all so utterly *ordinary* . . .

* * *

The previous several Chapters 5–9, if nothing else, provide mounting evidence that the Feferman objection can be met: a paraconsistent framework can sustain some long chains of (fairly) ordinary reasoning. Specifically, I laid out how the arguments *motivating* a paraconsistent approach can be established *within* a paraconsistent approach. I've argued, by direct demonstration, for the *independent viability* of the approach.

Along with this "proof of concept" demonstration of classical recapture/rehabilitation, I have tried to see something new. To that end, a recurring character in this study has been the Routley Set $\mathcal{Z}$, and Theorem 10 that may be summarized "as above, so below": everything is both a member of $\mathcal{Z}$ and not, and every nonempty set X has a(n inconsistent) proper part $\mathcal{Z}(X)$ that is *coextensive* (although not *identical*) with X, $X \equiv \mathcal{Z}(X)$. This is no small thing. Priest concluded his seminal paper on the logic of paradox by writing that "the discovery by Russell of a set which was both a member of itself and not a member of itself, is the greatest mathematical discovery since $\sqrt{2}$" [Priest, 1979, p. 240]. I believe the Russell set pales in comparison to Routley's discovery in 1977 of a set that is everywhere and nowhere.

In this concluding chapter, I will draw together some threads of the book, face some objections, and ask the main question – why are there paradoxes? – three last times.

10.1 Dividing the Universe

Why are there paradoxes?

Since the beginning of the Introduction, it has seemed that part of the answer is: because there are problems with, in Gödel's phrase, "dividing the totality of all existing things into two categories" exclusively and exhaustively. Whether it be truths or rainstorms, you can have *all* of them, or you can have *only* them, but you can't have both – at least, not classically.

That has been the story. But after the past several chapters, where many apparently familiar notions have been cast in a new light, it is time to revisit our original decisions and logical foundations, as seen from the far side of our paraconsistent travels.

10.1.0

You have to begin somewhere. Maybe in philosophy everything is eventually in question, *de omnibus dubitandum est*, but it can't all be in question all at once. If we are ever to arrive at (something like) a conclusion, then (something like) a starting point is needed. I have taken as a starting point some propositions, or propositional schemas, that I believe meet the standard of being axiomatic in the old, deep sense – propositions that cannot be deduced from anything more basic, the truth of which is self-evident. Along the way I have made several decisions about how to formalize these claims, and those decisions call for some scrutiny. This will be to take up questions of *negation*, via the *contraposability* of the implication connective in naive comprehension, and more generally of the conception of *truth, falsity, validity,* and *invalidity,* and their interrelations.

10.1.1

As captured in Axiom 1, the Frege–Cantor–naive set concept fixes both the extensions *and* antiextensions of sets: $x \in \{z : \varphi(z)\}$ iff $\varphi(x)$, and $x \notin \{z : \varphi(z)\}$ iff $\neg\varphi(x)$. Formally, this is a consequence of having phrased the "iff" in naive set comprehension using a relevant arrow $\rightarrow$, which obeys contraposition. Philosophically, my motivation for interpreting the Frege–Cantor view of extensions and antiextensions this way is as follows.

As an *alternative* to taking extensions to automatically fix antiextensions, one tradition in (classically) modeling paraconsistent set theory works by treating extensions and antiextensions as completely independent entities (e.g., [Libert, 2005]; cf. [Batens, 2020, p.905]). There is a germ of truth in this idea, since it is true that membership in set $\{z : \varphi(z)\}$ does not *rule out* membership in $\{z : \neg\varphi(z)\}$. Nevertheless, the idea of treating them as completely unrelated is quite classical; put a little uncharitably, it is a way of modeling paraconsistent negation as something other than real negation. Letting X be a set and X^+, X^- be its extension and antiextension, respectively, the idea is that X^+ and X^- are consistent ($x \in X^+$ is either "just" true or "just" false, and so too for X^-), and any "inconsistent" members of X end up in (the entirely classical again) intersection $X^+ \cap X^-$.

We already know from Chapter 3 that this won't work (as intended) if we are using a dialetheic paraconsistent metatheory. If the extensions and antiextensions are really supposed to be consistent, then somewhere along the way will be revenge (e.g., the extension of all extensions that are in their own antiextension or suchlike). Along with whatever other revenge-incoherence this "three-valued" picture might engender, if extensions are severed from their antiextensions in this way, it leads to a vicious infinite regress. For what about X^{++} and X^{+-}, or X^{-+} and X^{--}, the extension and antiextension of X^+ and X^- respectively? Either these fall back simply to X and $\overline{X}$ after all, or else we can keep breaking sets into further and further distinct entities:

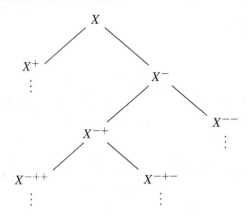

which reinstates a familiar hierarchical, and ultimately open-ended, picture; this is the sort of consistent-and-incomplete picture we have been at pains to avoid from the start. Part of the intuition behind Basic Law V is that a set just is its own extension, and its complement is its antiextension, given for free by negation – and that's that (cf. Priest [2006b, p. 29]). Basic Law V completely characterizes the nature, membership and nonmembership, of sets. As being φ just is to be one of the φs, so too the *non-φs* are just *not* among the φs. Put metaphysically, and more generally, every object always already divides the universe: into the part that is that object, and everything else.

These were our starting points, axioms. And this was our starting problematic: divisions of the universe, though intended to be exclusive and exhaustive, are not.

10.1.2

That is the opening statement of a case for using a contraposable conditional in Axiom 1, and for accepting the consequences of doing so; more to follow. But what does this logical choice, however philosophically motivated, already portend for *truth* and *validity*? – especially as the purpose of the exercise was ultimately to explain why there are paradoxes (i.e., certain sorts of valid arguments), by taking on a uniformly paraconsistent worldview. We have *derived* some of the motivating contradictions, but with that accomplished, what are we to say about the *status* the paradoxes themselves? For recall from the opening of the Introduction:

A *genuine* paradox is a valid argument with true premises and a false conclusion. Since validity preserves truth, then the conclusion of a genuine paradox must also be true; the conclusion is a true contradiction. So far so good (!). But by the semantics for conjunction and negation, all contradictions are false, so any contradictory conclusion of an argument is still false, even though sometimes also true. That means that a genuine paradox is an argument with true premises and a false conclusion. But an argument with true premises and a false conclusion is invalid (Section 0.2.2.3); so even a genuinely paradoxical argument is *invalid*. Ergo there are no sound paradoxical arguments, because there are no valid paradoxical arguments.

Why are there paradoxes? *There are no paradoxes.*

10.1.3

Of course, there are also paradoxes. This isn't an abnegation like Wittgenstein's "anyone who understands me eventually recognizes [my propositions] as nonsensical" [Wittgenstein, 1922, 6.54]; this rather is stone-cold rational dialetheism, of the kind that says dialetheism is false (as well as – because it is – true). Nevertheless, for a project about paradoxes and the revenge phenomenon, to end up saying that there are no paradoxes calls for some explanation. Let's look closer at how exactly *truth* plays in the formalism.

Box 0 of Section 0.2.2.3 informally presented some semantic conditions on extensional logical connectives, and by Chapter 3 I was recommending that a thoroughgoing dialetheic paraconsistentist use relational evaluations to formalize them. The semantic clauses when resituated into our paraconsistent set theory, homophonically given in terms of truth in an evaluation v from sentences to $\{t, f\}$, would look like this, say for negation:

$$v(\neg\varphi) \ni t \quad \text{iff} \quad v(\varphi) \ni f$$
$$v(\neg\varphi) \ni f \quad \text{iff} \quad v(\varphi) \ni t.$$

And then the question of contraposition resurfaces: what is the "iff"? Well, if truth in an-interpretation is to mirror truth ("truth in a model is" (or should be) "a model of truth," as in Chapter 3), then the biconditional here will work however it works in the T-schema. The T-schema is derived from naive set comprehension (Section 1.1.2), so if the biconditional $\leftrightarrow$ in naive comprehension contraposes, then the (derived) T-schema contraposes. That, in turn, means we need to think about how truth relates to falsity, via *negation*.

Adopting Priest's terminology [Priest, 2006b, p. 70], falsity is truth of negation, so φ is *false* iff $\neg\varphi$ is true; and φ is *untrue* iff φ is *not* true. Then directions of the truth schema include two salient conditions:

Exhaustive: Untrue implies false, $\neg T^\ulcorner\varphi\urcorner \to T^\ulcorner\neg\varphi\urcorner$.

Exclusive: False implies untrue, $T^\ulcorner\neg\varphi\urcorner \to \neg T^\ulcorner\varphi\urcorner$.

As we observed with Belnap and Dunn (Section 3.3.2), if the "not" in the definition of untrue is *not*, then falsity and untruth look to collapse. Both EXHAUSTIVE and EXCLUSIVE follow immediately, if the T-schema contraposes (whence $\neg T^\ulcorner\varphi\urcorner \leftrightarrow \neg\varphi$). So I think we are pushed to embrace, as I have (and will argue again in Section 10.1.7), both of EXHAUSTIVE and EXCLUSIVE, so that the symbols t, f relate in a very natural way: φ is true iff it is not false, and false iff it is not true:

$$\varphi \text{ is t} \quad \text{iff} \quad \neg(\varphi \text{ is f})$$
$$\text{iff} \quad \neg\varphi \text{ is f.}$$
$$\varphi \text{ is f} \quad \text{iff} \quad \neg(\varphi \text{ is t})$$
$$\text{iff} \quad \neg\varphi \text{ is t.}$$

Negation facts *inside* permeate *outside* and vice versa. In the case of contradiction, φ is both t and f, then also φ is *not* t and *not* f.[1]

This becomes important when we go back to the definition of validity, which I restate from the Introduction now in dialetheically appropriate terms:

Valid: An argument is valid iff on any assignment of truth values, when all the premises are true, the conclusion is true.

Invalid: An argument is *invalid* if there is an assignment of values where all the premises are true but the conclusion is untrue.

Now, if one were to *reject* the preceding EXCLUSIVE truth condition, then there is a distinction between "false" and "untrue"; and then the definition of invalidity is specifically about conclusions that are "just" false. One is not committed to saying that "there are no paradoxes" because the paradoxical arguments end in sentences which are, yes, false – but not (in any way) *untrue*.[2] If I on the other hand am *accepting* EXCLUSIVE, then it follows that all the paradoxical arguments are invalid, and so there are no paradoxes.

In fact, given EXCLUSIVITY, an even more drastic consequence follows: for the LP-fragment of the language, not only are all paradoxical arguments invalid, but *every* argument is invalid.[3] That is, just in terms of the vocabulary of LP, we have the somewhat prosaic but still disturbing discovery that *anything is possible*. I'll explain.

10.1.4

What is valid, and not, depends on what assignments of truth values exist. Well, we have that there is an assignment on which some sentence φ is true and φ is false; that's dialetheism. More jarringly, for these truth conditions over the LP fragment, there is even, as has been clear for a long time (cf. Priest, 2008, 8.10, problem 5), a trivial assignment where

---

[1] See also Priest and Routley [1989b, p. 516], which states that a closed "metalanguage" should include sentences of the form "*s* is true and *s* is not true." Cf. Field [2008, ch. 27].

[2] Priest "tentatively" rejects EXCLUSIVE in Priest [2006b, p. 71] on the grounds that "contradictions should not be multiplied beyond necessity" (but given the argument I am discussing here, his position has shifted to a less tentative one). Priest then asks what the difference between falsity and untruth could be, and answers that it is "surprisingly little," other than noting that truth and untruth are "more inconsistent" than truth and falsity (both sets of scare quotes in the original). Whether the T-schema contraposes is a point of difference between Priest and Beall; as noted in Chapter 4, Beall (like Field) ends up saying that validity is in some sense indefinable [Beall, 2009, p. 37].

[3] These issues were already noted in [Weber et al., 2016; Weber and Omori, 2019]. Similar problems are raised in [Young, 2005; Batens, 2019] and discussed in [Priest, 2020].

every φ is both true and false. In a three-valued presentation of the logic, every formula can be assigned the middle "b" value. And that's just because negation as theorized in LP can be inconsistent; by design, no LP connective forces "classicality." This trivial evaluation does not describe the way the actual world is, but the existence of such an evaluation is not ruled out *by logic alone*.

And this can be amplified, using a dialetheic paraconsistent metatheory, as follows (cf. [Weber et al., 2016]). Let v be any valuation, a two-place relation taking every proposition (in the language of LP) to at least one member of $\{t, f\}$. In the simplest case, take the trivial valuation

$$\langle p, x \rangle \in v \quad \text{iff} \quad x = t \vee x = f.$$

Since $t = t$ and $f = f$, this already gives

$$t \in v(p) \ni f$$

for every p, as promised. Then intersect with, yes, the Routley Set $\mathcal{Z}$, whence $v \cap \mathcal{Z} = \mathcal{Z}(v)$ is

$$\langle p, x \rangle \in \mathcal{Z}(v) \quad \text{iff} \quad (x = t \vee x = f) \ \& \ \langle p, x \rangle \in \mathcal{Z}.$$

Then $\langle p, t \rangle \notin \mathcal{Z}(v)$, so p is untrue. Also, $\langle p, f \rangle \notin \mathcal{Z}(v)$, so p is unfalse. This is a valuation making *every* sentence true, false, untrue, and unfalse. Then for any argument (in the language of LP), there is an assignment of values $\mathcal{Z}(v)$ where all the premises have some truth but the conclusion is untrue. Then every argument is invalid.[4]

10.1.5

This trivializes the theory of invalidity in LP. It is little more than the Routley Set showing through the surface. Modally, *everything is possible*, and nothing is necessary.[5] Our world is a possible impossible world.[6] Or so it would seem.

Now, I think there is more to truth, even logical truth, than what can be said or ruled out by an expressively very weak language like that of LP. I do not hold that every argument is invalid, and will show why in a moment. But even so, I would flag that nothing so far trivializes the theory of *validity* even in LP. It does not mean that all arguments are valid. It does not mean that we are prepared to accept as defeasibly reliable some arguments that fail to be deductively valid;[7] the theory of validity is *not* "quasivalidity" from Chapter 3.

---

[4] There might be some room for debate. One avenue is to go back to disputing, again, what the "iff" is in the semantics. If it is the DKQ-ish relevant arrow, then that arrow does contrapose, but $v \cap \mathcal{Z}$ is not a valuation, because relevance precludes introducing an irrelevant $x \in \mathcal{Z}$ in the consequents. (Specifically, the absurd valuation is not a valuation because $p \ \& \ (q \rightarrow r) \vdash q \rightarrow r \ \& \ p$ fails on relevance grounds.) If it is the BCK arrow, then $v \cap \mathcal{Z}$ is a valuation, but does not contrapose. (Thanks to Restall here.)

[5] For *possibilism* (or antinecessaritarianism), see [Mortensen, 1989]; cf. [Weber and Omori, 2019].

[6] As Priest and Routley write in 1983, "The only thing that would really surprise us about future work in these areas would be its failure to produce surprises" [Priest and Routley, 1983, p. 170].

[7] *Logical nihilism* is the position (independent of the concerns here) that there are no valid argument forms, though there are reliable instances thereof. See [Russell, 2018].

It does not mean that every sentence is true, or that every sentence is false. Some arguments are valid – *always* truth preserving, on every valuation (even the trivial one) – and in no way are *all* arguments valid. *Proofs* still matter, for establishing positively what holds, with deductive certainty. And some arguments are not valid in a very strong sense:

Strongly Invalid: An argument is *strongly invalid* if: if every assignment making the premises true makes the conclusion true, then ⊥.

The whole of Chapter 4 can be read as a demonstration of strong invalidity claims, such as the following:

- If every instance of disjunctive syllogism is valid, then ⊥.
- If every instance of contraction is valid, then ⊥.

And so forth. Arguments that are thusly annihilated – invalid, on pain of absurdity – are absolutely invalid, at least on pain of absurdity. None of the preceding considerations show that all arguments are strongly invalid. Some things are valid, not everything is valid, and some but not all things are strongly invalid. There is an enormous amount of *noise* around the theory of LP-invalidity; but nevertheless, there is a perfectly adequate way to exclude an argument from being valid, and to affirm others as (always) valid.

How much noise is too much noise? It depends on what kind of music you like.

10.1.6

Still, if everything is invalid, even just in LP, that's a lot of inconsistency. And if this result were somehow to extend from the LP-fragment to the *whole* language, it would throw into serious doubt what the purpose of the past several chapters has been, since it would mean that they were painstaking but necessarily invalid deductions (not just accidentally invalid due to mistakes). The inevitable question, unimpressed with my glib allusion to music, repeats (cf. Section 3.1.3.2): how much inconsistency is *too much*?

Whatever we conclude, I would say that we should reflect on the degree to which we are (even unintentionally) still holding up orthodoxy as the standard against which all else is measured. "Valid" means valid, even if its glutty version takes on some previously unexpected features. If we accept that dialetheism is a radically nonclassical thesis, and accept dialetheism, then we are going to get some results that diverge radically from classicality.

Nevertheless, the aim was always to understand the world as we find it, a world with sets and truth and copses of trees, and a world where not all arguments are invalid. In LP there is a sense in which "anything is possible." This sounds kind of sweet, but becomes a rather drastic problem because it has the effect that "everything has a counterexample," and so "no arguments are valid." LP as a logic invalidates everything and so effectively has no theory of invalidity. For critics who allege that dialetheism has given up the ability to express rejection, this would appear to be confirmation. Luckily, the discussion above about STRONG INVALIDITY points the way to a solution, via the ⊥ constant.

To see how a dialetheist can respond constructively to this situation, we get some perspective by asking: okay, so why isn't everything invalid according to *classical* logic? A classical valuation is a univocal function and assigns p exactly the value true when it assigns $\neg p$ exactly the value false. If there were a valuation that assigned p both true and false, then we would have that truth is *identical* to false (or $0 = 1$). The classicist reasons that, since in fact $0 = 1$ is false, there cannot be such a valuation.

When we move to a relational semantics, this line of reasoning isn't available. But the basic spirit of the argument still is. The bedrock of the argument isn't classical negation, but the conviction that truth and falsity are distinct, that there are at least two objects. And that looks like a precondition for almost anything. As I've said before, one can understand paraconsistency as a *thesis*, that *not everything is true, even if some contradictions are*. There is, indeed, no point in continuing rational inquiry if everything is true (and not true). As we saw in Section 2.3.1, some things are simply unacceptable, with no possibility of being true – and if you need formal *logic* to tell you that, then you are in more trouble than logic can help you with.

And we have, both from our underlying mathematics and as a basic conviction, that there is some constant $\bot$ representing such a thing – incoherence, absurdity. This was derived from the set theory, and used crucially in axioms for arithmetic and real analysis. We consider the theory of validity and invalidity, then, over the whole (closed) language, now including the absurdity constant; and if we make respecting the absurdity of $\bot$ an *adequacy condition* on any valuation, then this rules out "trivial" valuations. I endorse this option, and taking it in a forceful way – making it part of *logic*. We need to say – have said all along – more than LP alone can, that some things are absolutely false.[8]

So in Box 3, officially, is a restatement of the semantics for (propositional) logic, providing a *definition* of what it means for a relation v from propositions to $\{t, f\}$ to be an *evaluation*. A relation that fails to meet these conditions is not an evaluation. The conditions are given with the flexible biconditional $\Leftrightarrow$, but with the "contrapositives" stated explicitly too. The key thing is this: on these conditions, if $t \notin v(\top)$, then $\bot$. There is at least one proposition that is true and only true on pain of absurdity.[9]

Officially, an argument is *valid*, $\Gamma \vDash \varphi$, iff for every valuation v, if $t \in v(\psi)$ for all $\psi \in \Gamma$, then $t \in v(\varphi)$. An argument is invalid iff there is a valuation that counterexamples the argument, $t \in v(\psi)$ for all $\psi \in \Gamma$ but $t \notin v(\varphi)$. And then there is at least one argument that is valid, and not in any way invalid:

$$\top \vDash \top.$$

[8] How to phrase it? We could stipulate that that condition on $\bot$ is that any valuation v makes $\bot$ true whenever v makes everything true (as in [Weber and Omori, 2019]). But that's not enough: it would just say that, if there is a possibility that $\bot$ is true, then there is a possibility that everything is true.

[9] Thanks, once again, to Badia here, for the needed insight from [Badia, 2017] that a vertebrate language needs $\bot$.

Box 3 Truth in a model is a model of truth

$$t \in v(p) \quad \vee \quad f \in v(p)$$

$$t \in v(p) \quad \Leftrightarrow \quad f \notin v(p)$$
$$f \in v(p) \quad \Leftrightarrow \quad t \notin v(p)$$

$$t \in v(\neg p) \quad \Leftrightarrow \quad f \in v(p)$$
$$f \in v(\neg p) \quad \Leftrightarrow \quad t \in v(p)$$

$$t \in v(p \& q) \Leftrightarrow t \in v(p) \& t \in v(q) \qquad t \in v(p \vee q) \Leftrightarrow t \in v(p) \vee t \in v(q)$$
$$f \in v(p \& q) \Leftrightarrow f \in v(p) \vee f \in v(q) \qquad f \in v(p \vee q) \Leftrightarrow f \in v(p) \& f \in v(q)$$

$$t \in v(p \Rightarrow q) \quad \Leftrightarrow \quad t \in v(p) \Rightarrow t \in v(q)$$
$$f \in v(p \Rightarrow q) \quad \Leftarrow \quad t \in v(p) \& f \in v(q)$$

$$t \in v(\top) \quad \Leftrightarrow \quad \top$$
$$f \in v(\top) \quad \Leftrightarrow \quad \bot$$

For if this argument were invalid, then there would be some evaluation v such that $t \in v(\top) \not\ni t$ and that would imply $\bot$. Similarly,

$$\bot \models \bot$$

is valid, and not-invalid-or-else.

Does any such evaluation exist? For a moment, just consider a language with *only* $\top$. Then let $v = \{\langle \top, t \rangle\}$. This set exists by naive comprehension. If $\langle \top, f \rangle \in v$, then by the law of ordered pairs, $t = f$, and we assume that cannot be (for example, by letting $t = 1$ and $f = 0$). So this v is a consistent evaluation on $\top$. And then we can extend v with the other relational clauses to cover all atomic formulas.

What about the Routleyization of this evaluation, $\mathcal{Z}(v) \equiv v$? For each $\langle p, t \rangle \in v$, then $\langle p, t \rangle \in \mathcal{Z}(v)$ and $\langle p, t \rangle \notin \mathcal{Z}(v)$, but by that very fact, $\mathcal{Z}(v)$ is not a valuation, on pain of absurdity.

Dialetheic paraconsistency or no, there is at least one absolute truth, and at least one absolute falsity. We know, in fact, that beyond the LP vocabulary there are such things: $\top$ is a theorem, is in no way not a theorem, and $\neg \top$ cannot in any way be a theorem. And this stands in, arguably, for a great many of the results in this book, garden-variety theorems. Since $\bot$ is so fundamental, then, in making any sense of the world, this would be a good reason to take $\top$ and $\bot$ to be part of the logical vocabulary, part of deductive logic itself. Anything that counts as an evaluation must respect this condition, the nontriviality of the world. It's not "adding premises to reduce threat" or defaulting to classicality; it's taking the rejection of absurdity – absolute consistency, nontriviality – as a starting point for the *organon* of logic itself. Call it an appeal to the principle of sufficient reason, Premise 0, an appeal to the very possibility of reason itself.

10.1.7

With the threat of global invalidity apparently averted, we can return from logical technicalities to the more general philosophical issue of whether keeping the EXCLUSIVE condition is correct, or whether it "multiplies contradictions beyond necessity." For even if we have avoided committing ourselves to saying that *all* arguments are invalid, we still end up with the result that all *paradoxes* are invalid, and so that there are no paradoxes. I am prepared to accept this – there really are no paradoxes (and there really are) – but will spend some time discussing why, because it cuts to the bone of the whole glutty/dialetheic approach.

If we drop the EXCLUSIVE condition, and distinguish falsity from untruth, as Priest does, then even in just LP we avoid global invalidity, without adding $\bot$ to the logical vocabulary. So now I will say why I do not think dropping EXCLUSIVE is a good option, and why it is better to accept that, without placing some strong faith in $\top$ (or, equivalently, maintaining a very healthy aversion to $\bot$), "invalid" is trivialized.

One issue is pragmatic, shading in to a revenge issue. I've been at pains to recapture some fairly sophisticated paradoxes; a great many of the results in Chapter 5 and beyond turn on contraposition, starting from the non-self-identity of the Russell set. Without contraposition, you *can't even derive the liar paradox properly* (as it was derived in the Introduction)! Here's why. The diagonal lemma delivers $\ell \leftrightarrow \neg T\ell$; that's where the liar sentence originates. But without contraposition, that's not yet "this sentence is false," $\ell \leftrightarrow T\neg\ell$. Without contrapositon, one may put together $\ell \leftrightarrow \neg T\ell$ with the instance of the T schema $\ell \leftrightarrow T\ell$, to achieve $T\ell \leftrightarrow \neg T\ell$, and hence

$$T\ell \ \& \ \neg T\ell$$

[Priest, 2002b, p. 351] – but without contraposition, how are we to get the T-free liar contradiction $\ell \ \& \ \neg\ell$? It does not follow.[10] That seems very bad to me, in terms of paradox recapture: dialetheism, predicated on dealing better with the liar than any other position, had better be able to capture the liar paradox. Now, we have a connective that does not contrapose, $\Rightarrow$, but we also have one that contraposes, $\rightarrow$, the one that captures "genuine" absolute sufficiency, and this is the one in the naive set axiom. Without that, it is not clear to me that many dialetheic arguments can get off the ground. Revenge! More than mere utility, that suggests that a *uniformly* paraconsistent approach without contraposition will be very hard going.

Another issue is philosophical. Again, truth in a model should be a model of truth. What are the constraints on that? Lewis distinguishes between predicates like, on the one

---

[10] Priest supposes that the liar sentence is generated via functions, so that ℓ just is identical with $\neg T\ell$ as a given. (Cf. Priest 2006b, 3§3.5.) Assuming that much as already assumes classical mathematics (which Priest does (Section 3.1.2.6), but which we can't in the present "fundamentalist" setting.

hand, "in Australia" and "on the mountain," versus predicates like "according to the Bible" and "Fred said that." The former extensionally describe the way things are, in restricted domains, and so can commute with logical connectives without change of meaning or truth value: "in Australia, p or q," is the same as "in Australia, p, or in Australia, q"; and "in Australia, not p" is the same as "not: in Australia, p." But intensional predicates are not like this; either of "Fred says that not p" and "not: Fred says that p" could hold, or not, with or without the other. Lewis is chiefly concerned with the modifier "in a world," about which he says that "in world w, not p," is equivalent to "not: in world w, p." That is because

Worlds, as I understand them, are not like stories or story-tellers. They are like this world; and this world is no story, not even a true story. *[Lewis, 1986, p. 7]*

Put modally, dialetheism is the claim that there are true contradictions *in the actual world*; truth is truth-at-the-actual world, this world. This world is no story. And so "it is true that: not p," is not the rambling dreams of Fred; it is the claim that "not: it is true that p." Intensionality is good for sameness, but extensionality is good for difference. Negation commutes with truth.

10.1.8

Putting together these previous two reasons, pragmatic and philosophical, for accepting the EXCLUSIVE condition, we have what I take to be the most important point. It goes to the very statement of dialetheism itself: "there are true contradictions." What does it mean?

A contradiction is conjunction of a pair $p, \neg p$. A conjunction is true iff both conjuncts are; so some contradictions are true iff some true sentences have true negations. A negation is true iff its negand is false; so some sentences are both true and false. So much for the truth conditions on negation. What are the desiderata on falsity conditions, and do these constraints justify or motivate the conditions I've given?

The negation of p *contradicts* p. Otherwise, the conjunction of p and $\neg p$ is not a "contradiction." Is it enough to say that p is true iff $\neg p$ is false? For all that is asserted in the previous paragraph, there is nothing yet *oppositional* about truth and falsity (much as with the "classical" modeling of extensions and antiextensions X^+, X^-). To fill in this gap in the account, we also need EXCLUSIVE – to say that falsity excludes truth: what is false is not true. As Aristotle didn't quite say, truth is what is the case, and falsity is what is *not* the case. And *that* ties things up very neatly: a sentence is false iff it is not true, and so $\neg p$ is true iff p is not true. It means that the dialetheist is not just committed to sentence/negation pairs both being true, or even sentences being both true and false, but sentences being both true and *not* true. *That's* a true contradiction, when falsity is *inconsistent* with truth.

Now, "deviant" logicians have been dogged since the start by various forms of Quine's "change of logic, change of subject" argument.[11] Paraconsistent logicians (among others) have been told again and again that, whatever the merits of the operator they write as "¬,"

---

[11] And that argument (though "dogmatic and pointlessly controversial" [Williamson, 2014, p. 214] has, arguably, been widely accepted: logical pluralism today flourishes as a view that different logics are not rivals because they can be regarded as different subjects [Shapiro, 2014]. But that's not the point here.

it isn't *negation*. This is pretty annoying, so it is with some regret that I find myself saying something in the vicinity. But some changes of logic *are* changes of subject. If, according to a dialetheist, falsity does not contradict with truth, in the sense that false sentences are not true and true sentences are not false, then contradictions don't contradict. The dialetheist's defense that there are true contradictions is in danger of being like a theorist who makes some arresting statement ("Science explains why there is something rather than nothing!") only to concede that the defense is based on changing the meanings of some words ("nothing" means "a relativistic quantum field in a vacuum state").[12]

For me, as I've tried to emphasize throughout, the most attractive feature of dialetheism is (paradoxically?) its *honesty*. No one can even address the paradoxes without some sort of self-refutation; it just turns out that in life sometimes we are compelled to falsehoods. It is good to admit that the world is the way it is, rather than conjure up endlessly clever alternative stories. And I can only say honestly that there are true contradictions if they truly contradict.

Less sentimentally, we are newly in the same old situation as much of Chapter 3 was about. Either we fall back on classical consistency, and say that false does not mean untrue, severing a basic link between (un)truth and falsity and ultimately pegging our hopes to an "exclusive negation" from orthodox metatheory, or we finish the project: of describing the world as it is given to us, described by a completely substructural paraconsistent language, a world where some arguments are valid, some are absolutely not, no paradoxes are valid, some are, and outside the moon is beginning to rise just as before.

10.1.9

And so for these reasons I accept that there are no paradoxes (though there are some, too) and go (further) down the *ultraparaconsistent* rabbit hole. If a valid paradoxical argument is one where the premises cannot be true without the conclusion, then there are no valid paradoxical arguments (although there are a great many), not even "ℓ therefore ℓ." Propositions such as the liar or the Russell are theorems, but if a theorem is a statement that is necessarily true, then they are not theorems (though they are).

What can the ultraparaconsistent approach achieve? What can it tell us about the explanation for the paradoxes? It tells us to stop looking for answers in the wrong places. Since, on this ultraparaconsistent account, there are no paradoxes, then a fortiori it tells us that there is no special schema to isolate the paradoxes, nor is there any special status of "being paradoxical." It tells us that dialetheism is false and there are no true contradictions; ergo, there is no way to isolate the true contradictions, or the sound arguments that give rise to them. There is no special class of dialetheias, precisely demarcated from the nondialetheias, because there are true sentences, and false sentences, and that's it.[13] It tells us that, to think well about paradoxes, we must think paradoxically.

---

[12] See Albert's 2012 *New York Times* review of Krauss's *A Universe from Nothing*.
[13] Cf. Priest [2006b, p. 294].

And if one still wants to ask, "but of those false contradictions that are also true, then why are *they* (and not other false-only contradictions) true?" then the answer must be: the true contradictions are true because they are true. The lesson of revenge is: that's it. And if one still wants to ask, "okay, so how many theorems in this book are also not theorems?" then the answer must be: the false ones. That's it. Ultraparaconsistency tells us the answer to "how much inconsistency is too much?" It is simply whatever the answer is to how much truth is there.

10.1.10

The ultraparaconsistent approach recasts the central dilemma of the whole project. *The universe can be divided exclusively and exhaustively in two parts after all,* then: for any property φ whatsoever,

$$\mathcal{V} \equiv \{x : \varphi(x)\} \cup \{x : \neg\varphi(x)\}.$$

The repeating options we encountered in the opening chapters – the choice between all the truths (but not only) or only the truths (but not all) – is overcome, surpassed. With truth predicate $\mathsf{T}(x)$, the set of all p such that $\mathsf{T}\ulcorner p\urcorner$, paired with the set of all p such that $\neg\mathsf{T}\ulcorner p\urcorner$, is a perfect division of the propositions into all and only the truths, and all and only the falsities.

Even a *point* may be split in two directly, from the inside. This is, yes, to invoke the Routley Set, this time at the smallest scale imaginable. When a is a point,

$$\{a\}_L = \{x : x = a \text{ and } x \in \mathcal{Z}\} \qquad \{a\}_R = \{x : x = a \text{ and } x \notin \mathcal{Z}\}$$

is an exclusive, exhaustive division of the point into a left and right side.

10.2 The Last Horizon

Why are there paradoxes?

The ultraparaconsistent position would look to offer the type of solution "seen in the vanishing of the problem" [Wittgenstein, 1922, 6.521]. But that would at most be only half the story. Premise 0 of this study, stated in the Preface and invoked in the key concept for final theorem, is Hilbert's credo that all questions have answers. The principle of sufficient reason: everything has an *explanation*.[14] The limit of this principle is to demand an explanation of why there is some(paradoxical)thing rather than no(paradoxical)thing.

10.2.1

Paradoxes come at us from *above*. The last century of research especially has focused on this universal or global character of the paradoxes – that they arise at limits, overly large

---

[14] An explanation such that once we knew it, then, as Della Rocca puts it, "we could see it coming." A particularly strong, self-inverted version of this is: everything has an explanation *given in terms of explicability itself* [Della Rocca, 2008, p. 30].

domains, or by excessively expressive quantification. That's not wrong.[15] The fixed point Theorem 13 says that every mapping has a fixed point; some of the reason for this fact is that every map, regardless of its stated domain and range, is really from the universe to itself. That's how strange, isolated fixed point objects floating in the darkness of transconsistent space end up, unintended, as arguments of mundane schoolroom functions. This is hinted at too in Theorem 66. The universe in its totality is paradoxical.

Therefore, there is a temptation to attribute the paradoxes to the universe itself, as a grand unified whole object; and this makes it tempting to believe that paradox can be avoided by staying closer to home, around "smaller" objects. That *is* wrong. The universe is a paradoxical object, but it is not the only paradoxical object. Even if a paradox is a property of some "dark" object from beyond the rim of the void, the interesting question is: why *that* object in particular? Why *my* shoes? Paradoxes come at us from *below*, too.

The basic fact is that there are objects (the liar sentence, the Russell set, a baby growing older) such that when we describe their properties as best we can, we are led into contradiction. These objects, like the fixed point objects, just exist. The "universe" is responsible for the paradoxes insofar as it is where everything happens; it is not responsible for where specific heaps of sand, or sets of non-self-membered sets, fall. Some properties give rise to paradoxes, and there won't be an answer for why they exist – why there are *these* properties rather than *those* properties – because there is always already *every* property, out there in Aussersein or hanging from the coat rack or wherever properties are. There won't be a better answer for why these things exist than there is for why the universe itself exists. The explanation for the paradoxes, insofar as there is one, has to come from the objects, the properties, the rainstorms and ordinals and babies, themselves.

Paradoxes from above. Paradoxes from below. Paradoxes *between*.

10.2.2

Paradoxes arise at boundaries. In set (property) theory, there are contradictions at the "top edge" of the universe. For vague properties, there are contradictions at cutoff points. For material objects, there are contradictions at their boundaries. The edge of an object surrounds the edges of its complement, as in Proposition 59; the boundaries of some objects are *inside* those objects. And any attempt to dissolve the contradictions leads to revenge, in the form of *new* boundaries every bit as problematic as the original. This appeared to be nicely captured by Priest's inclosure schema where contradictions are at the "limits" – but in Chapter 2 we saw that the inclosure locates the paradoxes at only *extremal* boundaries, when paraconsistency allows the contradictions to be much more local. The sites for contradiction can be at the limits of big objects, like totalities, but they can also be *small*.

You are in the city and walking out of it. At some point, you are in the city; at some point, you are not; and for any two very nearby locations, no one is in the city while the other is

---

[15] As a children's song puts it: "What if you drew a giant circle / What if it went around all there is / Then would there still be such a thing as an outside / And does that question even make any sense?" (They Might Be Giants, "One Everything")

not. Somewhere there must be, quite literally, some stone, some *pebble* (a "calculus"), that is both in the city and not in the city. There is some spot *on* that pebble where the change is located, *in* that spot.

Cities, stones. These are not unusual objects. These are the objects found everywhere. There are paradoxes everywhere.

10.2.3

Astronomers use parallax to measure the distance to stars: observe the sky one night, then another, and compare the relative changes in position of the stars. The nearer stars will appear to have moved, while the farther away stars will appear to stay fixed. Parallax provides a way to measure where things are from a limited position. Using parallax, what would a star be that never changes position, no matter how much everything else moves? It would be *infinitely far away*.

Like many, I used to think that attaining a fixed point at infinity was the only way to make sense of the world, the only way to get an objective view of reality sub specie aeternitatis. I used to think, with Wittgenstein in the *Tractatus*, that "the sense of the world must lie outside the world" (6.41), that "the solution of the riddle of life in space and time must lie outside space and time" (6.4312). I thought, with Archimedes, that we need a place to stand – a fixed point – to move the world. Now, by definition there is no outside the universe, no such fixed point; but I thought that maybe dialetheism provided a way to do the impossible, a means to stand on the boundary of the universe, since, for example, it is coherently inconsistent to be on the boundary of the universal set $\{x : x = x\}$. But as in Chapter 2, for the comprehensive property of $\top$, there is no outside the universe. "All things that bear a property" is a class so comprehensive that the only "things" beyond it are utter absurdity. What lies beyond is not even out there in Aussersein, not even chaos.

Could there still be a star be that never changes position, no matter how much everything else moves, a star that is fixed without being infinitely far away? Could it be the fixed point that arises for every object, *on* every object? As in Theorem 53, as a force pulls a point toward truth and falsity both, then that point stays fixed. The absolute place from which to measure the ultimate value of any thing is right there, in the world, on the thing itself.

10.2.4

There is a fixed point for every mapping. I have shown how, properly understood, the universe *can* be divided exclusively and exhaustively, in a way that leaves room for fixed points. But all is not sanguine: questions remain; implosion looms ever large. The paradoxes are the gift that never stops giving: the "angst and awe ... in the face of contradiction" [Wittgenstein, 1956, p. 53]. The paradoxes are endlessly surprising; that's why they are paradoxes.

And so: what about implosion? Could the universe be a point? If it were, then we have seen ways in which it could be so without necessarily being a *simple* point – points can be complexes, with internal parts, space to live in. The classical view is that inconsistency has no structure; the ultraparaconsistent response is, not to make "inconsistency" somehow consistent,[16] reviving a simulation of inconsistency as a classical object, but rather to show that the transconsistent has genuine impossible structure – and maybe with it, some beauty – after all. The *ultimate* paradox recapture: we were worried that if the world is inconsistent, then it implodes – and Lemma 69 illustrates how it happens. Yet here we are. We have seen in the paraconsistent Brouwer Fixed Point Theorem 70 that even if all the points are turned to dust, broken on their insides, still there would be a fixed point, a place to stand.

If we are bounded in a nutshell, are we the kings of infinite space?[17] There is nothing, no explanation, to be found beyond the edge of the universe, however far away (or not) that edge is. There is no point in relative (absolute) consistency proofs. Thinking the world is not a matter of looking *up*, seeking the outside, but of looking *in*. In rejecting metalanguages, hierarchies, stratified type theories, we have the opposite – an imploded point. Is that not what the paraconsistent proof of Cantor's Theorem 15 shows? You don't chase the horizon; the horizon comes to you.

10.2.5

Paradoxes above, paradoxes below, paradoxes inside. Here's what the ultra-paraconsistent picture shows: in the naive universe, every object has a shadow, cast by and cast onto the Routley set. Everything is always one step away from oblivion. But Routley himself councils:

Being one world away from absurdity is very different from being in an absurd world. Being one step removed from disaster is often very different, and feels very different, from the disaster. *[Routley, 1983]*

That's our inconsistent world. There are contradictions at the liminal edges of coffee cups, and they are still just coffee cups, and they are impossible. Dialetheism is not a *solution* to the paradoxes. It is part of a *solvent*, as Routley puts it, a *resolution* via explanation, but also a way to continue to grapple with ever new paradoxes, endless surprise. Dialetheic paraconsistency is a motive and a method, a way to *accept* that the world is as we find it – and also *not* simply to accept it, but to wake up tomorrow and try again.

10.3 A Fixed Point Where None Can Be

Why are there paradoxes?

In this book I have done two things. First, I laid out an unrepentant case that dialetheism is true, that our world is inconsistent. Second, I showed how a substructural paraconsistent logic can provide a foundation for mathematical theories, by developing such theories at

---

[16] "If I were attempting to produce a consistent theory of the inconsistent, this would be fatal" [Priest, 2006b, p. 72].

[17] To paraphrase *Hamlet*. Less obliquely: "One must by all means stretch out one's fingers and make the attempt to grasp this amazing finesse, *that the value of life cannot be estimated*" [Nietzsche, 1976, Twilight of the Idols, Problem of Socrates, emphasis in the original text].

some length—sketching how our ordinary reasoning projects can persist in the presence of an absurdity, how things do not fall apart.

I have not provided a schematic form that all the genuine paradoxes fit, because there cannot be any such form; but in working out the details of the second point, a substructural picture of the world, we have narrowed in on the geometry, the structural locations, so to speak, of some of the contradictions – in the cracks, the boundaries, the *points* between things, and especially between a thing and itself. We have seen that core theorems are proved, that a chunk of the modern mathematical megalopolis holds. We have seen that a fixed point persists even in the most extreme circumstances. Truth does not need to be replaced or saved from contradiction. In the end, then, what does obsessing over the paradoxes forever teach?

10.3.1

Paradoxes are ordinary. Paradoxes are found at the boundary points of objects. But *every point is the boundary of some object*. Ordinary objects are host to paradoxes; paradoxes are ordinary.

This runs counter to a plausible intuition: that if a phenomenon occurs often, then the phenomenon cannot be strange. Or, it would be strange for something that is very common to be very strange. Can we be in a state of perpetual surprise? It is paradoxical in itself – surprising – that paradoxes be ordinary. "How could paradoxes be so ubiquitous? I ate toast this morning! And I didn't not eat toast!" Yes. When did the bread become toast? Was it still dark outside, or was there already light in the sky? Was the sun coming up over the horizon? Paradoxes are *ordinary*.

Gupta and Belnap say that, despite enormous effort, the liar paradox has not yet been solved because "insufficient attention has been paid to the ordinary, unproblematic uses of the concept of truth ... Often we come to understand the extraordinary only when we see it in terms of the ordinary" [Gupta and Belnap, 1993, p. 17]. Once we see that even ordinary uses of, e.g., truth are also paradoxical, then we can see that the liar is no more – or less – problematic than the morning weather report. The ordinary is extraordinary; *that is why the paradoxes are paradoxes*. Wittgenstein was almost right. "It is not *how* the world is that is [paradoxical], but *that* it exists at all" [Wittgenstein, 1922, 6.44].

10.3.2

Let us take stock of some of the more important results and themes from our efforts in substructural paraconsistent mathematics:

- Some things are not self-identical (Lemma 5).
- Sometimes things remain, even after everything is taken away ($\varnothing_X$, $\mathbf{0}_x$, etc.).
- Dividing a line leaves something in between; and in the cases where it rips a point, it leaves a fixed point (Theorem 53).
- There are boundaries inside boundaries (Proposition 59(iii)).

- There are points inside points (e.g., Section 8.3.1).
- The world can be subsumed into to any part of itself (Theorems 22 and 66).
- Spinning every point of a disc dissolves the disc (Theorem 70) – but even in that impossible vortex in which the smallest points are divided, broken, *some point does not move.*

The principle of sufficient reason itself leads to unreasonable contradictions. And that's part of what the paradoxes teach: to persist in making sense of the world even – especially – when the world seems senseless.

Perhaps most important – perhaps what is behind it all – is Theorem 10:

- Our world has the Routley set, above and below.

It is everywhere and it is nowhere, and it is everything and it is nothing, and when the ordinary familiar surface of reality shifts, sometimes you can see it there, in the cracks, in its cosmic and comic indifference. It is a dome over everything, of which every object is inconsistently a part. It is a ground beneath everything, from which every object has an inconsistent part. We are between. The earth meets sky at the horizon [ορος, *boundary*], and these together are our inconsistent world.[18]

10.3.3

Leibniz says that all objects "are generated by the continual flashes of silent lightning." Pick up a pebble. It is the boundary point between somewhere and somewhere else. You are holding a paradox in your hand. It doesn't hurt; it doesn't even look strange. It looks like a pebble. It *is* a pebble.

Look! There is for every thing and everything its fixed point. A continuous flash of silent lightning. A point, tearing itself apart. A fixed point, where none can be, and is.

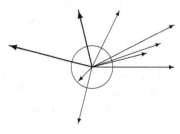

---

[18] "It's the strangest thing of all, / Stranger than all poets' dreams / And all philosophers' thoughts / That things are really what they seem to be ..." (Pessoa, XXIX).

Bibliography

Abad, J. (2008). The inclosure schema and the solution to the paradoxes of self-reference. *Synthese*, 160(2):183–202.

Aczel, P. (1988). *Non-Well-Founded Sets*. Number 14 in CSLI Lecture Notes. CSLI Publications.

Allwein, G. and Dunn, J. M. (1993). Kripke semantics for linear logic. *Journal of Symbolic Logic*, 58(2):514–545.

Anderson, A. R. and Belnap, N. D. (1975). *Entailment: The Logic of Relevance and Necessity*, volume 1. Princeton University Press.

Anderson, A. R., Belnap, N. D., and Dunn, J. (1992). *Entailment: The Logic of Relevance and Necessity*, volume 2. Princeton University Press.

Armstrong, M. (1979). *Basic Topology*. Springer.

Arntzenius, F. (2012). *Space, Time, and Stuff*. Oxford University Press.

Arruda, A. I. and Batens, D. (1982). Russell's set versus the universal set in paraconsistent set theory. *Logique et Analyse*, 25(8):121–133.

Asenjo, F. (1966). A calculus of antinomies. *Notre Dame Journal of Formal Logic*, 7:103–105.

Asenjo, F. (1975). Logic of antinomies. *Notre Dame Journal of Formal Logic*, 16:17–44.

Asenjo, F. G. (1989). Toward an antinomic mathematics. In [Priest et al., 1989], pages 394–414.

Austin, D. F., editor (1988). *Philosophical Analysis*. Kluwer.

Azzouni, J. (2006). *Tracking Reason: Proof, Consequence, and Truth*. Oxford University Press.

Bacon, A. (2013). Non-classical metatheory for non-classical logics. *Journal of Philosophical Logic*, (42):335–355.

Badia, G. (2017). *The Languages of Relevant Logic*. PhD thesis, University of Otago.

Badia, G. and Weber, Z. (2019). A substructural logic for inconsistent mathematics. In Reiger, A. and Young, G., editors, *Dialetheism and Its Applications*, pp. 155–176. Springer.

Badici, E. (2008). The liar paradox and the inclosure schema. *Australasian Journal of Philosophy*, 86(4):583–596.

Barendregt, H. (1984). *The Lambda Calculus: Its Syntax and Semantics*. North Holland.

Barnes, E. (2010). Ontic vagueness: a guide for the perplexed. *Nous*, 44(4):601–627.

Barwise, J. and Moss, L. (1996). *Vicious Circles*. CSLI Publications.

Batens, D. (2019). Looting liars masking models. In Baskent, C. and Ferguson, T., editors, *Graham Priest on Dialetheism and Paraconsistency*, pages 139–164. Springer.

Batens, D. (2020). Adaptive fregean set theory. *Studia Logica*, 108:903–939.

Batens, D., Mortensen, C., Priest, G., and van Bendegem, J.-P., editors (2000). *Frontiers of Paraconsistent Logic*. Research Studies Press.

Beall, J. (1999). From full blooded platonism to really full blooded platonism. *Philosophia Mathematica*, 7(3):322–325.

Beall, J., editor (2003). *Liars and Heaps*. Oxford University Press.

Beall, J., editor (2007). *Revenge of the Liar*. Oxford University Press.

Beall, J. (2009). *Spandrels of Truth*. Oxford University Press.

Beall, J. (2011). Multiple-conclusion LP and default classicality. *Review of Symbolic Logic*, 4(2):326–336.

Beall, J. (2013a). Free of detachment: logic, rationality, and gluts. *Noûs*, 49(2):410–423.

Beall, J. (2013b). Shrieking against gluts: the solution to the "just true" problem. *Analysis*, 73(3):438–445.

Beall, J. (2014a). End of inclosure. *Mind*, 123(491):829–849.

Beall, J. (2014b). Finding tolerance without gluts. *Mind*, 123(491):791–811.

Beall, J., Brady, R. T., Hazen, A., Priest, G., and Restall, G. (2006). Relevant restricted quantification. *Journal of Philosophical Logic*, 35:587–598.

Beall, J. and Colyvan, M. (2001a). Heaps of gluts and hyde-ing the sorites. *Mind*, 110:401–408.

Beall, J. and Colyvan, M. (2001b). Looking for contradictions. *Australasian Journal of Philosophy*, 79:564–9.

Beall, J., Glanzberg, M., and Ripley, D. (2018). *Formal Theories of Truth*. Oxford University Press.

Beall, J. and Murzi, J. (2013). Two flavors of curry paradox. *Journal of Philosophy*, 110(3):143–165.

Beall, J., Priest, G., and Weber, Z. (2011). Can u do that? *Analysis*, 71(2):280–285.

Beall, J. and Restall, G. (2006). *Logical Pluralism*. Oxford University Press.

Beardon, A. F. (2012). *Algebra and Geometry*. Cambridge University Press.

Běhounek, L. and Cintula, P. (2005). Fuzzy class theory. *Fuzzy Sets and Systems*, 154(1):34–55.

Běhounek, L. and Haniková, Z. (2015). Set theory and arithmetic in fuzzy logic. In Montagna, F., editor, *Petr Hajek on Mathematical Fuzzy Logic*, pages 63–89. Springer. Outstanding Contributions to Logic Series, volume 6.

Bell, J. L. (2008). *A Primer of Infinitesimal Analysis*. Cambridge University Press. Second Edition.

Bell, J. L. (2014). *Intuitionistic Set Theory*. College Publications. Studies in Logic 50.

Belnap, N. D. and Dunn, J. M. (1981). Entailment and the disjunctive syllogism. *Contemporary Philosophy: A New Survey*, 1:337–366.

Berberain, S. K. (1977). *A First Course in Real Analysis*. Springer.

Berto, F. (2008). Adunaton and material exclusion. *Australasian Journal of Philosophy*, 86(2):165–190.

Birkhoff, G. and Maclane, S. (1999). *Algebra*, Third Edition. American Mathematical Society. First Edition 1967.

Bishop, E. (1967). *Foundations of Constructive Analysis*. McGraw-Hill.

Bishop, E. and Bridges, D. S. (1985). *Constructive Analysis*. Springer.

Black, M. (1951). Achilles and the tortoise. *Analysis*, 11:91–101.

Blizard, W. (1989). Multiset theory. *Notre Dame Journal of Formal Logic*, 30(1):36–66.

Bolzano, B. (1950 [1851]). *Paradoxes of the Infinite*. Routledge and Keegan Paul. Translated by F. Prihonsky.

Bolzano, B. (1973). *Theory of Science*. D. Reidel Publishing. Edited, with an introduction, by J. Berg. Translated by B. Terrell.

Boolos, G. (1971). The iterative conception of set. *Journal of Philosophy*, 68(8):215–231. Reprinted in in [Boolos, 1998].

Boolos, G. (1998). *Logic, Logic and Logic*. Harvard University Press.

Boos, W. (1983). A self-referential "cogito." *Philosophical Studies*, 44(2):269–290.

Boyer, C. (1959). *The History of the Calculus and Its Conceptual Development*. Dover Publications.

Braddon-Mitchell, D. and Miller, K. (2006). The physics of extended simples. *Analysis*, 66(3):222–226.

Brady, R. (1971). The consistency of the axioms of the axioms of abstraction and extensionality in a three-valued logic. *Notre Dame Journal of Formal Logic*, 12: 447–453.

Brady, R., editor (2003). *Relevant Logics and Their Rivals, Volume II: A Continuation of the Work of Richard Sylvan, Robert Meyer, Val Plumwood and Ross Brady*. Ashgate.

Brady, R. (2006). *Universal Logic*. CSLI Publications.

Brady, R. and Mortensen, C. (2014). Logic. In Oppy, G. and Trakakis, N., editors, *History of Philosophy in Australia and New Zealand*, pp. 679–705. Springer, Dordrecht.

Brady, R. T. (1989). The non-triviality of dialectical set theory. In [Priest et al., 1989], pages 437–470.

Brady, R. T. and Routley, R. (1989). The non-triviality of extensional dialectical set theory. In [Priest et al., 1989], pages 415–436.

Brentano, F. (1988). *Philosophical Investigations on Space, Time and the Continuum*. Croom Helm. B. Smith (trans).

Bridges, D. and Vita, L. S. (2011). *Apartness and Uniformity: A Constructive Development*. Springer.

Brown, B. and Priest, G. (2004). Chunk and permeate i: the infinitesimal calculus. *Journal of Philosophical Logic*, 33:379–388.

Brown, J. (2008). *Philosophy of Mathematics*. Routledge. Second Edition.

Bueno, O. and Colyvan, M. (2012). Just what is vagueness? *Ratio*, 25:19–33.

Burgess, J. P. (2005). No requirement of relevance. In Shapiro, S., editor, *The Oxford Handbook of Philosophy of Mathematics and Logic*, pages 727–750. Oxford University Press.

Burgess, J. P. and Woods, J. (2015). *Review of the Boundary Stones of Thought* by Ian Rumfitt. Notre Dame Philosophical Reviews, October 24, 2015.

Buskes, G. and van Rooij, A. (1997). *Topological Spaces: From Distance to Neighborhood*. Springer.

Cantini, A. (2003). The undecidability of Grišin's set theory. *Studia Logica*, 74(3):345–368.

Cantor, G. (1895). Beiträge zur begründung der transfiniten mengenlehre (zweiter artikel). *Mathematische Annalen*, 49:207–246.

Cantor, G. (1915). *Contributions to the Founding of the Theory of Transfinite Numbers*. Dover. Edited, translated, and introduced by P. E. B. Jourdain, from [Cantor, 1895].

Cantor, G. (1967). Letter to Dedekind. In [van Heijenoort, 1967], pages 113–117. 1899.

Caret, C. R. and Hjortland, O. T., editors (2015). *Foundations of Logical Consequence*. Oxford University Press.

Caret, C. R. and Weber, Z. (2015). A note on contraction-free logic for validity. *Topoi*, 31(1):63–74.

Carnielli, W. and Coniglio, M. E. (2016). *Paraconsistent Logic: Consistency, Contradiction and Negation*. Springer.

Carnielli, W., Coniglio, M., and Marcos, J. (2007). Logics of formal inconsistency. In Gabbay, D. and Guenthner, F., editors, *Handbook of Philosphical Logic*, volume 14, pages 1–93. Springer-Verlag.

Carr, M. H. (2007). *The Surface of Mars*. Cambridge University Press.

Casati, F. and Fujikawa, N. (2019). Nothingness, meinongianism, and inconsistent mereology. *Synthese*, 196:3739–3772.

Casati, R. (2003). *The Shadow Club*. Knopf.

Casati, R. and Varzi, A. C. (1999). *Parts and Places: The Structures of Spatial Representation*. MIT Press.

Chang, C. (1963). The axiom of comprehension in infinite valued logic. *Mathematica Scandinavia*, 13:9–30.

Chisholm, R. M. (1984). Boundaries as dependent particulars. *Grazer philosophische Studien*, 10:87–95.

Chvalovský, K. and Cintula, P. (2012). Note on deduction theorems in contraction-free logics. *Mathematical Logic Quarterly*, 58(3):236–243.

Cintula, P. and Paoli, F. (2016). Is mutliset consequence trivial? *Synthese*, https://doi.org/10.1007/s11229-016-1209-7.

Clarke, B. L. (1985). Individuals and points. *Notre Dame Journal of Formal Logic*, 26:61–75.

Cogburn, J. (2017). *Garcian Meditations*. Edinburgh University Press.

Cohen, P. J. (1966). *Set Theory and the Continuum Hypothesis*. W. A. Benjamin.

Cohen, S. M., Curd, P., and Reeve, C., editors (2000). *Readings in Ancient Greek Philosophy: From Thales to Aristotle*. Hackett.

Cohn, P. (2003). *Basic Algebra: Groups, Rings, and Fields*. Springer.

Colyvan, M. (2008a). The ontological commitments of inconsistent theories. *Philosophical Studies*, 141(1):115–123.

Colyvan, M. (2008b). Vagueness and truth. In Dyke, H., editor, *From Truth to Reality: New Essays in Logic and Metaphysics*, pp. 29–40. Routledge.

Colyvan, M. (2009). Applying inconsistent mathematics. In *New Waves in Philosophy of Mathematics*, pp. 160–172. Palgrave Macmillan.

Colyvan, M. (2012). *An Introduction to the Philosophy of Mathematics*. Cambridge University Press.

Conway, J. (1976). *On Numbers and Games*. Academic Press.

Cook, R. T. (2014). There is no paradox of logical validity. *Logical Universalis*, 8(3–4):447–467.

Cotnoir, A. and Baxter, D., editors (2014). *Composition as Identity*. Oxford University Press.

Curry, H. B. (1942). The inconsistency of certain formal logics. *Journal of Symbolic Logic*, 7:115–117.

da Costa, N. (2000). Paraconsistent mathematics. In [Batens et al., 2000], pages 165–180.

da Costa, N. C., Krause, D., and Bueno, O. (2007). Paraconsistent logics and paraconsistency. In [Jacquette, 2007], pages 791–912.

da Costa, N. C. A. (1974). On the theory of inconsistent formal systems. *Notre Dame Journal of Formal Logic*, 15:497–510.

Dainton, B. (2010). *Time and Space*. McGill-Queen's University Press. Second Edition.

Dauben, J. W. (1979). *Georg Cantor: His Mathematics and Philosophy of the Infinite*. Princeton. University Press.

Dedekind, R. (1901). *Essays on the Theory of Numbers*. Dover. Ed. and trans. by W. W. Beman, 1901. Includes *Stetigkeit und irrationale Zahlen* [1872] and *Was sind und was sollen die Zahlen?* [1888]. Dover Edition 1963.

Della Rocca, M. (2008). *Spinoza*. Routledge.

Deloup, F. (2005). The fundamental group of the circle is trivial. *American Mathematical Monthly*, 112(5):417–425.

Devlin, K. J. (1979). *Fundamentals of Contemporary Set Theory*. Springer. Second Edition 1993, retitled "The Joy of Sets."

Dicher, B. and Paoli, F. (2019). ST, LP, and tolerant metainferences. In Baskent, C. and Ferguson, T., editors, *Graham Priest on Dialetheism and Paraconsistency*, pages 383–408. Springer.

Dietrich, E. (2015). *Excellent Beauty*. Columbia University Press, New York.

Dillon, M. (1997). *Merleau-Ponty's Ontology*. Northwestern University Press.

Drake, F. (1974). *Set Theory: An Introduction to Large Cardinals*. North Holland.

Dummett, M. (1978). *Truth and Other Enigmas*. Oxford University Press.

Dummett, M. (1991). *Frege: Philosophy of Mathematics*. Duckworth.

Dümont, J. and Mau, F. (1998). Are there true contradictions? A critical discussion of Graham Priest's *Beyond the Limits of Thought*. *Journal for General Philosophy of Science*, 29(2):289–299.

Dunn, J. (1987). Relevant predication 1: The formal theory. *Journal of Philosophical Logic*, 16:347–381.

Dunn, J. and Restall, G. (2002). Relevance logic. In Gabbay, D. M. and Günthner, F., editors, *Handbook of Philosophical Logic*, volume 6, pages 1–128. Kluwer. Second Edition.

Dunn, J. M. (1979). A theorem in 3-valued model theory with connections to number theory, type theory, and relevant logic. *Studia Logica*, 38:149–169.

Dunn, J. M. (1980). Relevant Robinson's arithmetic. *Studia Logica*, 38:407–418.

Dunn, J. M. (1988). The impossibility of certain higher-order non-classical logics with extensionality. In [Austin, 1988], pages 261–280.

Dzhafarov, E. N. and Dzhafarov, D. (2010). Sorites without vagueness I: classificatory sorites. *Theoria*, 76(1):4–24.

Estrada-González, L. (2016). Prospects for triviality. In Anreas, H. and Verdée, P., editors, *Logical Studies of Paraconsistent Reasoning in Science and Mathematics*. Springer.

Etchemendy, J. (1990). *The Concept of Logical Consequence*. Harvard University Press.

Euclid (1956). *The Thirteen Books of the Elements*. Dover. Edited, translated, and annotated by T. L. Heath.

Feferman, S. (1984). Toward useful type-free theories, I. *Journal of Symbolic Logic*, 49(1):75–111.

Field, H. (2008). *Saving Truth from Paradox*. Oxford University Press.

Field, H. (2020). Properties, propositions, and conditionals. *Australasian Philosophical Review*, forthcoming.

Field, H., Lederman, H., and Øgaard, T. F. (2017). Prospects for a naive theory of classes. *Notre Dame Journal of Formal Logic*, 58(4):461–506.

Fitch, F. (1964). Universal metalanguages for philosophy. *The Review of Metaphysics*, 17(3):396–402.

Forster, T. (1982). Axiomatising set theory with a universal set. Typeset 1997.

Forster, T. (1995). *Set Theory with a Universal Set*. Clarendon Press.

Fraenkel, A. (1953). *Abstract Set Theory*. Amsterdam.

Fraenkel, A. Bar-Hillel, Y., and Levy, A. (1958). *Foundations of Set Theory*. North Holland.

Frankel, C. (2005). *Worlds on Fire: Volcanoes on the Earth, Moon, Mars, Venus, and Io*. Cambridge University Press.

Franks, C. (2009). *The Autonomy of Mathematical Knowledge*. Cambridge University Press.

Frege, G. (1885). Über formale Theorien der Arithmetik. *Sitzungsberichte der Jenaischen Gesellschaft für Medizin und Naturwissenschaft*, 2:94–104. Reprinted in *Collected Papers on Mathematics, Logic, and Philosophy*. Basil Blackwell, 1984.

Frege, G. (1893/1903a). *The Basic Laws of Arithmetic*. Oxford. Translated and edited by P. A. Ebert and M. Rossberg, with C. Wright; appendix by R. T. Cook.

Frege, G. (1893/1903b). *Grundgesetze der Arithmetik, Begriffsschriftlich abgeleitet*. Verlag Hermann Pohle.

French, R. (2016). Structural reflexivity and the paradoxes of self-reference. *Ergo*, 3(5):113–131.

French, R. and Ripley, D. (2015). Contractions of noncontractive consequence relations. *Review of Symbolic Logic*, 8(3):506–528.

Friedman, H. and Meyer, R. (1992). Whither relevant arithmetic? *Journal of Symbolic Logic*, 57:824–31.

Galatos, N., Jipsen, P., Kowalski, T., and Ono, H. (2007). *Residuated Lattices*. Elsevier.

Geach, P. (1955). On insolubilia. *Analysis*, 15:71–72.

Gilles, D., editor (1992). *Revolutions in Mathematics*. Oxford Unviersity Press.

Gilmore, P. C. (1974). The consistency of partial set theory without extensionality. In [Jech, 1974], pages 147–153.

Girard, J.-Y. (1998). Light linear logic. *Information and Computation*, 143:175–204.

Girard, P. and Weber, Z. (2019). Modal logic without contraction in a metatheory without contraction. *Review of Symbolic Logic*, 4(12):685–701.

Goddard, L. and Routley, R. (1973). *The Logic of Significance and Context*. Scottish Academic Press.

Gödel, K. (1964). What is Cantor's continuum problem? In Benacerraf, P. and Putnam, H., editors, *Philosophy of Mathematics*, pages 258–273. Cambridge University Press. Paper originally written in 1947.

Goldblatt, R. (1998). *Lectures on the Hyperreals*. Springer.

Goodship, L. (1996). On dialethism. *Australasian Journal of Philosophy*, 74(1):153–161.

Grey, J. (2008). *Plato's Ghost: The Modernist Transformation of Mathematics*. Princeton University Press.

Grimeisen, G. (1972). Continuous relations. *Mathematische Zeitschrift*, 127:35–44.

Grišin, V. (1982). Predicate and set theoretic calculi based on logic without contraction rules. *Mathematics of USSR–Izvestija*, 18(1):41–59.

Gupta, A. and Belnap, N. (1993). *The Revision Theory of Truth*. MIT Press.

Hájek, A. (2016). Philosophical heuristics and philosophical methodology. In Cappelen, H., Gendler, T., and Hawthorne, J., editors, *Oxford Handbook of Philosophical Methodology*. Oxford University Press.

Halbach, V. (2014). *Axiomatic Theories of Truth*. Cambridge University Press. Second Revised Edition.

Hallett, M. (1984). *Cantorian Set Theory and Limitation of Size*. Clarendon Press.

Halmos, P. (1974). *Naive Set Theory*. Springer.

Hamkins, J. D. and Kikuchi, M. (2016). Set-theoretic mereology. *Logic and Logical Philosophy*, 25(3):285–308.

Hatcher, A. (2001). *Algebraic Topology*. Cambridge University Press.

Hausdorff, F. (1957). *Set Theory*. Chelsea Publishing Co.. Third Edition. First Edition 1914.

Hausdorff, F. (2005). *Hausdorff on Ordered Sets*. American Mathematical Society. Edited, with notes, by J. M. Plotkin.

Hellman, G. and Shapiro, S. (2018). *Varieties of Continua*. Oxford University Press.

Heyting, A. (1956). *Intuitionism: An Introduction*. North Holland.

Hilbert, D. (1902a). *Foundations of Geometry*. Open Court Classics.

Hilbert, D. (1902b). Mathematical problems. *Bulletin of the American Mathematical Society*, 8:437–479. Translated from the 1900 address by Mary Winston Newton.

Hilbert, D. (1925). On the infinite. In van Heijenoort 1967, pp. 367–392.

Hindley, J. R. and Seldin, J. P. (2008). *Lambda-Calculus and Combinators*. Cambridge University Press.

Hinkis, A. (2013). *Proofs of the Cantor–Bernstein Theorem: A Mathematical Excursion*. Birkhäuser.

Hinnion, R. and Libert, T. (2003). Positive abstraction and extensionality. *Journal of Symbolic Logic*, 68(3):828–836.

Hocking, J. G. and Young, G. S. (1961). *Topology*. Dover.

Hoffman, B. (1966). *About Vectors*. Dover.

Holmes, M. R., Forster, T., and Libert, T. (2012). Alternative set theories. In Gabbay, D., Kanamori, A., and Woods, J., editors, *Handbook of the History of Logic, Vol. 6, Sets and Extensions in the Twentieth Century*, pages 559–632. Elsevier/North-Holland.

Horgan, T. (1994). Robust vagueness and the forced-march sorites paradox. *Philosophical Perspectives*, 8:159–188.

Hudson, H. (2005). *The Metaphysics of Hyperspace*. Oxford University Press.

Humberstone, L. (2006). Variations on a theme of Curry. *Notre Dame Journal of Formal Logic*, 47(1):101–131.

Humberstone, L. (2011). *The Connectives*. MIT.

Hyde, D. (1994). Why higher-order vagueness is a pseudo-problem. *Mind*, 103:35–41.

Hyde, D. (1997). From heaps and gaps to heaps of gluts. *Mind*, 106:440–460.

Hyde, D. (2008). *Vagueness, Logic, and Ontology*. Ashgate.

Hyde, D. and Raffman, D. (2018). Sorites paradox. In Zalta, E., editor, *Stanford Encyclopedia of Philosophy*. http://plato.stanford.edu/entries/sorites-paradox/.

Hyde, D. and Sylvan, R. (1993). Ubiquitous vagueness without embarrassment. *Acta Analytica*, 10(1):7–29. Reprinted In Hyde, D. and Priest, G., editors, *Sociative Logics and Their Applications*, pages 189–1199. Ashgate.

Istre, E. (2017). *Normalized Naive Set Theory*. PhD Thesis, University of Canterbury.

Jacquette, D., editor (2007). *Philosophy of Logic*. Elsevier.

Jané, I. (1995). The role of the absolute infinite in Cantor's conception of set. *Erkenntnis*, 42:375–402.

Jänich, K. (1984). *Topology*. Springer-Verlag.

Jänich, K. (2010). *Vector Analysis*. Springer.

Jaśkowski, S. (1969). Propositional calculus for contradictory deductive systems. *Studia Logica*, 24:143–57. Originally published in Polish in 1948.

Jech, T., editor (1974). *Axiomatic Set Theory*. American Mathematical Society.

Jech, T. (1978). *Set Theory*. Academic Press.

Jones, V. (1998). A credo of sorts. In Dales, H. and Oliveri, G., editors, *Truth in Mathematics*, pages 203–214. Clarendon Press.

Kahan, W. (1987). Branch cuts for complex elementary functions, or much ado about nothing's sign bit. In Iserles, A. and Powell, M. J. D., editors, *The State of the Art in Numerical Analysis*. Clarendon Press, Oxford.

Kanamori, A. (1994). *The Higher Infinite: Large Cardinals in Set Theory from Their Beginnings*. Springer-Verlag.

Keefe, R. (2000). *Theories of Vagueness*. Cambridge University Press.

Kelley, J. L. (1955). *General Topology*. Springer-Verlag.

Kitcher, P. (1989). Explanatory unification and the causal structure of the world. In Kitcher, P. and Salmon, W., editors, *Scientific Explanation*, pages 410–505. University of Minnesota Press.

Kleene, S. C. (1952). *Introduction to Metamathematics*. North-Holland.

Komori, Y. (1989). Illiative combinatory logic based on BCK-logic. *Mathematica Japonica*, 34:585–596.

Koslicki, K. (2008). *The Structure of Objects*. Oxford University Press.

Kripke, S. (1975). Outline of a theory of truth. *Journal of Philosophy*, 72:690–716.

Kroon, F. (2004). Realism and dialetheism. In [Priest et al., 2004].

Kunen, K. (1980). *Set Theory: An Introduction to Independence Proofs*. North Holland.

Kuratowski, K. (1962). *Introduction to Set Theory and Topology*. Pergamon. Trans. by L. Boron.

Lakatos, I. (1976). *Proofs and Refutations: The Logic of Mathematical Discovery*. Cambridge University Press.

Lamb, E. (2014). The saddest thing I know about the integers. *Scientific American*. Roots of Unity blog, http://blogs.scientificamerican.com/roots-of-unity/the-saddest-thing-i-know-about-the-integers/.

Landini, G. (2009). Russell's schema, not Priest's inclosure. *History and Philosophy of Logic*, 30(2):105–139.

Lavine, S. (1994). *Understanding the Infinite*. Harvard University Press.

Lawvere, W. (1969). Diagonal arguments and Cartesian closed categories. *Lecture Notes in Mathematics*, 92:134–145.

Lawvere, W. and Schanuel, S. (2009). *Conceptual Mathematics: A First Introduction to Categories*. Cambridge University Press. First appeared from Buffalo Workshop Press 1991.

Lee, J. M. (1997). *Riemannian Manifolds: An Introduction to Curvature*. Springer.

Leibniz, G. (1714). *The Monadology*. Early Modern Texts. Trans. J. Bennett, 2017.

Leibniz, G. (1951). *Selections*. Scribners. Edited by P. Wiener.

Leibniz, G. (2002). *The Labyrinth of the Continuum: Writings on the Continuum Problem, 1672–1686*. Yale University Press. Translated and edited by R. T. W. Arthur.

Levy, A. (1979). *Basic Set Theory*. Springer Verlag. Reprinted by Dover in 2002.

Lewis, D. (1991). *Parts of Classes*. Basil Blackwell.

Lewis, D. (1999). Many, but almost one. In *Papers in Metaphysics and Epistemology*, pages 164–182. Cambridge University Press.

Lewis, D. K. (1986). *On the Plurality of Worlds*. Blackwell.

Libert, T. (2005). Models for paraconsistent set theory. *Journal of Applied Logic*, 3:15–41.

Lycan, W. (2010). What is a paradox? *Analysis*, 70(4):615–622.

MacCaull, W. (1996). Kripke semantics for logics with BCK implication. *Bulletin of the Section of Logic*, 25:41–51.

Maddy, P. (1983). Proper classes. *Journal of Symbolic Logic*, 48:113–139.

Mancosu, P., editor (2008). *The Philosophy of Mathematical Practice*. Oxford University Press.

Mancosu, P. (2011). Explanation in mathematics. In Zalta, E., editor, *Stanford Encyclopedia of Philosophy*. https://plato.stanford.edu/.

Mares, E. (2004a). *Relevant Logic*. Cambridge University Press.

Mares, E. (2004b). Semantic dialetheism. In [Priest et al., 2004], pages 264–275.

Mares, E. (2019). The universality of relevance. In Weber, Z., editor, *Ultralogic as Universal? by Richard Routley*. Synthese Library, Springer.

Mares, E. and Paoli, F. (2014). Logical consequence and the paradoxes. *Journal of Philosophical Logic*, 43:439–469.

Martin, B. (2014). Dialetheism and the impossibility of the world. *Australasian Journal of Philosophy*, 93(1):61–75.

Martin, N. and Pollard, S. (1996). *Closure Spaces and Logic*. Kluwer.

Mates, B. (1981). *Skeptical Essays*. University of Chicago Press.

McCarty, C. (2008). Completeness and incompleteness for intuitionistic logic. *Journal of Symbolic Logic*, 73(4):1315–1327.

McDaniel, K. (2007). Extended simples. *Philosophical Studies*, 133(1):131–141.

McGee, V. (1990). *Truth, Vagueness, and Paradox*. Hackett.

McKubre-Jordens, M. and Weber, Z. (2012). Real analysis in paraconsistent logic. *Journal of Philosophical Logic*, 41(5):901–922.

McKubre-Jordens, M. and Weber, Z. (2016). Paraconsistent measurement of the circle. *Australasian Journal of Logic*, 14(1):268–280.

Meadows, T. (2015). Unpicking Priest's bootstraps. *Thought: A Journal of Philosophy*, 4(3):181–188.

Meadows, T. and Weber, Z. (2016). Computation in non-classical foundations? *Philosophers' Imprint*, 16:1–17.

Mendelson, B. (1975). *Introduction to Topology*. Allyn and Bacon, Inc.

Meyer, R. K. (1975). Arithmetic formulated relevantly. Typescript, *Australasian Journal of Logic*, 18(5):154–288.

Meyer, R. K. (1976). The consistency of arithmetic. typescript, *Australasian Journal of Logic*, 18(5):289–379.

Meyer, R. K. (1985a). A farewell to entailment. In Dorn, G., et al., editors, *Foundations of Logic and Linguistics*, pages 577–636. Springer Science and Business Media.

Meyer, R. K. (1985b). Proving semantical completeness "relevantly" for R. *Logic Research Paper* (23). RSSS Australian National University.

Meyer, R. K. (1996). Kurt Gödel and the consistency of $R^{\#\#}$. In Hajek, P., editor, *Gödel '96: Logical Foundations of Mathematics, Computer Science and Physics – Kurt Gödel's Legacy, Brno, Czech Republic, August 1996, Proceedings*, pages 247–256. Springer-Verlag.

Meyer, R. K. (1998). $\supset$-E is admissible in "true" relevant arithmetic. *Journal of Philosophical Logic*, 27:327–351.

Meyer, R. K. and Mortensen, C. (1984). Inconsistent models for relevant arithmetics. *Journal of Symbolic Logic*, 49:917–929.

Meyer, R. K. and Mortensen, C. (1987). Alien intruders in relevant arithmetic. *Technical Report*, TR-ARP(9/87).

Meyer, R. K. and Ono, H. (1994). The finite model property for BCK and BCIW. *Studia Logica*, 53:107–118.

Meyer, R. K. and Restall, G. (1999). "Strenge" arithmetics. *Logique et Analyse*, 42:205–220.

Meyer, R. K. and Routley, R. (1977). Extensional reduction (I). *The Monist*, 60:355–369.

Meyer, R. K., Routley, R., and Dunn, J. M. (1978). Curry's paradox. *Analysis*, 39:124–128.

Meyer, R. K. and Urbas, I. (1986). Conservative extension in relevant arithmetic. *Mathematical Logic Quarterly*, 32(1-5):45–50.

Miller, K. (2009). Stuff. *American Philosophical Quarterly*, 46(1):1–18.

Moore, G. (1993). *G. E. Moore: Selected Writings*. Routledge Baldwin, T., editor.

Moore, G. H. (1982). *Zermelo's Axiom of Choice*. Springer Verlag.

Mortensen, C. (1988). Inconsistent number systems. *Notre Dame Journal of Formal Logic*, 29:45–60.

Mortensen, C. (1989). Anything is possible. *Erkenntnis*, 30:319–337.

Mortensen, C. (1995). *Inconsistent Mathematics*. Kluwer Academic Publishers.

Mortensen, C. (2010). *Inconsistent Geometry*. College Publications.

Moschovakis, Y. (2010). Kleene's amazing second recursion theorem. *Bulletin of Symbolic Logic*, 16(2):189–239.

Munkres, J. R. (2000). *Topology*. Prentice Hall.

Murzi, J. and Rossi, L. (2020). Generalized revenge. *Australasian Journal of Philosophy*, 98:153–177.

Myhill, J. (1960). Some remarks on the notion of proof. *Journal of Philosophy*, 57:461–471.

Myhill, J. (1975). Levels of implication. In Anderson, A. R., Marcus, R. C., and Martin, R. M., editors, *The Logical Enterprise*, pages 179–85. Yale University Press.

Myhill, J. (1984). Paradoxes. *Synthese*, 60:129–143.

Nietzsche, F. (1976). *The Portable Nietzsche*. Penguin. Translated, edited, and with an introduction by W. Kaufmann.

Odifreddi, P. (1989). *Classical Recursion Theory*. North Holland.

Øgaard, T. F. (2016). Paths to triviality. *Journal of Philosophical Logic*, 45(3):237–276.

Øgaard, T. F. (2017). Skolem functions in non-classical logics. *Australasian Journal of Logic*, 14(1):181–225.

Omori, H. (2015). Remarks on naive set theory based on LP. *Review of Symbolic Logic*, 8(2):279–295.

Omori, H. and Weber, Z. (2019). Just true? On the metatheory for paraconsistent truth. *Logique et Analyse*, 248:415–433.

Ono, H. (2010). Logics without the contraction rule and residuated lattices. *Australasian Journal of Logic*, 8:50–81.

Ono, H. and Komori, Y. (1985). Logic without the contraction rule. *Journal of Symbolic Logic*, 50:169–201.

Paul, L. (2006). Coincidence as overlap. *Noûs*, 40(4):623–659.

Pavlović, D. (1992). On the structure of paradoxes. *Archive for Mathematical Logic*, 31(6):397–406.

Petersen, U. (2000). Logic without contraction as based on inclusion and unrestriced abstraction. *Studia Logica*, 64:365–403.

Pitts, A. M. and Taylor, P. (1989). A note on Russell's paradox in locally Cartesian closed categories. *Studia Logica*, 48(3):377–387.

Potter, M. (2004). *Set Theory and Its Philosophy*. Clarendon Press.

Pratt-Harmon, I. (2007). First order mereotopology. In Aiello, M., Pratt-Harmon, I. E., and van Benthem, J., editors, *Handbook of Spatial Logics*, pages 13–98. Springer.

Priest, G. (1979). The logic of paradox. *Journal of Philosophical Logic*, 8:219–241.

Priest, G. (1980). Sense, entailment and *modus ponens*. *Journal of Philosophical Logic*, 9:415–435.

Priest, G. (1989). Reductio ad absurdum et modus tollendo ponens. In [Priest et al., 1989], pages 613–626.

Priest, G. (1990). Boolean negation and all that. *Journal of Philosophical Logic*, 19:201–215.

Priest, G. (1991a). The limits of thought – and beyond. *Mind*, 100(3):361–370.

Priest, G. (1991b). Minimally inconsistent LP. *Studia Logica*, 50:321–331.

Priest, G. (1994a). Is arithmetic consistent? *Mind*, 103(411):337–349.

Priest, G. (1994b). What could the least inconsistent number be? *Logique et Analyse*, 145:3–12.

Priest, G. (1997a). Inconsistent models of arithmetic part I: finite models. *Journal of Philosophical Logic*, 26:223–35.

Priest, G. (1997b). On a paradox of Hilbert and Bernays. *Journal of Philosophical Logic*, 26:45–56.

Priest, G. (2000). Inconsistent models of arithmetic, II: the general case. *Journal of Symbolic Logic*, 65:1519–1529.

Priest, G. (2002a). *Beyond the Limits of Thought*. Oxford University Press. Second Edition. First Edition 1995, Cambridge University Press.

Priest, G. (2002b). Paraconsistent logic. In Gabbay, D. M. and Günthner, F., editors, *Handbook of Philosophical Logic*, volume 6, pages 287–394. Kluwer. Second Edition.

Priest, G. (2003). A site for sorites. In [Beall, 2003], pages 9–24.

Priest, G. (2005). *Towards Non-Being*. Oxford University Press.

Priest, G. (2006a). *Doubt Truth Be a Liar*. Oxford University Press.

Priest, G. (2006b). *In Contradiction: A Study of the Transconsistent*. Oxford University Press. Second Edition.

Priest, G. (2006c). Spiking the field artillery. In Beall, J. and Armour-Garb, B., editors, *Deflationism and Paradox*, pp. 41–52. Oxford University Press.

Priest, G. (2007). Revenge, field, and ZF. In [Beall, 2007], pp. 225–233.

Priest, G. (2008). *An Introduction to Non-Classical Logic*. Cambridge University Press. Second Edition.

Priest, G. (2010). Inclosures, vagueness, and self-reference. *Notre Dame Journal of Formal Logic*, 51(1):69–84.

Priest, G. (2013a). Indefinite extensibility – dialetheic style. *Studia Logica*, 101:1263–1275.

Priest, G. (2013b). Mathematical pluralism. *Journal of the IGPL*, 21:4–13.

Priest, G. (2014). *One*. Oxford University Press.

Priest, G. (2015). Fusion and confusion. *Topoi*, 34(1):55–61.

Priest, G. (2016). Thinking the impossible. *Philosophical Studies*, 173:2649–2662.

Priest, G. (2017a). A note on the axiom of countability. *IfCoLog Journal of Logics and Their Applications*, 4:1351–1356. First published in *Al-Mukhatabat*, 2012, 1(23–32).

Priest, G. (2017b). What if? The exploration of an idea. *Australasian Journal of Logic*, 14(1). Special issue: Non-Classicality, edited by Z. Weber, P. Girard, and M. McKubre-Jordens, M.

Priest, G. (2019a). Dialetheism and the sorites paradox. In Ohms, S. and Zardini, E., editors, *The Sorites Paradox*, pages 135–150. Cambridge University Press.

Priest, G. (2019b). Some comments and replies. In Baskent, C. and Ferguson, T., editors, *Graham Priest on Dialetheism and Paraconsistency*, pages 575–675. Springer.

Priest, G. (2020), Metatheory and dialetheism. *Logical Investigations* 26 (2020):48–59.

Priest, G., Beall, J., and Armour-Garb, B., editors (2004). *The Law of Non-Contradiction*. Clarendon Press.

Priest, G. and Routley, R. (1983). *On Paraconsistency*. Research School of Social Sciences, Australian National University. Reprinted as introductory chapters in [Priest et al., 1989].

Priest, G. and Routley, R. (1989a). Applications of paraconsistent logic. In [Priest et al., 1989], pages 367–393.

Priest, G. and Routley, R. (1989b). The philosophical significance and inevitability of paraconsistency. In [Priest et al., 1989], pages 483–537.

Priest, G., Routley, R., and Norman, J., editors (1989). *Paraconsistent Logic: Essays on the Inconsistent*. Philosophia Verlag.

Putnam, H. (1990). *Realism with a Human Face*. Harvard University Press.

Putnam, H. (1994). Peirce's continuum. In Ketner, K., editor, *Peirce and Contemporary Thought: Philosophical Inquiries*. Fordham University Press.

Quine, W. (1969). *Set Theory and Its logic*. Harvard University Press.

Rayo, A. and Uzquiano, G., editors (2006). *Absolute Generality*. Oxford University Press.

Reinhard, W. N. (1974). Remarks on reflection principles, large cardinals, and elementary embeddings. In [Jech, 1974].

Restall, G. (1992). A note on naïve set theory in *LP*. *Notre Dame Journal of Formal Logic*, 33:422–432.

Restall, G. (1994). *On Logics without Contraction*. PhD Thesis, University of Queensland.

Restall, G. (2010a). Models for substructural arithmetics. *Australasian Journal of Logic*, 8:82–99.

Restall, G. (2010b). On t and u and what they can do. *Analysis*, 70(4):673–676.

Restall, G. (2013). Assertion, denial and non-classical theories. In Tanaka, K., Berto, F., Mares, E., and Paoli, F., editors, *Paraconsistency: Logic and Applications*, pages 81–99. Springer.

Ripley, D. (2013). Paradoxes and failures of cut. *Australasian Journal of Philosophy*, 91(1):139–164.

Ripley, D. (2015a). Comparing substructural theories of truth. *Ergo*, 2(13):299–328.

Ripley, D. (2015b). Contraction and closure. *Thought*, 4:131–138.

Ripley, D. (2015c). Embedding denial. In [Caret and Hjortland, 2015], pages 289–309.

Ripley, D. (2015d). Naive set theory and nontransitive logic. *Review of Symbolic Logic*, 8(3):553–571.

Rizza, D. (2013). Deconstructing a topological sorites. *Philosophia Mathematica*, 21(3):361–364.

Robinson, A. (1966). *Non-Standard Analysis*. North Holland.

Rogers, J. (1995). *The Giant Planet Jupiter*. Cambridge University Press.

Rosenblatt, L. (2021). On structural contraction and why it fails. *Synthese* 198, pages 2695–2720.

Rossberg, M. (2013). Too good to be "just true". *Thought*, 2(1):1–8.

Routley, R. (1975). Universal semantics. *Journal of Philosophical Logic*, 4:327–356.

Routley, R. (1977). *Ultralogic as Universal?* First appeared in two parts in *The Relevance Logic Newsletter* 2(1):51–90, January 1977 and 2(2):138-175, May 1977; reprinted as appendix to [Routley, 1980b], pp. 892–962; new edition as *The Sylvan Jungle*, volume 4, edited by Z. Weber, Synthese Library, 2019.

Routley, R. (1979). Dialectical logic, semantics and metamathematics. *Erkenntnis*, 14:301–331.

Routley, R. (1980a). The choice of logical foundations: non-classical choices and the ultralogical choice. *Studia Logica*, 39(1):77–98.

Routley, R. (1980b). *Exploring Meinong's Jungle and Beyond*. Philosophy Department, RSSS, Australian National University, Canberra. Departmental Monograph number 3.

Routley, R. (1983). Nihilism and nihilist logics. *Discussion Papers in Environmental Philosophy*, no. 4, Department of Philosophy at the Australian National University.

Routley, R. and Meyer, R. K. (1976). Dialectical logic, classical logic and the consistency of the world. *Studies in Soviet Thought*, 16:1–25.

Routley, R., Plumwood, V., Meyer, R. K., and Brady, R. T. (1982). *Relevant Logics and Their Rivals*. Ridgeview.

Routley, R. and Routley, V. (1985). Negation and contradiction. *Revista Colombiana de Matemáticas*, 19:201–231.

Rubin, H. and Rubin, J. E. (1985 [1963]). *Equivalents of the Axiom of Choice*. North Holland.

Rudin, W. (1953). *Principles of Mathematical Analysis*. McGraw-Hill. Third Edition 1976.

Russell, B. (1905a). On denoting. *Mind*, 14(56):479–493.

Russell, B. (1905b). On some difficulties in the theory of transfinite numbers and order types. *Proceedings of the London Mathematical Society*, 4:29–53.

Russell, B. (1959). *My Philosophical Development*. Allen and Unwin.

Russell, G. (2018). Logical nihilism: could there be no logic? *Philosophical Issues*, 28(1):308–324.

Sainsbury, R. M. (1995). *Paradoxes*. Cambridge University Press. Second Edition.

Sartre, J.-P. (1958). *Being and Nothingness*. Methuen. Trans. H. Barnes.

Scharp, K. (2013). *Replacing Truth*. Oxford University Press.

Schröder, B. S. (2003). *Ordered Sets: An Introduction*. Birkhaüser.

Shapiro, L. (2011). Deflating logical consequence. *Philosophical Quarterly*, 61(243):320–342.

Shapiro, S. (1997). *Philosophy of Mathematics: Structure and Ontology*. Oxford University Press.

Shapiro, S. (2002). Incompleteness and inconsistency. *Mind*, 111:817–832.

Shapiro, S., editor (2005). *The Oxford Handbook of Philosophy of Mathematics and Logic*. Oxford University Press.

Shapiro, S. (2006). *Vagueness in Context*. Oxford University Press.

Shapiro, S. (2014). *Varieties of Logic*. Oxford University Press.

Shapiro, S. and Wright, C. (2006). All things indefinitely extensible. In [Rayo and Uzquiano, 2006], pages 255–304.

Shirahata, M. (1994). *Linear Set Theory*. PhD Thesis, Stanford University.

Shirahata, M. (1999). Fixpoint theorem in linear set theory. Typescript.

Sider, T. (2000). Simply possible. *Philosophy and Phenomenological Research*, 60(3):585–590.

Sider, T. (2001). *Four-Dimensionalism*. Oxford University Press.

Simons, P. (2004). Extended simples: a third way between atoms and gunk. *The Monist*, 87:371–85.

Skolem, T. (1963). Studies on the axiom of comprehension. *Notre Dame Journal of Formal Logic*, 4:162–170.

Slaney, J. (1982). The square root of two is irrational (and no funny business). Typescript.

Slaney, J. K. (1989). RWX is not Curry-paraconsistent. In [Priest et al., 1989], pages 472–480.

Slaney, J. K., Meyer, R. K., and Restall, G. (1996). Linear arithmetic desecsed. Technical Report TR-ARP-2-96, Automated Reasoning Project, Australian National University.

Slote, M. (2011). *The Impossibility of Perfection*. Oxford University Press.

Smith, B. (1997). Boundaries: an essay in mereotopology. In Hahn, L., editor, *The Philosophy of Roderick Chisholm*, Library of Living Philosophers, pages 534–561. Open Court.

Smith, N. (2012). *Logic: The Laws of Truth*. Princeton University Press.

Smith, N. J. (2000). The principle of uniform solution (of the paradoxes of self-reference). *Mind*, 109:117–22.

Smith, N. J. (2008). *Vagueness and Degrees of Truth*. Oxford University Press.

Smith, P. (2007). *An Introduction to Gödel's Theorems*. Cambridge University Press.

Smullyan, R. and Fitting, M. (1996). *Set Theory and the Continuum Problem*. Clarendon Press.

Smullyan, R. M. (1991). Some unifying fixed point principles. *Studia Logica*, 50(1):129–141.

Sorensen, R. (1985). An argument for the vagueness of "vague." *Analysis*, 27:34–37.

Sorensen, R. (1994). A thousand clones. *Mind*, 103:47–54.

Sorensen, R. (2001). *Vagueness and Contradiction*. Oxford University Press.

Sorensen, R. (2008). *Seeing Dark Things: The Philosophy of Shadows*. Oxford University Press.

Spivak, M. (2006). *Calculus*. Cambridge University Press. Third Edition.

Steen, L. A. and Jr, J. A. S. (1978). *Counterexamples in Topology*. Springer-Verlag.

Stillwell, J. (1998). *Numbers and Geometry*. Springer.

Sylvan, R. (1992). On interpreting truth tables and relevant truth table logic. *Notre Dame Journal of Formal Logic*, 33(2):207–215.

Sylvan, R. and Copeland, J. (2000). Computability is logic relative. In Hyde, D. and Priest, G., editors, *Sociative Logics and Their Applications*, pages 189–199. Ashgate.

Takeuti, G. and Zaring, W. M. (1971). *Introduction to Axiomatic Set Theory*. Springer-Verlag.

Tappenden, J. (2002). The liar and sorites paradoxes: toward a unified treatment. In Williamson, T. and Graff, D., editors, *Vagueness*. Ashgate. Reprinted from *Journal of Philosophy*, 90(11):551–577.

Tarski, A. (1944). The semantic conception of truth and the foundations of semantics. *Philosophy and Phenomenological Research*, 4:341–376.

Tarski, A. (1956a). The concept of truth in formalized languages. In [Tarski, 1956b], pages 152–278. Polish original 1933.

Tarski, A. (1956b). *Logic, Semantics, Metamathematics: Papers from 1923 to 1938*. Clarendon Press. Translated by J. H. Woodger.

Tedder, A. (2015). Axioms for finite collapse models of arithmetic. *Review of Symbolic Logic*, 3:529–539.

Terui, K. (2004). Light affine set theory: a naive set theory of polynomial time. *Studia Logica*, 77:9–40.

Terui, K. (2014). Open problems: Brouwer's fixed point theorem and Cantor's naive set theory in substructural and fuzzy logics. Typescript.

Thomasson, A. L. (2007). *Ordinary Objects*. Oxford University Press.

Thomas, M. (2014). A conjecture about the interpretation of classical mathematics in naive set theory, presented at the LMU Paraconsistent Reasoning in Science and Mathematics Conference in June 2014, at http://sites.google.com/a/uconn.edu/morgan-thomas/ (accessed May 11, 2020).

Tieszen, R. (2005). *Phenomenology, Logic, and the Philosophy of Mathematics*. Cambridge University Press.

Tits, J. (1957). Sur les analogues algébriques des groupes semi-simples complexes. In *Colloque d'algebre superieure*, pages 261–289. Librairie Gauthier-Villars.

Unger, P. (1979). There are no ordinary things. *Synthese*, 41:117–154.

Unger, P. (1980). The problem of the many. *Midwest Studies in Philosophy*, 5:411–67.

van Aken, J. (1986). Axioms for the set theoretic hierarchy. *Journal of Symbolic Logic*, 51(4):992–1004.

van Atten, M. (2007). *Brouwer Meets Husserl: On the Phenomenology of Choice Sequences*. Synthese Library, Springer.

van Bendegem, J. P. (2003). Classical arithmetic is quite unnatural. *Logic and Logical Philosophy*, 11:231–249.

van Dalen, D. (1997). How connected is the intuitionistic continuum? *Journal of Symbolic Logic*, 62(4):1147–1150.

van Heijenoort, J., editor (1967). *From Frege to Gödel: A Source Book in Mathematical Logic, 1879–1931*. Harvard University Press.

van Inwagen, P. (1990). *Material Beings*. Cornell University Press.

Varzi, A. (1997). Boundaries, continuity and contact. *Noûs*, 31:26–58.

Varzi, A. (2014). Counting and countenancing. In Cotnoir, A. and Baxter, D., editors, *Composition as Identity*, pages 47–69. Oxford University Press.

Varzi, A. (2004). Boundary. In Zalta, E. N., editor, *The Stanford Encyclopedia of Philosophy*. https://plato.stanford.edu/.

Wansing, H. and Priest, G. (2015). External curries. *Journal of Philosophical Logic*, 44(4):453–471.

Weber, Z. (2009). *Paradox and Foundation*. PhD Thesis, University of Melbourne.

Weber, Z. (2010a). Explanation and solution in the inclosure argument. *Australasian Journal of Philosophy*, 88(2):353–357.

Weber, Z. (2010b). Extensionality and restriction in naive set theory. *Studia Logica*, 94(1):87 – 104.

Weber, Z. (2010c). A paraconsistent model of vagueness. *Mind*, 119(476):1025–1045.

Weber, Z. (2010d). Transfinite numbers in paraconsistent set theory. *Review of Symbolic Logic*, 3(1):71–92.

Weber, Z. (2012). Transfinite cardinals in paraconsistent set theory. *Review of Symbolic Logic*, 5(2):269–293.

Weber, Z. (2014). Naive validity. *Philosophical Quarterly*, 64(254):99–114.

Weber, Z. (2016a). On paraconsistent downward Löwenheim–Skolem theorems. In Arazim, P. and Dancak, M., editors, *Logica Yearbook 2015*. College Publications.

Weber, Z. (2016b). On closure and truth in substructural theories of truth. *Synthese*, https://doi.org/10.1007/s11229-016-1226-6.

Weber, Z. (2019). At the limits of thought. In Baskent, C. and Ferguson, T., editors, *Graham Priest on Dialetheism and Paraconsistency*, pages 555–574. Springer (Outstanding Contributions to Logic).

Weber, Z. (2020). Property identity and relevant conditionals. *Australasian Philosophical Review*, (forthcoming).

Weber, Z., Badia, G., and Girard, P. (2016). What is an inconsistent truth table? *Australasian Journal of Philosophy*, 94(3):533–548.

Weber, Z. and Colyvan, M. (2010). A topological sorites. *Journal of Philosophy*, 107(6):311–325.

Weber, Z. and Cotnoir, A. (2015). Inconsistent boundaries. *Synthese*, 192(5):1267–1294.

Weber, Z. and Omori, H. (2019). Observations on the trivial world. *Erkenntnis*, 5(84):975–994.

Weber, Z., Ripley, D., Priest, G., Hyde, D., and Colyvan, M. (2014). Tolerating gluts. *Mind*, 123(491):791–811.

Weir, A. (1998). Naive set theory is innocent! *Mind*, 107:763–798.

Weir, A. (2004). There are no true contradictions. In Priest, G., Beall, J. C., and Armour-Garb, B., editors, *The Law of Non-Contradiction*, pages 385–417. Clarendon Press.

Weyl, H. (1919). *The Continuum*. Dover.

White, R. (1979). The consistency of the axiom of comprehension in the infinite valued predicate logic of łukasiewicz. *Journal of Philosophical Logic*, 8:503–534.

White, R. B. (1993). A consistent theory of attributes in a logic without contraction. *Studia Logica*, 52:113–142.

Whitehead, A. N. and Russell, B. (1910). *Principia Mathematica*. Cambridge University Press. In three volumes, 1910–1913.

Whittle, B. (2004). Dialetheism, logical consequence and hierarchy. *Analysis*, 64(4):318–326.

Willard, S. (1970). *General Topology*. Addison-Wesley.

Williamson, T. (1994). *Vagueness*. Routledge.

Williamson, T. (2014). Logic, metalogic and neutrality. *Erkenntnis*, 79(2):211–231.

Wilson, W. (1931). On semi-metric spaces. *American Journal of Mathematics*, 53:361–373.

Wittgenstein, L. (1922). *Tractatus Logico-Philosophicus*. Routledge. Translated by D. E. Pears and B. F. McGuinness.

Wittgenstein, L. (1953). *Philosophical Investigations*. Blackwell. Trans. G. E. M. Anscome.

Wittgenstein, L. (1956). *Remarks on the Foundations of Mathematics*. MIT Press. Edited by G. H. von Wright, R. Rees, and G. E. M. Anscome.

Woods, J. (2003). *Paradox and Paraconsistency*. Cambridge University Press.

Woods, J. (2019). Logical partisanhood. *Philosophical Studies*, 176:1203–1224.

Wright, C. (1976). Language mastery and the sorites paradox. In Evans, G. and McDowell, J., editors, *Truth and Meaning*. Oxford University Press.

Yanofsky, N. S. (2003). A universal approach to self-referential paradoxes, incompleteness and fixed points. *Bulletin of Symbolic Logic*, 9(3):362–386.

Young, G. (2005). *Revenge: Dialetheism and Its Expressive Limitations*. PhD Thesis, University of Glasgow.

Zalta, E. (2007). Frege's theorem. *Stanford Encyclopedia of Philosophy*. https://plato .stanford.edu/.

Zardini, E. (2011). Truth without contra(di)ction. *Review of Symbolic Logic*, 4:498–535.

Zermelo, E. (1967). Investigations in the foundations of set theory. In [van Heijenoort, 1967], pages 200–215.

Zhong, H. (2012). Definability and the structure of logical paradoxes. *Australasian Journal of Philosophy*, 90(4):779–788.

Zimmerman, D. (1996). Could extended objects be made out of simple parts? An argument for atomless gunk. *Philosophy and Phenomenological Research*, 56:1–29.

Index

$1 + 1 = 2$, 193

absolute consistency, *see also* nontriviality, 14, 294
absolute sufficiency, 126, 151
absolutely separated, 277
absorption, *see also* contraction
absurdity, 14, 18, 79–80, 111, 141, 164, 166, 292
Ackermann constants, 142
addition, 193–195
adherent points, 245
adjunction, 116, 133
∀, 18
Anderson, Alan Ross, 101
annihilation, *see also* negation
antecedent strengthening, 135
antilogism, 141
antisymmetry, 153
Archimedes, 23, 83, 281, 299
Archimedes' axiom, 199
argument by cases, 134
Aristotle, 5, 51, 182, 238, 295
arithmetic
    classical, *see also* PA
    inconsistent, 96, 97, 189, 199
    relevant, 97
    Robinson, 193
Arruda, Ayda Ignez, 162
assertion, 114
asymmetry, 156
Aussersein, 95, 168, 298, 299
axiom of abstraction, 28, 151
axiom of choice, 185
axiom of comprehension, 29, 95, 152
    unresricted, 29
axiom of extensionality, 28, 151
axiom of infinity, 169

Badia, Guillermo, 84, 130, 132, 144, 188, 193, 292
Basic Law V, 29, 30, 35, 128, 152, 287
Beall, J. C., x, 6, 35, 59, 74, 78, 86–94, 111, 114, 119, 192, 289
Bell, John L., 230
Belnap, Nuel, 88, 101, 105, 236, 288
Berkeley, George, 213
Bernstein, Felix, 36
Berry's paradox, 33
Bishop, Errett, 257
bivalence, 4, 7, 18, 20
blip function, 24, 176
Bolzano, Bernard, 34, 253
Boole, George, 100
Boolos, George, 38
Borges, Jorge Luis, 80
⊥, 14
boundary, 50–51, 77, 108, 266–269
    eliminativism, 55
Brady, Ross, 29, 90, 114, 125, 126, 144, 170, 176
Brouwer's fixed point theorem, 25, 108, 281
Brouwer, L. E. J., 239
Burali-Forti paradox, 32, 39, 74, 151, 180, 183
Burgess, John P., 85, 87, 93
Butchardt, Sam, 48

cancellation, 195, 219, 222–224, 226, 231, 235
Cantini, Andrea, 122
Cantor, Georg, 31, 34, 39, 66, 80, 178, 180, 182, 183, 185, 245, 247
Cantor–Bernstein theorem, 187
Cantor's theorem, 31, 35, 68, 73, 167, 179, 185
cardinality, 31, 72, 73, 180, 187
Caret, Colin, 110
Cartesian product, 170
Casati, Roberto, 50, 269
category theory, 66

Chase, James, 44
Church, Alonzo, 23
classical recapture, 96–99
closed (set/object), 52, 257
closed theory, 16, 69, 93, 101, 115, 117, 119, 129, 143, 145, 289, 292
closure operator, 256
  as logical consequence, 264
closure space, 258
cogito, 23, 83
collapsing lemma, 92
Colyvan, Mark, 43, 44, 72, 100
complement
  exclusive, 160
  relative, 160
  self-, 160
  unrestricted, 159
$\overline{X}$, 159
&, 18
conjunction (as intensional), 134
connectedness, 44, 53, 245, 270
⊢, 132
consistency operators, 95
continuity, 6, 52, 53, 66, 230, 239, 251
  continuous mappings, 250–253, 273
contraction, 112, 116–119, 129, 132, 134, 137, 158, 161, 163, 181, 186, 190, 195, 197, 199, 200, 202, 204, 228, 232, 237, 258, 260, 264, 267, 291
  on identity, 121, 154, 202
  on theorems, 142
contradiction, 295–296
  in the world, x
contraposition, 5, 20, 73, 76, 126, 129, 151, 160, 185, 201, 263, 286–288, 294
Conway, John, 210
Cook, Roy, 103
Cotnoir, Aaron, 50, 257
counterexample, 21, 291
counterexample (principle), 21
Curry's paradox, 24, 78, 103, 110
  for operators, 110–114
  for validity, 116
Curry, Haskell, 111
cutoff, 48

∂, 266
da Costa, Newton, 90, 95
de Morgan laws, 18, 20
Dedekind, Richard, 8–9, 39, 52, 54, 169, 172, 178, 190, 228, 234, 239, 253
Dedekind cuts, 52, 239–243, 247
definite descriptions, 48, 78, 119
density, 44, 234
derivable, 131
Descartes, René, 83
diagonal lemma, 175
dialetheism, x–xi, 14

fictionalism, 59
  and the inclosure schema, 82
  semantic, 59
∨, 18
disjunction (as extensional), 134
disjunctive syllogism, 15, 19, 41, 47, 48, 75, 76, 78, 86, 87, 94, 98, 100, 102, 106, 111, 113, 119, 141, 144, 160, 193, 204, 210, 241, 270, 291
distribution, 138
division, 202–203
  algorithm, 202
doppelgänger problem, 128, 165
double negation, 18
Dummett, Michael, 38, 73
Dunn, J Michael, 88, 105, 114, 193, 214, 236, 288
Dunn–Mortensen problem, 213–215, 226, 235

empty, 164, 165, 237, 278
$\varnothing$, 164
$\varnothing_X$, 160
epistemicism, 45
∈, 29
equivalence relation, 152, 186
Escher, M. C., 81
Euclid, 37, 50, 100, 101, 201, 255
Euclid's lemma, 205
Euclid's theorem, *see also* Infinitude of Primes
Euler, Leonhard, 100
existence, 157, 172
∃, 18
explanation, 26, 66, 69, 83, 296–300
explosion, 14, 97, 99, 108, 161, 166, 215, 224, 252
extensionality, 120, 123, 126–128
  and intensionality, 34–35, 135

f'', 171
factor (principle), 137, 139, 142, 153, 161, 166, 171, 175, 200, 258, 262, 266, 278
Feferman objection, 87, 88, 90, 96, 99, 105, 108, 211, 285
field, 227
Field, Hartry, 85, 87–90, 104, 119, 144, 289
Fitch, Fredric, 11, 16
fixed point property, 24, 178, 274
fixed point theorem, 23, 174
Forster, Thomas, 39
foundations of mathematics, x, 28, 35, 89, 94, 100, 183, 185
  crisis in, 30
Fréchet space, 259
Frege, Gottlob, 21, 29, 94, 97, 189
French, Rohan, 113
functions, 171
  and skolemization, 171, 178
  versus mappings, 171–173, 177, 250
  versus relations, 102, 104
Fundamental Theorem of Arithmetic, 205

Gödel, Kurt, *see also* incompleteness theorems, 7, 285
gamma problem, 98
gaps, 12, 43, 46, 55, 176
Gentzen, Gerhard, 117
Girard, Jean Yves, 124
Girard, Patrick, 69, 84, 130
gluts, x, 3, 14, 46, 56, 83, 103, 271
God, 35, 59
    Nyarlathotep, 80
Goodship, Laura, 111
googolplex, 208
greatest common divisor, 203
greatest lower bound principle, 238
Grelling's paradox, 33
Grišin's paradox, 122
group, 221
gunk, 240

Hallett, Michael, 80
Halmos, Paul, 11
Harman, Gilbert, 87
Hausdorff, Felix, 32, 36, 185, 257
Hausdorff space, 237, 259
hierarchy, 10, 13, 36, 82, 115, 180, 287, 300
Hilbert, David, xi, 16, 33, 37, 101
Hinnion–Libert paradox, 122
⊃, 19
Humberstone, Lloyd, 43
Husserl, Edmund, 8

idempotence, 113, 117, 121, 140, 163
identity, 126–128
    and contraction, *see also* contraction
=, 152
≡, 154
image (of a relation), 171
⇒, 124
→, 125
implosion, 23, 25, 163, 176, 184, 228, 279, 299
inclosure schema, 69–79, 111, 298
incompleteness theorems, 16–17, 100
inconsistent mathematics, xi, 82, 90, 96, 98–102, 108,
    129, 171, 185, 214, 253
induction (mathematical), 189, 191, 201
    complete, 201
    transfinite recursion, 184
inexpressibility, 12
inf, 238
infinite descent, 32, 163, 201, 210, 211
infinitesimals, 212, 215, 247–249
Infinitude of Primes, 206
infinity, 9, 169, 180, 234, 239, 299
    actual, potential, absolute, 183
injection, 173, 187
    soft, 184
instrumentalism, 89
interior operator, 263

intermediate value theorem, 25, 46, 253
intersection, 158
    of closed sets, 262
∩, 158
interval, 244
invalidity, 19, 21, 289, 290
inverse element, 221
Istre, Erik, 135, 178

Kierkegaard, Søren, 61
Kleene, Stephen, 23
Knaster's theorem, 173
König's paradox, 33
Kripke, Saul, 13, 85
Kroon, Fredrick, 59
Kuratowski, Kazimierz, 170, 257

λ-calculus, 23, 177
large cardinals, 36, 185
law of excluded middle, 12, 18, 73, 76, 79, 134, 141,
    144, 167, 176, 207, 278
law of noncontradiction, 20
Lawvere's diagonal theorem, 66, 82
least number principle, 107, 201, 207, 211
least upper bound principle, 108, 237
Leibniz continuity condition, 44, 243
Leibniz's law, *see also* substitution
Leibniz, G .W., 20
Leibniz, G. W., xi, 52, 212, 230, 239, 247, 249,
    272, 281
⩽ in ℕ, 197
⩽ in ℝ, 234
Lewis, C. I., 19
Lewis, David, 89, 119, 216, 294
liar paradox, 5, 30, 59, 68, 82, 83, 103, 104, 127, 176,
    180, 294, 296, 298, 301
    and the inclosure schema, 71
limit on size, 38
linearity, 200
logic
    BCK, 124, 146
    classical, 12
    DKQ, 97, 125, 146
    of formal inconsistency, 95
    fuzzy, 46, 124
    intuitionistic, 12, 46, 85, 90, 239, 257
    is not a game, 96
    is the laws of truth, 21
    LP, 17, 80, 92, 102, 117, 156, 229,
        289–291
    paracomplete, 12, 111
    paraconsistent, ix, 14, 111
    plural, 35
    R, 97, 229
    relevant, 97, 125
    subDLQ, 131
    substructural, 117, 117

logic (Cont.)
  ultra-, 90, 97, 126
  universal, *see also* universality
logical pluralism, 84, 91, 295
Lucretius, 70

mapping, 172
  composition of, 173, 252, 273
  ultracontinuous, 281
Mares, Edwin, 59, 162
material conditional, 19, 41, 47–49, 78, 111, 129, 144,
    173, 181, 193, 204, 243
McKubre-Jordens, Maarten, 204, 230
Meadows, Toby, 84, 92
mereological nihilism, 4, 45
mereology, 35, 188, 238
metalanguage, 10, 13, 85, 101, 115, 132
metatheory, 84, 90, 96
Meyer, Robert, 16, 59, 90, 97, 98, 108, 114, 132, 178,
    191, 250
Mirimanoff's paradox, 32, 181
modus ponens, 5, 19, 78, 111, 112, 114, 117, 118,
    121, 131, 133, 278
  pseudo, 113–114, 191
monad, 249
monoid, 226
Moore, G. E., 7
Mortensen, Chris, 97, 101, 178, 191, 212, 214, 243,
    256, 257, 290
multiplication, 195–197
multisets, 132
Myhill, John, 115

negation, 105, 130, 142, 192, 288–289, 295
  annihilation, 112, 125, 130, 134, 141, 192,
    193, 291
¬, 18
nested intervals theorem, 245
Newton, Isaac, 212
Nietzsche, Friedrich, 17, 71, 110, 300
no retraction theorem, 253, 277
non-self-identical, 157, 205, 278
non-self-identity lemma, 157
nonstandard analysis, 247
nontriviality, 14, 124, 125, 144–145, 176, 294
numbers
  cardinal, *see also* cardinality
  even and odd (not both), 208
  hyperreal, 247
  irrational, 189, 207, 209
  natural, 189, 231
  natural, axioms for, 191
  ordinal, *see also* ordinals
  prime, 204
  real, 213, 230
  real, axioms for, 230
  surreal, 247

objects, ordinary, 51, 61, 94, 95, 298, 301
Omori, Hitoshi, 84
On (set of all ordinals), 32, 182
1_x, 216
Ono, Hiroakira, 124
open (set/object, 52, 264, 269
$\langle x, y \rangle$, 170
ordinals, 32, 37, 74, 80, 89, 182, 189, 280

PA, 97–99, 190, 193
paraconsistent mathematics, *see also* inconsistent
    mathematics
paradox
  definition, 3, 287
  options, 9, 13, 21, 41, 45, 54, 84, 94, 104, 286, 297
  solution, 22, 26, 297, 300
parallax, 299
Parmenides, 25
partial order, 153
  strict, 156
partition, 185
Peano postulates, *see also* PA
Peano, Giuseppe, 29, 94, 97, 100, 190
permutation, 114
Pessoa, Fernando, 9, 119, 283, 302
Petersen, Uwe, 7, 29, 176
phenomenology, 8, 50
Plato, 5, 17, 34
Poincaré, Henri, 36
point, 7, 25, 52, 218, 238–239, 255, 270, 277–281,
    297
possibilism, 290
powerset, 162
$\mathscr{P}$, 162
Priest, Graham, x, 13, 18, 45, 69–74, 86, 92, 94, 96,
    103, 111, 114, 144, 285, 288, 289, 294, 298
principle of sufficient reason, xi, 51, 52, 83, 281, 294,
    297, 302
principle of uniform solution, 65, 69, 70
proper classes, 38
Pythagoras, 127, 209

quantification, 139
  restricted, 143, 263
  universal, 11, 18, 82, 143, 178
quantifier duality, 21, 139, 156
quasivalid, 88, 94, 96, 290
Quesada, Miró, x
Quine, W.V.O., 3, 86, 157

$\mathbb{R}$, 230
r, 157
rationality, 80, 94
reductio, 5, 12, 71, 73, 76, 77, 140–141, 189, 197,
    199, 209, 211, 234, 277
relation, 170
  relational semantics, 17, 102, 292

residuation, 114, 136

Restall, Greg, 70, 91, 122, 156, 165, 189, 191, 290

retraction, 253, 273
   soft, 276

revenge, 4, 11, 13, 42, 45–47, 55, 56, 59–61, 72, 89, 103, 106, 143, 286, 291, 294, 297, 298

revisionism in mathematics, 98–100

Richard's paradox, 33

ring, 225

Ripley, David, 14, 113, 134, 189

Robinson, Abraham, 247

Rosenblatt, Lucas, 113

Rossberg, Marcus, 55, 103

Routley, Richard, x, 6, 16, 29, 85, 90, 93, 98, 114, 126, 151, 185, 285, 300

Routley Set, 95, 167–168, 179, 186, 236, 272, 276, 285, 290, 293, 297, 300, 302

Russell, Bertrand, 10, 48, 65, 70, 73, 94, 101, 157, 180

Russell paradox, 31, 35, 60, 73, 80, 296
   and the inclosure schema, 75–78

Russell set, 31, 68, 73, 76–77, 80, 95, 107, 111, 157, 159, 160, 162, 167, 169, 285, 294, 298

s, 190

Sartre, Jean Paul, 161, 215

Schönflies, Arthur, 36

Scott, Dana, 37, 178

self-reference, 10, 23

semantic closure, 10, 12, *see also* universality

separation, 53, 270

separation axioms, 259

set concept
   iterative, 36–38
   naive, 28, 33–35

set theory, 28–40
   classical, *see also* ZFC
   naive, 28
   non-well-founded, 155
   versus property theory, 38, 125, 128

shadows, 50

Shapiro, Stewart, 15, 38, 39, 89, 96, 239

Shirahata, Masaru, 121

shrieking, 88, 192

simple consistency, 14

Skolem, Thoralf, 40

Slaney, John, 39, 78, 102, 113, 189, 197, 210

Smith, Nicholas J.J., 21, 69

smooth infinitesimal analysis, 101, 247

Socrates, 17, 26, 29

Sorensen, Roy, 42

sorites paradox, 40–43
   and the inclosure schema, 72–73, 78–79
   continuous, 43, 108, 243
   discrete, 41, 107, 207
   inductive, 41
   line drawing, 42
   topological, 57, 108, 272

Spinoza, Baruch, 52

$\subseteq$, 153

$\sqsubseteq$, 154

substitution, 120–122, 124, 126, 129, 131, 258

subvaluationism, 46

sup, 237

supervaluationism, 46

suppression, 136

surjection, 173

Sylvan, Richard, *see also* Routley, Richard

symmetry (principle), 52, 239, 240, 269, 281

T-schema, 5, 10, 22, 112, 176, 288, 289, 294

Tappenden, James, 59

Tarski, Alfred, 5, 85, 103

Tarski's theorem, 10, 104

Tennant, Neil, 85

Terui, Kazushige, 26, 87, 121

theorem, 19, 132

theory, 15

They Might Be Giants, 298

tolerance (principle), 41–44, 53, 54, 57, 72, 78, 81, 107, 208, 243

$\top$, 165

topology, 262

totally separated space, 277

transconsistent, 93, 96, 109, 135, 298

transfinite, 32, 182

transfinite recursion, *see also* induction (mathematical)

transitivity, 5, 14, 20, 85, 113, 117, 132

trichotomy, 200, 234

triviality, 14, *see also* implosion, 122

truth
   deflationism, 115
   inconsistentcy theory of, 10
   in a model, 89, 288, 292
   naive theory of, 5, 17, 30
   values, 17, 102–103

truth tables, 104

Turing, Alan, 23

ultraparaconsistent, 296–297, 300

union, 158
   of closed sets, 261

$\cup$, 158

uniqueness, 48, 49, 78, 119, 155, 171, 201, 206, 220, 241, 250

unit element
   absolute, 221
   relative, 221

universal set, 11, 24, 31, 35, 39, 71, 77, 79, 121, 128, 164–167, 179, 268, 275, 276, 297–300

universality, 16, 23, 93–94, 115, 178

$\mathcal{V}$, 164

vagueness, 43

Val, 115
validity, 19, 115, 132, 287, 289
   predicate, 115
van Dalen, Dirk, 239
Varzi, Achille, 51, 269
vectors, 216–221
von Neumann, John, 32, 36, 69, 102, 182

Wang, Hao, 73
weakening, 20, 125, 129
Weir, Alan, 14, 38
well-founded, 181
well-ordered, 182
Weyl, Hermann, 35, 37, 239
Whitehead, Alfred North, 55
Williamson, Timothy, 49, 240

Wittgenstein, Ludwig, 12, 69, 83, 157, 288, 301
Woods, Jack, 85
world
   actual, 15
   inconsistent, x, 15
   is not a story, 295
Wright, Crispin, 38, 39, 41

$\mathcal{Z}$, *see also* Routley Set
Zardini, Elia, 117
Zeno, 238
Zermelo, Ernst, 36, 168
zero, 190
0_x, 216
zeros, laws of, 232
ZFC, 36, 38, 39, 87, 89, 90, 92, 98, 145, 168, 193

I saw a man pursuing the horizon;
Round and round they sped.
I was disturbed at this;
I accosted the man.
"It is futile," I said,
"You can never—"
"You lie," he cried,
and ran on.
Stephen Crane

Printed in the United States
by Baker & Taylor Publisher Services